"This volume provides the reader with an exquisite amalgam of classical and radically new wireless solutions, striking a compelling balance between the underlying theory and the associated practical issues."

Lajos Hanzo, *University of Southampton*

"This book provides a comprehensive introduction to green and software-defined wireless networks. The authors are recognized experts in these areas, and provide theoretical foundations as well as examples of real-world applications. By connecting theory with practice, this book will be very useful for academic researchers and practicing engineers."

Leonard Cimini, *University of Delaware*

"This book provides a comprehensive yet focused description of green and software-defined wireless networks with rich real-world examples of employment."

Zhisheng Niu, *Tsinghua University*

Green and Software-Defined Wireless Networks

Understand the fundamental theory and practical design aspects of green and soft wireless communications networks with this expert text. It provides comprehensive and unified coverage of fifth-generation (5G) physical layer design, as well as design of higher and radio access layers and the core network, drawing on viewpoints from both academia and industry. Get to grips with the theory through authoritative discussion of information-theoretical results, and learn about fundamental green design trade-offs, software-defined network architectures, and energy-efficient radio resource management strategies. Applications of wireless big data and artificial intelligence to wireless network design are included, providing an excellent design reference, and real-world examples of employment in software-defined 5G networks, and energy-saving solutions from wireless communications companies and cellular operators help to connect theory with practice. This is an essential text for graduate students, professionals, and researchers.

Chih-Lin I is the chief scientist of wireless technologies at the China Mobile Research Institute, having previously held senior positions at AT&T and the Industrial Technology Research Institute, Taiwan.

Guanding Yu is a professor at the College of Information Science and Electronics Engineering at Zhejiang University, China.

Shuangfeng Han is a senior project manager in the Green Communication Research Center and Fellow at the China Mobile Research Institute, and vice chair of the wireless technology work group in China's IMT-2020 (5G) Promotion Group.

Geoffrey Ye Li is a professor in the School of Electrical and Computer Engineering at the Georgia Institute of Technology, and a fellow of the IEEE.

Green and Software-Defined Wireless Networks

From Theory to Practice

CHIH-LIN I
China Mobile Research Institute

GUANDING YU
Zhejiang University

SHUANGFENG HAN
China Mobile Research Institute

GEOFFREY YE LI
Georgia Institute of Technology

CAMBRIDGE
UNIVERSITY PRESS

CAMBRIDGE
UNIVERSITY PRESS

University Printing House, Cambridge CB2 8BS, United Kingdom

One Liberty Plaza, 20th Floor, New York, NY 10006, USA

477 Williamstown Road, Port Melbourne, VIC 3207, Australia

314–321, 3rd Floor, Plot 3, Splendor Forum, Jasola District Centre, New Delhi – 110025, India

79 Anson Road, #06–04/06, Singapore 079906

Cambridge University Press is part of the University of Cambridge.

It furthers the University's mission by disseminating knowledge in the pursuit of
education, learning, and research at the highest international levels of excellence.

www.cambridge.org
Information on this title: www.cambridge.org/9781108417327
DOI: 10.1017/9781108277389

© Cambridge University Press 2019

First published 2019

Printed in the United Kingdom by TJ International Ltd. Padstow Cornwall

A catalogue record for this publication is available from the British Library.

Library of Congress Cataloging-in-Publication Data
Names: I, Chih-Lin, author. | Yu, Guanding, author. | Han, Shuangfeng, author. |
    Li, Ye (Geoffrey), author.
Title: Green and software-defined wireless networks : from theory to practice / Chih-Lin I,
    Guanding Yu, Shuangfeng Han, Geoffrey Li.
Description: Cambridge, United Kingdom ; New York, NY : Cambridge University Press, 2019. |
    Includes bibliographical references and index
Identifiers: LCCN 2018041844 | ISBN 9781108417327 (hardback : alk. paper)
Subjects: LCSH: Software-defined networking (Computer network technology) |
    Wireless communication systems–Energy conservation. | Wireless communication
    systems–Energy consumption. | Radio resource management (Wireless communications)
Classification: LCC TK5105.5833 .I2 2019 | DDC 004.6–dc23
LC record available at https://lccn.loc.gov/2018041844

ISBN 978-1-108-41732-7 Hardback

Contents

Preface

The past several decades have witnessed revolutionary progress in wireless networks, from the early first-generation systems to the current fourth-generation systems. Since the 1980s, the coverage, capacity, and capability of wireless networks have been marvelously improved, in correspondence with the dramatically increasing number of mobile subscribers. The market is now demanding fifth-generation (5G) systems, to be deployed in the coming years, that can support higher capacity, higher mobility, lower latency, lower cost, and better service. In addition to these requirements, 5G cellular networks are also anticipated to be more environmentally friendly to tackle the global warming crisis. As a by-product, the rapid growth of the worldwide information and communication industry also contributes a large amount of carbon emissions, which is comparable to the aviation industry and marks an undesirable increasing trend. Therefore, reducing carbon emissions and operating expenditure costs becomes more and more important goals for network infrastructure design. Meanwhile, from the perspective of user devices, how to reduce their battery consumption is also a major concern for network evolution. Faced with these challenges, energy efficiency has become an important metric for 5G cellular networks, and so-called green communications have been a growing trend.

On the other hand, by leveraging programmable control, management, and data planes, software-defined networking (SDN) makes it possible to enable flexible, scalable, configurable, and reliable mobile networks. Therefore, the industry and academia have launched several research initiatives on SDN-enabled 5G cellular networks worldwide. The soft design of cellular networks can be implemented in many aspects, such as cloud radio access network (C-RAN), baseband processing virtualization, and the software-defined air interface (SDAI).

Aiming at providing green and soft network architecture, infrastructure, and protocol, the traditional design philosophy of cellular networks should be revisited. For instance, the classical Shannon theory suggested a monotonic relationship between the energy efficiency (EE) and spectral efficiency (SE). However, with circuit power being considered, the SE–EE relation is no longer monotonic, therefore optimal SE and EE may be achieved simultaneously. Moreover, the idea of cell-centric design has been adopted for every generation of cellular networks. However, the soft concept enables a user-centric network architecture that no longer relies on the traditional cellular topology. Other design philosophies that can also be revisited include the decoupling of data and control planes, the reconfiguring of spectrum and air interference, and the application of big data analytics for wireless communications, just to name a few.

This book aims to provide a comprehensive overview of green and software-defined wireless networks, covering both theory and practice aspects. We start from the theoretical framework of green communications, including the information theoretical analysis of energy-efficient design and some fundamental trade-offs in green radio networks. Then, we introduce several design principles and practical architectures to facilitate green networks, with emphases on C-RAN and big-data-enabled networks. After that, several strategies on energy-efficient signaling design and resource management for next-generation networks are presented. The framework of SDAI for SDN-enabled 5G networks and some key design issues, are also provided. The book ends with some practices for energy saving solutions for different cellular networks, most of which are from China Mobile, the largest cellular operator in the world.

This book serves as an important reference for both academic students and engineers in this area. It includes both theoretical and practical results on future green and soft wireless networks, aiming at providing a compressive overview from both academic and industrial viewpoints. The theoretical part of this book is mainly contributed by Professor G. Yu and Professor G. Y. Li, while the practical part is mainly contributed by Dr. C.-L. I and Dr. S. Han. Although we try to include recent progresses on green and soft wireless networks as much as possible, we cannot cover all important results in this field due to the rapid development of 5G wireless techniques. The authors would like to thank the experts in the Green Communication Research Center of the China Mobile Research Institute for their contribution to the book, particularly Qi Sun, Jinri Huang, Sen Wang, Jiqing Ni, Gang Li, Siming Zhang, Guozhen Xu, Wei Zhou, Jun Zuo, Zhiming Liu, Yami Chen, Ailing Wang, Kai Yan, Zhiming Fan, Guizhen Wang, Junshuai Sun, Xueyan Huang, Xingyu Han, Sen Bian, Ke Li, Xidong Wang, Yaxing Qiu, Zecai Shao, Tian Xie, Ran Duan, and Chunfeng Cui.

List of Abbreviations

(I)FFT	(Inverse) Fast Fourier Transformation
2G/3G/4G	Second/Third/Fourth Generation
3GPP	3rd Generation Partnership Project
5G	Fifth Generation
AAS	Active Antenna System
ACK	Acknowledgment
ACLR	Adjacent Channel Leakage Ratio
ADC	Analog-to-Digital Converter
AI	Artificial Intelligence
AM	Acknowledged Mode
ANDSF	Access Network Discovery Support Functions
AOA	Angle of Arrival
AP	Access Point
AP	Application Protocol
API	Application Programmable Interface
AR	Augmented Reality
ARQ	Automatic Repeat Request
ATCA	Advanced Telecom Computing Architecture
BBU	Baseband Unit
BC	Broadcast Channel
BD	Big Data
BOM	Bill of Material
BP	Back-Propagation
BS	Base Station
CA	Carrier Aggregation
Caffe	Convolutional Architecture for Feature Extraction
CAPEX	Capital Expenditure
CDMA	Code Division Multiple Access
CM	Channel Measurement
CMCC	China Mobile Communications Corporation
CN	Core Network
CN-GW	Core Network Gateway
CNN	Convolution Neural Network
CoMP	Coordinated Multipoint

CP	Control Plane
CP-OFDM	Cyclic Prefix Orthogonal Frequency-Division Multiplexing
CPRI	Common Public Radio Interface
CQI	Channel Quality Indicator
CRS	Cell-Specific Reference Signal
CSI	Channel State Information
CSIR	CSI at the Receiver
CSIT	CSI at the Transmitter
CU/DU	Central Unit/Distributed Unit
CU-C	Central Unit – Control
CU-U	Central Unit – User Plane
D2D	Device-to-Device
D2I	Device-to-Infrastructure
DAC	Digital-to-Analog Converter
DAQ	Data Acquisition
DAS	Distributed Antenna System
DC	Dual Connectivity
DFT-S-OFDM	Discrete Fourier Transform-Spread-OFDM
DL	Downlink
DMRS	Demodulation Reference Signal
DNN	Depth Neural Network
DOA	Direction of Arrival
DRB	Data Radio Bearer
DS-CDMA	Direct-Sequence Code Division Multiple Access
DSP	Digital Signal Processing (*or* Processor)
DT	Data Technology
DTX	Discontinuous Transmission
DwPTS	Downlink Pilot Time Slot
E2E	End-to-End
EE	Energy Efficiency
eMBB	Enhanced Mobile Broadband
eNodeB/eNB	Enhanced Node B
EPC	Evolved Packet Core
E-RAB	Evolved Radio Access Bearer
ERP	Effective Radiated Power
e-UTRAN	Evolved UMTS Terrestrial Radio Access Network
EVM	Error Vector Magnitude
FBMC	Filter Bank Multi-Carrier
FDD	Frequency Division Duplex
FH	Fronthaul
f-OFDM	Filtered-OFDM
FPGA	Field-Programmable Gate Array
GBSCM	Geometry-Based Stochastic Channel Model
GFDM	Generalized Frequency Division Multiplexing

GFS	Google File System
GMSK	Gaussian Filtered Minimum Shift-Keying
gNB	gNodeB
GPS	Global Positioning System
GSM	Global System for Mobile Communications
GTX	Gigabit Transceiver
HARQ	Hybrid ARQ
HPBW	Half-Power Beam Width
HSR	High-Speed Railway
HSS	Home Subscriber Server
Hys	Handover Hysterias Value
I2I	Indoor-to-Indoor
IF	Intermediate Frequency
IoT	Internet of Things
IP	Internet Protocol
IS-95	Interim Standard 95
IT	Information Technology
JT	Joint Transmission
KPI	Key Performance Indicator
L1/L2	Layer1/Layer2
LNA	Low-Noise Amplifier
LO	Local Oscillator
LSAS	Large-Scale Antenna System
LTE	Long-Term Evolution
LVDS	Low-Voltage Differential Signaling
MA	Multiple Access
MAC	Media Access Control
MANO	Management and Orchestration
MAP	Maximum A Posteriori Probability
MBSFN	Multicast Broadcast Single Frequency Network
MCD	Multilevel Centralized and Distributed
MCES	Multi-RAT Cooperation Energy-Saving System
MCPA	Multi-Carrier Power Amplification
MCS	Modulation and Coding Schemes
MCU	Microcontroller Unit
MEC	Mobile Edge Computing
MeNB	Master eNB
MIB	Main Information Block
MIMO	Multi-Input Multi-Output
MLP	Multiple Layer Perception
MME	Mobility Management Entity
MMSE	Minimum Mean Square Error
mMTC	Massive Machine-Type Communication
mmWave	Millimeter Wave

MPA	Message-Passing Algorithm
MPC	Multipath Components
MPM	Mobile Platform Monitor
MR	Measure Report
MRS	Mobile Relay Station
MSK	Minimum Shift-Keying
MU-MIMO	Multi-User MIMO
MUSA	Multi-User Shared Access
NACK	Negative Acknowledgement
NEF	Network Exposure Function
NF	Network Functions
NFV	Network Function Virtualization
NG Core	Next-Generation Core
NGFI	Next-Generation Front-Haul Interface
NIST	National Institute of Standards and Technology
NLOS	Non-Line-of-Sight
NLP	Natural Language Processing
NoMA	Non-Orthogonal Multiple Access
NR	New Radio
NSSF	Network Slice Selection Function
NWD	Network Data Analytic
O2I	Outdoor-to-Indoor
O2O	Outdoor-to-Outdoor
OAM	Operation Administration and Maintenance
OBSAI	Open Base Station Architecture Initiative
OFDM	Orthogonal Frequency-Division Multiplexing
OFDMA	Orthogonal Frequency-Division Multiple Access
OMA	Orthogonal Multiple Access
OMC	Operating and Maintenance Center
OMC-R	Operation and Maintenance Center-Radio
ONU	Optical Network Unit
OOB	Out-of-Band
OPEX	Operational Expenditure
OQAM	Offset Quadrature Amplitude Modulation
OTA	Over the Air
OTFS	Orthogonal Time Frequency Space
OTN	Optical Transport Networks
OTT	Over the Top
PA	Power Amplifier
PAN	Personal Area Network
PAPR	Peak-to-Average Power Ratio
PAS	Power Angular Spectrum
PBCH	Physical Broadcast Channel
PCF	Policy Control Function

PCFICH	Physical Control Format Indicator Channel
PCRF	Policy and Charging Rules Function
PDCCH	Physical Downlink Control Channel
PDCP	Packet Data Convergence Protocol
PDMA	Pattern Based Division Multiple Access
PDSCH	Physical Downlink Shared Channel
PDU	Protocol Data Unit
PER	Packet Error Rate
PGW	Packet Gateway
PHICH	Physical Hybrid-ARQ Indicator Channel
PHY	Physical Layer
PoE	Power Over Optical Network Unit Ethernet
PRACH	Physical Random Access Channel
PS	Phase Shifter
PSD	Power Spectrum Density
PSS	Primary Synchronization Signals
PUSCH	Physical Uplink Shared Channel
QoE	Quality of Experience
QoS	Quality of Service
QPSK	Quadrature Phase Shift-Keying
QSFP	Quad Small Form-Factor Pluggable
RAN	Radio Access Networks
RAN1	Radio Access Network Layer 1
RAT	Radio Access Technology
RB	Resource Blocks
RE	Resource Element
RF	Radio Frequency
RFIC	Radio Frequency Integrated Circuits
RLC	Radio Link Control
RNC	Radio Network Controller
RNN	Recurrent Neural Network
ROHC	Robust Header Compression Mechanism
RRC	Radio Resource Control
RRM	Radio Resource Management
RRU	Remote Radio Unit
RS	Reference Signal
RSMA	Resource Spread Multiple Access
RSRP	Reference Signal Receiving Power
RSRQ	Reference Signal Receiving Quality
RT	Real Time
SA	Standalone
SC-FDMA	Single-Carrier FDMA
SCMA	Sparse Code Multiple Access
SCPA	Single-Carrier Power Amplification

SDAI	Software-Defined Air Interface
SDN	Software-Defined Network
SDU	Service Data Unit
SE	Spectral Efficiency
SeNB	Secondary eNB
SERDES	Serializer-Deserializer
SFP	Small Form-Factor Pluggable
SGW	Serving Gateway
SIB	System Information Block
SIC	Successive Interference Cancellation
SINR	Signal-to-Interference Plus Noise Ratio
SISO	Single Input Single Output
SLA	Service Level Agreement
SN	Sequence Number
SNR	Signal-to-Noise Ratio
SOA	Service Oriented Architecture
SON	Self-Organized Network
SRB	Signaling Radio Bearer
SRS	Sounding Reference Signal
SVD	Singular-Value Decomposition
TB	Transport Block
TCO	Total Cost of Owner
TCP	Transmission Control Protocol
TCSL	Time Cluster-Spatial Lobe
TDD	Time Division Duplexing
TDMA	Time Division Multiple Access
TD-SCDMA	Time Division-Synchronous Code Division Multiple Access
TTI	Transmission Time Interval
TTT	Time to Trigger
TXRU	Transmit and Receive Unit
UCN	User-Centric Network
UDN	Ultradense Network
UE	User Equipment
UFMC	Universal Filtered Multi-Carrier
UL	Uplink
UM	Unacknowledged Mode
UpPTS	Uplink Pilot Time Slot
UP	user plane
URLLC	Ultra-Reliable Low Latency Communications
UW	Unique Word
V2V	Vehicle-to-Vehicle
V2X	Vehicle-to-Everything
VCR	Virtual Channel Representation
WBD	Wireless Big Data

WDM	Wavelength Division Multiplexing
WLAN	Wireless Local Area Networks
w-OFDM	Windowed OFDM
WOLA	Weighted Overlap and Add
WPAN	Wireless Personal Area Network
WSSUS	Wide-Sense Stationary Uncorrelated Scatter
ZFBF	Zeroforcing Beamforming
ZIF	Zero Intermediate Frequency

1 Introduction

1.1 Why Green and Soft?

For the past forty years, mobile communication systems have been undergoing a revolutionary change from the first-generation (1G) analog cellular systems to the fourth-generation (4G) long-term evolution (LTE) systems. 4G could provide high-speed data service, including internet access, high-definition video broadcasts, and so on. With the development of mobile internet and service diversity, wireless traffic is growing rapidly, especially in developing countries, as shown in Fig. 1.1.

From Fig. 1.1, we can predict that the data explosion will continue in the future, driven by the vigorous development of mobile internet and internet of things (IoT). More and more mobile internet applications have emerged to meet the diverse demands of subscribers. The fifth-generation (5G) wireless networks will touch many aspects of our daily life in the future, such as home, work, leisure, and transportation. As a consequence, a consistent service experience should be supported in various scenarios, including dense residential areas, office buildings, stadiums, open-air gatherings, subways, highways, high-speed trains, and wide-area coverage scenarios. IoT is focused on communications between things and things, and between things and people, involving not only individual users, but also a large number of various vertical industrial customers. IoT applications are usually complex and diverse, therefore 5G should be more flexible and more scalable, to support massive device connections and meet diverse requirements.

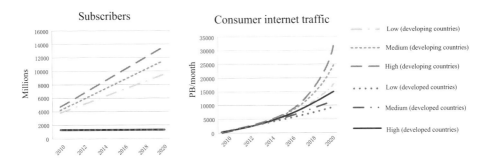

Figure 1.1 The growth of subscribers and consumer internet traffic.

The very challenging requirements from mobile traffic in these scenarios are characterized by ultra-high traffic volume density, mobility, or connection density, etc. The key performance indicators (KPIs), as defined by the International Telecommunications Union (ITU) [1], include peak data rates of 20 Gb/s, user experienced data rates of 100 Mb/s, support for up to 500 km/h mobility, 1 ms latency, a connection density of 10^6 devices/Km^2, a network energy efficiency improvement of 100X, and an area traffic capacity of 10 Mb/s/m^2. Though all the requirements need not be met simultaneously, the design of 5G networks and radio access should provide flexibility to more efficiently support various applications in diversified scenarios. In addition to the technical requirements on 5G, the operators are also faced with the requirement of taking social responsibilities when deploying the 5G networks. The first one is how to tackle the global warming issue. As it is becoming more and more serious, the impact of wireless communication networks on the environment has drawn extensive concerns. It has been reported that worldwide information and communication technology (ICT) contributes around 2 percent of the global carbon dioxide emissions (comparable to the global aviation industry), and it is estimated to grow to 4 percent by year 2020 [2]. In addition, ICT accounts for about 10 percent of global electricity consumption, and mobile network is one of the significant components of ICT energy consumption. Therefore, 5G mobile networks should be more energy-efficient than ever before to reduce both the operational costs and carbon dioxide emission. Motivated by this, network operators, regulatory bodies, such as the 3rd Generation Partnership Project (3GPP) and ITU, have conducted several research activities aiming at improving the network energy efficiency (EE). A lot of work related to green communication has been done, such as the Mobile Virtual Centre of Excellence (VCE) Green Radio project, EARTH project, and the international Greentouch Consortium [3, 4].

Another issue is environmental pollution from the outdated infrastructure equipment. The transition of mobile communication systems from one generation to another occurs generally at the expense of abandoning huge amounts of equipment in either core networks or radio access networks of the previous generations, which may pollute the environment if not handled properly. Is there a possibility that such generation transition can be conveniently and efficiently achieved via software upgrade, without abandoning old hardware or replacing it with newly manufactured equipment? The 5G network is therefore motivated to be reconfigurable with software-defined networking (SDN) [5] and air interface agility in implementation.

In the past several decades, high capacity and spectral efficiency (SE) are the primary design goals of mobile network, but now we need to pursue a SE–EE codesign network. Besides satisfying diverse user demands, future mobile communication systems should be able to support lower power consumption to build a greener mobile communication network and achieve greater sustainability. Meanwhile, the 5G network needs to facilitate a converged network synergistic with information and communication technology convergence, multiple radio access technology (RAT) convergence, radio access network (RAN) and core network convergence, content convergence, and spectrum convergence.

1.2   Green: From UE to Infrastructure

Reducing carbon emissions and operating expenditure (OPEX) costs are important goals for wireless cellular networks. The profound meaning of green is to heighten efficiency in utilization of any resources supporting wireless communications from the network side to the user terminal (UE) side.

For the UE side, the required energy in the UE's battery is increasing with the development of mobile internet. How to optimize the battery life of mobile users is still a challenging task. To solve this problem, several methods have been proposed. For example, power-saving mode (PSM), such as discontinuous reception (DRX) mechanism, has been introduced in LTE for power saving at the UE. DRX enables the UE to switch from an active state to a short or long sleep state without sacrificing the quality of service (QoS). In the sleep state, a terminal remains attached with the network. However, it is not accessible because it does not check for paging.

In 5G, a new user mode, called RRC_INACTIVE mode [6], is introduced. In the RRC_INACTIVE mode, the terminal can return to communication state without RRC connection setup procedure, and hence the energy consumption can be reduced further.

To cope with the limited battery energy problem, radio frequency (RF) energy-harvesting technology has garnered extensive attention recently [7–9]. Since the RF signals radiated by transmitter carry both information and power at the same time, it is natural to think that the devices can be powered by the energy from the received electromagnetic waves. An example of a simultaneous wireless information and power transfer (SWIPT) system is shown in Fig. 1.2. In this system, the UE intends to decode the information and harvest energy from the received signal simultaneously. The power of the received signal at the UE is split into two parts, for decoding the information and for energy harvesting. With the energy-harvesting mechanism, the UE is expected to be not only environmentally friendly but also self-sustainable.

Recently, many mobile application developers have also shown interest in how to prolong the battery life. For applications involved with heavy network usage, such as online video, an appropriate mobile network mechanism can be designed to avoid sustained mobile network signaling interaction, frequent small data transaction, and so

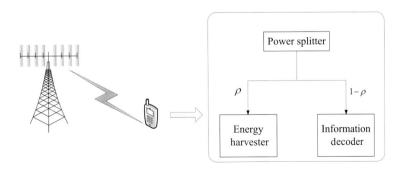

Figure 1.2 SWIPT system.

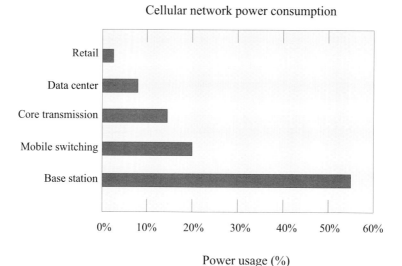

Figure 1.3 Cellular network power consumption.

on, and hence reduce power consumption. For applications requiring short interaction time and less data transaction, such as game play, the power consumption is mainly from an application processor (AP) module. Therefore, the optimization of an AP part should be considered.

As shown in Fig. 1.3 [10], based on the research of the Green Radio program, a base station (BS) takes up more than 50 percent of the cellular network power consumption. Therefore, most of the projects mainly focus on energy-saving issues at the BS and put a lot of effort into developing innovative techniques for reducing the energy to operate a RAN and to design appropriate radio architectures for energy saving.

To improve the energy efficiency of wireless networks, the Mobile VCE Green Radio Project [8] has proposed several techniques and concepts, including higher-efficiency antennas, power amplifiers (PA), multi-hop relaying techniques, BS cooperation, interference cancellation techniques, as well as energy-aware packet transmission and scheduling protocols. The results suggest that as much as 90 percent of the overall energy can be reduced under high-traffic conditions when combining high-efficiency antenna, PA, coordinated multipoint (CoMP) techniques, and shifting the network topology to heterogeneous networks (HetNets). The EARTH project [9] has investigated the energy-saving problem from both network aspects and radio components level. For the network level, new deployment strategies of RANs, self-organizing management mechanisms, and signaling protocols for energy efficiency optimization have been considered. For the component level, the power-efficient transceiver that can adapt to traffic load has been studied. For example, the supply voltages can be changed with estimated average signal power of incoming baseband signals, and some baseband boards can also be switched off. The GreenTouch project has also conducted a lot of studies to improve the energy efficiency from different aspects, including the mobile networks, the fixed access networks, and the core networks [11]. To make the mobile access networks more energy efficient, several schemes have been suggested. By decoupling the control and data plane functionalities, the small cells can be turned on and off based on the traffic load to save energy consumption. Antenna-sleeping technology is utilized in dynamic multiple-input and multiple-output (MIMO) systems to enable the BSs to switch between single-user and multiuser mode with the optimal number of active antennas. For fixed-access networks, a Gigabit Passive Optical Network (GPON)-based fiber-to-the-premises (FTTP) solution and redesigned low-power optical transceiver for optical access are brought together. To burst the energy efficiency of a core network, power-saving network components have been investigated, including energy-efficient routers, transponders, and improved digital signal processors. In addition, the relationship between the traffic demand and the power consumption of routers and transponders is used as guidance to find the optimal combination of routers and transponders. The study of GreenTouch has demonstrated that the net energy consumption in end-to-end communication networks can possibly be reduced by up to 98 percent by 2020 while taking into account the traffic growth between 2010 and 2020.

In the chip level, many new materials have been utilized for energy saving. In PA design, GaAs (gallium arsenide) transistors are usually used because of their ability to operate at high frequency, and they can generate signals with lesser noise. Currently, the GaN transistor has attracted great attention of manufacturers due to its superior characteristics of high output power, high breakdown voltage, and high temperature stability [12]. It has been reported that using GaN technology allows more than six times the output power of existing PA, using GaAs transistors. Besides, how to improve the cooling capacity without increasing energy consumption is crucial for both UE and network infrastructure. It has been proved that graphene-based materials have better thermal dissipation ability, therefore the graphite sheet has been widely used as thermal-averaging material in terminal design to delete the extra high temperature point.

1.3   Soft: From Core Network to RAN

Unlike software upgrade, it usually takes a long time to evolve to a new communication system, since the launch cycle of new standards is long, and new equipment needs to be developed and integrated. Therefore, it is necessary to make the network more flexible and reconfigurable. Soft design is the key to achieve these goals, and it will bring agility to the implementation of network elements from both core networks and access networks, as well as the building blocks of air interface. In a soft network, computing, storage, and radio resources are virtualized and centralized in order to reach dynamic and user-centric resource management.

The soft idea in communication networks can be traced back to the 1990s. In early 1990, the communication industry realized that it was difficult to define a unique standard for future mobile systems, and hence software-defined radio (SDR) emerged [13]. Some components of radio systems are implemented using software on a programmable platform instead of implementing on the hardware, so that modulation, coding scheme,

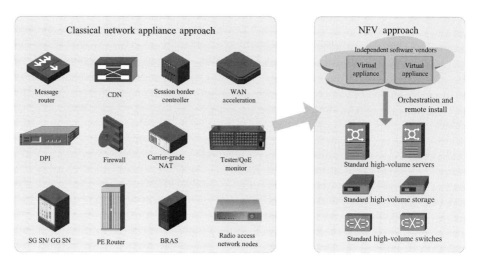

Figure 1.4 NFV vision.

resource management, and so on can be adaptive to various scenarios. Another famous example is SDN [5, 14]. SDN decouples the control plane and data plane, where the central controller is responsible for path selection based on the global network state. This configuration has many benefits, including cost reduction on routers, fast deployment of network services, capable of supporting UE from various applications, and so on. Nowadays, network function virtualization (NFV) technology enables operators to manage the infrastructure more efficiently [15, 16]. The goal of NFV is to transform the infrastructure from fragmented non-commodity hardware into platform building on common internet technology (IT) servers and storages, as shown in Fig. 1.4 [15]. To maximize the benefit from IT, end-to-end reconfigurable designs should be considered. Besides the NFV utilized in core network, virtualization of RAN, known as cloud radio access network (C-RAN), has been proposed [17]. A C-RAN network brings the baseband units (BBUs) processing resource to a centralized pool so that the resource can be managed and allocated on demand. The centralization of the baseband processing provides more energy-efficient cooling. By virtualizing the baseband processing, new features can be added to the network within months or even a few days, as opposed to years in the traditional infrastructure.

The soft concept should be extended to the air interface as well. The traditional air interface design focused on a "one-size-fits-all" approach to achieve a global optimization or trade-off. However, when it comes to 5G, soft physical layer air interface design is needed to assist with SDN and NFV technologies for providing users with diversified services and consistent quality of experience. To address this issue, the concept of software-defined air interface (SDAI) has been proposed to provide a configurable mechanism to customize air interface design to support different services under different conditions [18]. SDAI involves an intelligent controller to make air interface service oriented, and the multiple fundamental building blocks, such as multiple access, waveforms, modulation and coding, and spatial processing schemes, are case-specific configurable but with unified architecture. The unified architecture and the maximum sharing of foundational functionalities should be utilized as much as possible to ensure high energy and computational efficiency.

1.4   Green vs. Soft: An Unsolvable Contradiction?

For several decades, it has been widely accepted that dedicated hardware like an application-specific integrated circuit (ASIC) is required in communication systems to achieve highly efficient BS operation. The BSs were designed following this approach. If these functions are to be realized using field-programmable gate array (FPGA), which is capable of flexible configuration of baseband algorithms according to different scenarios and radio access technology protocols (e.g., LTE and WiFi), the power consumption would be much higher and green communication is thus difficult to achieve. Therefore, the contradiction between green communication and soft implementation has puzzled both IT and communication technology (CT) engineers worldwide, such that green plus soft design for wireless communication system has been regarded impossible for a long time and has not been investigated seriously.

Generally the contradiction between green and soft is true for a single BS implementation. The softer the implementation of algorithms, the more energy consumption it incurs. If we take the whole communication network into consideration, is the contradiction still valid? Is it possible to achieve green plus soft design if we pool the baseband computation in a central processing unit and conduct soft implementation using FPGA or general-purpose CPU with smart workload allocation based on the temporal and spatial mobile traffic variations? Later on in this book, we will show how green and soft network design can be achieved simultaneously.

1.5   Rethinking Green and Soft 5G Network Design

Characterized by a mixed set of KPIs, such as data rates, latency, mobility, energy efficiency, and traffic density, 5G services demand a fundamental revolution on the end-to-end network architecture and key technologies design. Toward a "Green and Soft" 5G, eight innovative 5G research and development themes have been proposed by China Mobile, including:

1. Rethinking Shannon to start a green journey on wireless systems;
2. Rethinking Ring and Young for no more "cells";
3. Rethinking signaling and control to make network applications- and load-aware;
4. Rethinking antennas to make BSs invisible via SmarTiles;
5. Rethinking spectrum and air interface to enable wireless signals to "dress for the occasion";

6. Rethinking fronthaul to enable soft RAN via next-generation fronthaul interface (NGFI);
7. Rethinking the protocol stack for flexible configurations of diversified access points and optimal baseband function split between the BBU pool and the remote radio systems;
8. Rethinking big data (BD) analytics in wireless communication systems to facilitate globally optimized resource allocation and scheduling via big-data-enabled network architecture and signaling procedure.

1.5.1   Rethink Shannon

After decades of high-speed development, the scale of ICT, particularly communication networks, is huge enough such that its power consumption is no longer a negligible factor in global energy consumption. Considering a 1,000-times capacity increase by 2020, the power consumption of future networks is not affordable if the network is designed with the current energy-scaling rule.

Classical Shannon theory, a "bible" in the technical communications domain, has been leading the development of communication systems for more than half a century. The extension of Shannon theory from scalar to vector in the early 1990s triggered the invention of the MIMO technology, which brought another golden twenty years of wireless communication systems. The SE and EE relationship is recently explored by rethinking Shannon theory, with a simple mathematical manipulation, for guidance on development of future green communication systems in the next decade. By only considering transmit power over the air that traditional Shannon theory dealt with, a monotonic trade-off between SE and EE always exists, which means that increasing SE will induce an EE reduction. That would not have been very interesting, nor useful. However, in any realistic network operations, the circuit power consumed by the equipment also takes an important part besides the transmit power. This power accounts for a greater and greater share of the total power as the cell becomes smaller and smaller. If taking into account the circuit power, the relationship between SE and EE is no longer monotonic. There is actually a win-win region for EE and SE, which presents a broad R&D field for joint SE and EE optimization [19–21]. It applies in future networks from each individual component technology to network-wide performance evaluation, ranging from the equipment level to the network level.

Diverse traffic fluctuation in the temporal and spatial domains provides another opportunity to rethink Shannon theory, and different scales of traffic characteristics can be well exploited to improve both SE and EE. Network architecture and deployment can be smartly optimized by taking advantage of spatial correlation properties. Resources can be more efficiently managed and allocated by using the small-scale variations of traffic volume. Transmission technology can be adaptively selected or combined in different scenarios to implement EE–SE codesign. In 5G standardization, many new technologies are being studied, e.g., hybrid beamforming for higher frequency bands, non-orthogonal multiplex access schemes, and new waveforms. The EE–SE codesign needs to be taken into consideration.

1.5.2   Rethink Ring and Young

The concept of cellular systems was proposed in 1947 by two researchers from Bell Labs, Douglas H. Ring, and W. Rae Young. Since the first generation of cellular standards, this cell-centric design has been maintained through every new generation of standards including 4G. Toward the timeline of 2020, with the introduction of HetNets and ultra-dense networks (UDNs) [22], multiple layers of radio networks have come into being. Energy consumption, interference, and mobility issues are becoming more serious due to smaller inter-site distance. Diverse types of BSs with different coverage, transmit power, frequency bands, among others are introduced. Traffic fluctuation is more significant than before, taking into account emerging millions of mobile data applications. Therefore, in practical deployment, it is clear that the traditional homogeneous cell-centric design of mobile networks does not match the anticipated traffic variations and diverse radio environments.

The design of user-centric 5G radio networks should start with the principle of "no more cells" (NMC), departing from cell-based coverage, resource management, and signal processing. It should be predicated on the spatial and temporal variation of user demand, rather than a fixed cell-bounded configuration.

Given a great deal of overlapped coverage in a UDN, to alleviate interference, more radio channel information between radio access points nearby should be shared in real-time, and more joint collaboration between neighboring cells is required. Dynamically for each user, the available radio resources from multiple access points could be jointly scheduled, and the selection of control plane (CP)/user plane (UP) and UL/DL channels, respectively could be done separately.

Centralized mobility control across different cells is also essential to reduce handover interruption delay. Besides that, multi-connectivity is viewed as a promising way to realize high throughput, ultra reliability, and seamless mobility. Multi-connectivity control and user plane anchor require centralized processing across multiple cells.

In addition, enabled by SDN and NFV, multi-RAT convergence and centralized BD processing are also motivating centralized processing across multiple cells.

A macro BS, utilizing LTE evolution or a new RAT at lower frequency, provides wide coverage and serves as a signaling BS while small cells at higher frequency, such as millimeter wave (mmWave), aim at boosting throughput and offloading traffic. Furthermore, to reduce the CAPEX/OPEX of small cells, by considering smaller coverage, supporting fewer users with low mobility, more relaxed synchronization requirement, and smaller time and frequency selective fading, "data-only carrier," with on-demand system information without broadcasting overhead, can be implemented to reduce interference and energy consumption. Macro cells can also help small ones in discovery, synchronization, measurement, etc.

With the emergence of C-RAN, many technologies toward realization of the concept of NMC can be facilitated. By taking into account differentiated fronthaul conditions, RAN can be split into a central unit (CU) and a distributed unit (DU), where the CU is in charge of centralized collaboration and user plane anchor, and the DU is responsible for radio scheduling and transmission.

1.5.3   Rethink Signaling and Control

As the proliferation of mobile internet continues, new services and applications appear at a fast pace. Some have exhibited orders of magnitude higher overhead over-the-air than more traditional services, since signaling/control of current networks is "connection-oriented."

In the 5G era, the user and traffic characteristics will be much more diversified, and the resource-contending environment will be more complex [23]. Therefore, more intelligent and adaptive signaling/control mechanisms are desired for 5G networks to achieve low-cost transmission with high signaling efficiency. Thus, 5G over-the-air signaling/control should be an intelligent combination of both connection-oriented and connectionless mechanisms. It should be also application aware, load condition aware, and user status (e.g., mobility) aware.

A new lightweight state, besides IDLE and CONNECTED, should be introduced to support "connectionless" mode. Under such state, the end-to-end (E2E) connection shall be resumed quickly without starting from scratch, so that access delay can be reduced significantly, and signaling overhead can also be reduced accordingly.

In addition, RAN signaling and control should not be limited to RAN protocol layers. Cross-layer optimization between RAN and high-layer applications seems to be a promising technology trend. RAN could be enhanced to "smart RAN" with service awareness without impairment of user privacy to improve users' quality-of-experience (QoE), for example, application level adjustment with radio channel fluctuation, and differentiated RAN L2/L3 treatment with service awareness.

Furthermore, the signaling and control should be slice aware and tailored for service requirements of diversified slices. The mobile networks shall be able to provide differentiated slices with customized signaling/control, where the differential access control, network entities, mobility management, security control, etc. are totally on demand. For example, during the low load period, a slim air interface can be configured to achieve low cost [24]. Customized signaling/control for differentiated network slices shall be designed for different contexts (user, service, network circumstance). More importantly, a network framework is required for signaling/control allocation and network function orchestration. SDN is extremely suitable for such a signaling/control framework. It provides a flexible and centralized control framework, and its open programmable interfaces also make it scalable to support new services. Moreover, with the centralized SDN framework, more contexts should be collected, and the big-data-enabled processing will be performed better.

1.5.4   Rethink Antenna

Targeting significant capacity enhancement in 2020, the 5G network is expected to be ultradense with massive antennas deployed either in a distributed or centralized manner. Theoretically, massive MIMO or large-scale antenna is expected to significantly improve network capacity and reduce the inter-cell and intra-cell interference, hence they may enhance both the SE and EE. However, to accommodate a few hundred antennas and transceiver chains all on one infrastructure in a traditional cell site is very hard, unless moving up to the mmWave band. For massive MIMO in the current cellular frequency bands, we can make BS invisible, by configuring the active antenna arrays in a flexible manner on the walls of city buildings and town houses.

Traditional multiple antenna transmission schemes, signaling protocol, and network structure may not be sufficient and efficient in 5G, thus mandating fundamental rethinking. The key considerations include, e.g., theoretical and practical algorithms of massive MIMO, practical implementation with a low-power and low-cost massive MIMO system (especially the transceiver design), flexible and adaptive installation of antenna arrays with irregular antenna configurations, and distributed or centralized signal processing. In the aspect of standardization, dramatic changes may be needed in reference signals design, transmit and receive scheme design, RF path calibration, channel estimation, and feedback. Proper beamforming structures need to be carefully investigated to identify the optimum digital, analog, or hybrid beamforming to best meet the requirements. The much-reduced power in each RF chain may bring novel RF chain design, e.g., making use of low-power, low-cost terminal-grade RFIC. It would be desirable to provide "SmarTile," a 2×2 or 8×8 active antenna module, as the building blocks of centralized massive MIMO. The global optimal utilization of system resources with distributed massive MIMO, on the other hand, would be greatly facilitated via C-RAN architectures.

1.5.5 Rethink Spectrum and Air Interface

To provide high data rates with the capability for all spectrum access, 5G air interface should provide flexible configuration according to the diverse service requirements. The traditional "one-size-fits-all" air interface paradigm needs to undergo a fundamental change as well. MmWave spectrum is considered crucially important as a choice of new spectrum because of its significantly greater bandwidth. 3GPP has already probed into the intelligent joint utilization of the existing licensed and unlicensed bands. Unified duplexing and full duplex provide another efficient solution to the utilization of the existing symmetric and asymmetric spectrum. New spectrum regulations are contemplating licensed shared access.

To support the diverse scenarios in 5G, the next-generation air interface will need to access all available spectrum, be scalable to deliver massive capacity and massive connectivity, and be adaptable to support new and existing services and applications with extreme requirements. The software-defined concept is expected to be one cornerstone of the 5G air interface framework. SDAI will meet the diverse demands in 5G by reconfiguring among an EE–SE co-optimized set of combinations of physical layer building blocks, including frame structure [25], duplex mode [26], waveforms [27–29], multiple access scheme [30], modulation and coding, spatial processing scheme [21], signaling, control [24], and protocol stack processing, etc.

1.5.6 Rethink Fronthaul

Due to the synchronous digital hierarchy (SDH)-based transmission nature, traditional fronthaul (FH) solutions, e.g., common public radio interface (CPRI), fell short both in required bandwidth and architecture flexibility. It is therefore well viewed that CPRI is inapplicable to support diverse 5G requirements as well as 5G technologies such as C-RAN, NFV, and large-scale antenna systems (LSAS). As a result, the fronthaul is needed to be rethought to accommodate 5G. In 2014, the concept of NGFI is proposed [31]. The essence of NGFI is to make the fronthaul traffic be traffic-dependent, independent of the number of antennas and thus to exploit statistical multiplexing. NGFI requires redesign of the baseband unit-remote radio head (BBU-RRH) function split and packetization on fronthaul data. By decoupling the FH bandwidth from the antenna number, NGFI can better support large antenna technologies. With decoupling of cell and UE processing, the NGFI data rate varies with traffic change, which enables exploiting the statistical multiplexing gain, which improves efficiency. The use of Ethernet for NGFI transmission brings the benefits of improved reliability and flexibility due to the packet-switching nature of Ethernet.

1.5.7   Rethink Protocol Stack

In traditional LTE/LTE-Advanced, the basic element of a communication network is the "cell," which is in total charge of the radio resources and the connected UEs. As a result, the UE context can only be established based on a specific cell, which works well in LTE, since the rate of inter-cell handover is not so often, and the signaling procedure for inter-cell handover is comparatively slow. But in 5G NR, the introduction of UDN leads to possible frequent handover so that reusing signaling procedure in 4G will degrade the quality of services. Therefore, the multilevel centralized and distributed (MCD) design logic is proposed, which could enable the protocol stack of 5G NR to inherit the advantages of the traditional cell-centric structure, especially on radio resource management, while adapting to the explosive growth of data traffic and the increasing density of the deployed BSs.

With the application of MCD, the UE becomes a basic element of RAN besides the cell. As a result, the UE is responsible for the UE context management while the cell is responsible for the radio resources management. By decoupling logical channels from cells, the MCD structure decreases the frequency and the complexity of the handover, and reduces the delay of signaling procedure. As an innovative concept, the introduction of MCD design logic in 5G NR is expected to enhance the multicell cooperation and increase the flexibility of the radio resource management while ensuring the stability of the system, which perfectly matches the characteristics of CU/DU architecture and the corresponding 5G protocol stack adopted by 3GPP.

1.5.8   Rethink Big Data Analytics in Wireless Communications

In the past 50 years, we have also seen the success of the business model of "IT+CT" in applications, such as targeted advertisements and Internet Credits. The next 50 years will be the era of the "IT+CT+DT" (data technology), where BD and artificial intelligence (AI) bring new momentum to the cellular industry development in terms of network optimization, capacity improvement, customized services, and better user experience, giving rise to more innovative and disruptive technologies.

The concept of using BD for wireless network optimization is no longer new. However, previous work has been primarily focused on long-term policies in the network, such as network planning and management. Besides, the source of the data collected for analysis/model training is mostly limited to the core network. It is the right time to rethink how the mobile network is affected by BD. BD capability available throughout wireless network is expected to deliver enhanced system performance and bring profound impacts on the design and standardization of the next-generation network architecture, protocol stack, signaling procedure, and physical layer processing [32].

1.6 Skeleton of This Book

In view of the great importance and fundamental impacts of green and soft design methodology on the efficiency of wireless communication networks, this book endeavors to elaborate designs and practices of green and soft technologies. The skeleton of this book is as follows.

In Chapter 2, the theoretical framework of green communications is presented. It starts with an introduction of energy efficiency, some important metrics, and information theoretical analysis. Then some fundamental trade-offs in green radio networks are investigated, especially the EE–SE trade-off in orthogonal frequency division multiple access (OFDMA) networks. The EE design from the perspective of optimization theory is also discussed.

In Chapter 3, some green communications techniques from the network-layer perspective are investigated, including mainly sleep control and cell zooming strategies for BSs, joint downlink and uplink energy-efficient resource allocation algorithms, and energy-efficient design issues in homogeneous networks and heterogeneous networks.

In Chapter 4, the exploration of green and soft wireless networks is discussed. An end-to-end 5G network architecture design is given, with highly efficient design of network functions, interfaces, and protocols. The inevitable transition from fixed network entities and deployment to NFV and SDN is analyzed, with introduction of C-RAN as an enabling technology toward green and soft 5G network. BD-enabled wireless network design is also addressed.

In Chapter 5, the framework of SDAI for 5G and beyond is presented. Instead of a global optimized air interface, which is a trade-off among many factors, the SDAI will be highly motivated to meet the diverse demands by reconfiguring combinations of the physical layer building blocks, including frame structure, waveforms, multiple access scheme (orthogonal and non-orthogonal), duplex mode (time division duplex (TDD), frequency division duplex (FDD), and flexible duplex), modulation and coding, and spatial processing scheme.

In Chapter 6, the existing energy-saving practices in 2G, 3G, and 4G are examined. The analysis includes global system for mobile communications (GSM), time division-synchronous code division multiple access (TD-SCDMA), time division long term evolution (TD-LTE), wireless local area networks (WLAN) energy saving, inter-RAT energy-saving platform multi-RAT cooperation energy-saving system (MCES), C-RAN application and field trial, green applications, and energy-efficient, large-scale antenna systems (invisible BSs).

References

[1] ITU-R, "IMT-Vision – Framework and overall objectives of the future development of IMT for 2020 and beyond," June 2015. www.itu.int/rec/R-REC-M.2083.

[2] European Commission, "Digital agenda: Global tech sector measures its carbon footprint," Release, Brussels, Belgium, Mar. 2013. http://europa.eu/rapid/press-release_IP-13-231_en.htm.

[3] Energy aware radio and network technology (EARTH) project. https://ieeexplore.ieee.org/document/5449938/.

[4] "GreenTouch" Consortium, Technical report. www.greentouch.org.

[5] B. A. A. Nunes, M. Mendonca, X.-N. Nguyen, K. Obraczka, and T. Turletti, "A survey of software-defined networking: Past, present, and future of programmable networks," *IEEE Commun. Surveys Tuts.*, vol. 16, no. 3, pp. 1617–1634, Sept. 2014.

[6] 3GPP TR 38.802, "Study on new radio (NR) access technology – Physical layer Aspects," 2016.

[7] J. Zhang et al., "Large system secrecy rate analysis for SWIPT MIMO wiretap channels," *IEEE Trans. on Inf. Forensics and Security*, vol. 11, no. 1, pp. 74–85, Jan. 2016.

[8] J. Thompson et al., "Overview of green radio research outcomes," Invited paper, *Workshop on Smart and Green Commun. & Networks*, 2012.

[9] M. Gruber et al., "EARTH – Energy Aware Radio and Network Technologies," *IEEE 20th Int. Symposium on Personal, Indoor and Mobile Radio Commun. (PIMRC)*, 2009.

[10] C. Han, T. Harrold et al., "Green Radio: radio techniques to enable energy-efficient wireless networks," *IEEE Commun. Mag.*, vol. 49, no. 6, pp. 46–54, June 2011.

[11] GreenTouch Final Results from Green Meter Research Study, June 2015 www.greentouch.org.

[12] Z. Popovic, "Amping up the PA for 5G: Efficient GaN power amplifiers with dynamic supplies," *IEEE Microwave Mag.*, vol. 18, no. 3, pp. 137–149, May 2017.

[13] E. Buracchini, "The software radio concept," *IEEE Commun. Mag.*, vol. 38, no. 9, pp. 138–143, Sept. 2000.

[14] D. Kreutz, F. M. V. Ramos, P. E. Verssimo et al. "Software-defined networking: A comprehensive survey," *Proceedings of the IEEE*, vol. 103, no. 1, pp. 14–76, Jan. 2015.

[15] C. H. Park, "VNF management method using VNF group table in network function virtualization," ICACT, 2017.

[16] R. Mijumbi, J. Serrat, J. L. Gorricho et al., "Network function virtualization: State-of-the-art and research challenges," *IEEE Commun. Surv. Tuts.*, vol. 18, no. 1, pp. 236–262, 1st quart. 2016.

[17] J. Huang et al., "Overview of cloud RAN," *URSI General Assembly and Scientific Symposium (URSI GASS)*, Beijng, 2014.

[18] Q. Sun et al., "Software defined air interface: A framework of 5G air interface," *2015 IEEE Wireless Commun. and Networking Conf. Workshops (WCNCW)*, 2015.

[19] Z. Xu, Z. Pan, and Chih-Lin, I, "Fundamental properties of the EE–SE relationship," *2014 IEEE Wireless Commun. and Networking Conf. (WCNC)*, pp. 1115–1120.

[20] G. Li et al., "Energy-efficient wireless communications: Tutorial, survey, and open issues," *IEEE Wireless Commun.*, vol. 18, no. 6, pp. 28–35, 2011.

[21] S. Han, C.-L. I, Z. Xu, and C. Rowell, "Large-scale antenna systems with hybrid analog and digital beamforming for millimeter wave 5G," *IEEE Commun. Mag.*, vol. 53, no. 1, pp. 186–194, 2015.

[22] N. Bhushan et al., "Network densification: The dominant theme for wireless evolution into 5G," *IEEE Commun. Mag.*, vol. 52, no. 2, pp. 82–89, 2014.

[23] NGMN Alliance, "NGMN KPIs and deployment scenarios for consideration for IMT2020, v. 1.0," Dec. 2015.

[24] Y. Chen, G. Li, Z. Pan, and C.-L. I, "Small data optimized radio access network signaling/control design," ICCWS, pp. 49–54, 2014.

[25] G. Wunder et al., "5GNOW: Intermediate frame structure and transceiver concepts," *IEEE GLOBECOM Workshop*," pp. 565–570, 2014.

[26] DUPLO Deliverable D4.1.1, "Performance of full duplex systems," Jan. 2014.

[27] G. Wunder et al., "5GNOW: Non-orthogonal, asynchronous waveforms for future mobile applications," *IEEE Commun. Mag.*, vol. 52, no. 2, pp. 97–105, 2014.

[28] PHYDAYS, "FBMC physical layer: A primer," June 2010.

[29] J. Abdoli, M. Jia, and J. Ma, "Filtered OFDM: A new waveform for future wireless systems," *IEEE Signal Processing Advances in Wireless Communication (SPAWC)*, pp. 66–70, Jul. 2015.

[30] CMCC, "On unified framework for multiple access schemes," 3GPP TSG RAN WG1 Meeting, R1-162870, Apr. 2016.

[31] C.-L. I, J. Huang, Y. Yuan et al., "Rethink fronthaul for soft RAN," *IEEE Commun. Mag.*, vol. 53, no. 9, pp. 82–88, Sept. 2015.

[32] S. Han, C.-L. I, G. Li, S. Wang, and Q. Sun, "Big data enabled mobile network design for 5G and beyond," *IEEE Commun. Mag.*, vol. 55, no. 9, pp. 150–157, Sept. 2017.

2 Theoretical Framework toward Green Networks

Traditional designs in cellular networks focuses on spectrum efficiency, which is defined as the amount of bits transmitted by each unit of bandwidth. Since the first-generation cellular system, spectrum efficiency improvement, along with network coverage enhancement, has been the most important issue in network design. Many appealing technologies, such as orthogonal frequency-division multiplexing (OFDM), multiple-input multiple-output (MIMO), small-cell networking, and full-duplex communications, have been proposed in this regard.

With the explosion of wireless data applications in recent years, energy consumption of wireless networks has aroused much interest in the 5G era. The motivation of so-called energy-efficient communications or green networks is to save the energy consumption of the whole cellular network [1]. This chapter focuses on the theoretical framework toward green radio networks. In this chapter, we will first introduce the definition of energy efficiency and some important metrics for green network design. Following that, the study of energy efficiency from information theoretical aspects will be outlined. Then some fundamental trade-offs in green radio networks will be introduced, especially the energy efficiency (EE)–spectral efficiency (SE) trade-off in orthogonal frequency-division multiple access (OFDMA) networks. The EE design from the perspective of optimization theory will be also introduced in this chapter. We will finally present the EE-oriented radio resource allocation algorithms for both orthogonal and non-orthogonal systems.

2.1 Metrics for Green Radio

There are various definitions of green metrics [2], which can be roughly classified as two kinds: energy efficiency metrics and energy consumption metrics. Table 2.1 depicts some typical definitions from link level, access level, and network level.

Link-Level Metric for Green Radio
The energy consumption of a point-to-point communication link presents the required energy for transmitting one bit information (Joules/bit). Minimizing the energy consumption has been considered for a long time. Radio resource allocation aims to

Table 2.1 Green metrics for different levels.

	Energy efficiency	Energy consumption
Link level	b/Joule; b/s/W; b/s/W; (b · m)/s/Hz/W	Joule/bit
Access level	GEE; WSEE; WMEE	
Network level	m^2/W; user/W	W/m^2 W/user

minimize the average required transmit power for a given average data rate requirement. The corresponding energy efficiency metric could be defined as the amount of transmitted bit for each Joule of energy (bits/Joule), as

$$EE = \frac{R}{\zeta P_t + P_c},$$ (2.1)

where ζ is the inverse of power amplifier efficiency, R is the transmit data rate, P_t relates to the transmit power, and P_c corresponds to the circuit power consumption, which can be modeled as a linear function of data rate

$$P_c = P_s + \xi R.$$ (2.2)

Here, P_s is the static circuit power in the transmit mode and ξ is a constant denoting dynamic power consumption per unit data rate. In some literatures, the dynamic power consumption is ignored, which corresponds to a special case that $\xi = 0$ of the mode just discussed. From (2.2), this energy efficiency metric can also be interpreted as the achievable data rate for a given transmit power (b/s/W).

Some other metrics have also been often used in certain circumstances. Corresponding to the spectral efficiency (b/s/Hz), power efficiency can be defined as the achievable spectral efficiency for a given supplied power resources (b/s/Hz/W). More generally, the radio efficiency ((b·m)/s/Hz/W) provides a more thorough definition, which takes into account the transmission distance.

These metrics mainly focus on a single point-to-point data link and can be applied in modulation and coding design, SE and EE trade-off, fundamental energy efficiency analysis, and other related topics. Link-level energy efficiency largely depends on the data rate of communication channels and the energy consumption in the transmitter and the receiver. Therefore, for a wireless channel, link-level energy efficiency is greatly impacted by the channel fading, such as path loss, shadowing, and fast fading. Also, given a point-to-point data link, energy consumption mainly consists of transmit power and circuit power. Link-level energy efficiency is important for the fundamental and theoretical study of green radios. For example, the fundamental limits of power consumption for each bit information is $\ln 2 \cdot N_0$ for an additive white Gaussian noise (AWGN) channel with a noise power density of N_0.

Access-Level Metric for Green Radio

The point-to-point link-level metric can be extended into a multiple-user link, e.g., multiple access network, multiuser interference, and device-to-device communications. In these scenarios, each communication node/user has its own energy efficiency, which should be considered for the EE metric. There are mainly three well-established metrics to aggregate the different EEs. The global energy efficiency (GEE) is defined as

$$\text{GEE} = \frac{\sum_{k=1}^{K} R_k}{\sum_{k=1}^{K} \zeta P_{t,k} + P_{c,k}}, \tag{2.3}$$

in which K is the number of total links, and $R_k, P_{t,k}$, and $P_{c,k}$ refer to the data rate, transmit power, and circuit power of user k, respectively. This GEE is actually the overall energy efficiency of the whole network consisting of K links.

The GEE represents the ratio between the overall amount of bit and the overall energy consumption, and thus can be interpreted as the benefit–cost ratio of the entire network. However, the GEE does not consider the individual energy efficiency, leading to potential unfairness among different users' EE. The maximization of GEE does not always provide the maximum EE for each user.

On the other hand, weighted sum energy efficiency (WSEE) is defined as the weighted summation of all EEs, as

$$\text{WSEE} = \sum_{k=1}^{K} \omega_k \frac{R_k}{\zeta P_{t,k} + P_{c,k}}, \tag{2.4}$$

where ω_k is the weight for link k. Moreover, weighted minimum energy efficiency (WMEE) is defined as

$$\text{WMEE} = \min_{k=1,\ldots,K} \omega_k \frac{R_k}{\zeta P_{t,k} + P_{c,k}}. \tag{2.5}$$

The WSEE and the WMEE are capable of characterizing the complete Pareto-optimal EE region of users, as it can prioritize individual EE by varying the weight value, ω_k. In regard, the maximization of WSEE or WMEE makes a good balance, or trade-off, for the EE of each individual link.

Network-Level Metric for Green Radio

While considering the network level, there are some other green metrics that characterize network energy efficiency performance, and that involve the overall consumed power of the whole network and the system-level performance. Again, there are two kinds of metrics: energy consumption metrics and energy efficiency metrics.

The area energy consumption is defined as the average network power consumption divided by the network coverage (W/m^2). This metric takes into account all network power consumptions, including radio transmission power, fixed circuit power related to the operation system, cooling system, etc. Thus, it is more related to carbon dioxide emissions and the carbon footprint. A counterpart area energy efficiency (AEE) metric can be defined as the average network coverage per consumed power (m^2/W), as

$$\text{AEE} = \frac{A}{P} \quad m^2/W, \tag{2.6}$$

where A is the coverage size and P is the overall consumed power. The AEE is more suitable for the rural environment than for the urban environment, since the network in the rural environment is usually coverage-limited.

In the urban environment where the network is capacity-limited, a more useful metric defined as the number of supported users per power unit can be used, which has a unit of users/W or W/user. With the densification of 5G small-cell base stations where the number of access points could be comparable to that of the associated users, the average power consumption per user becomes an important metric related to the operation cost, particularly the electricity bill, of cellular operators.

2.2   EE Study from Information Theory

Energy efficiency has been investigated from information theory since the very beginning. The capacity of a band-limited AWGN channel is given as

$$R = B \log_2 \left(1 + \frac{P_t}{N_0 B} \right), \tag{2.7}$$

where B is the bandwidth, P_t is the transmit power, and N_0 is the noise power density. In this section, without loss of generality, we assume that $N_0 = 1$. Spectral efficiency (SE) is defined as $\eta = R/B$, which means the amount of transmitted bits for a given unit bandwidth.

On the other hand, EE can be defined as the transmitted bit for one Joule energy, as $\epsilon = R/P_t$. Thus, the unit of EE is bit/Joule. Then, without considering circuit power consumption, the capacity region in (2.7) can be rewritten as

$$\epsilon = \frac{\eta}{2^\eta - 1}. \tag{2.8}$$

The above equation illustrates the SE–EE trade-off for point-to-point AWGN channel [3]. According to (2.8), EE monotonically decreases with SE, indicating that high SE and high EE cannot be simultaneously achieved in general. Moreover, we can also analyze that the maximum value of EE can be achieved at $(\ln 2)^{-1}$ when η approaches 0.

The relation of SE–EE trade-off can also be extended into multiuser channel scenario [4]. Multiple access channel (MAC) is one of the important multiuser channel models, which corresponds to the multiuser uplink communication in cellular networks. MAC is also the only channel model whose capacity region is already known.

The capacity region for Gaussian MAC with K users can be expressed as a convex region restricted by the following inequalities [3]

$$\sum_{i \in \mathcal{S}} R_i \leq B \log_2 \left(1 + \frac{\sum_{i \in \mathcal{S}} P_i}{N_0 B} \right), \forall \mathcal{S} \subset \{1, 2, \dots, K\}, \qquad (2.9)$$

where P_i and R_i are the transmit power and data rate of user i, respectively.

We now investigate the EE of Gaussian MAC, as well as the EE–SE trade-off. The above inequalities enclose the SE region, which characterizes the achievable data rate trade-off among different users. That is, each user cannot increase its own SE without degrading the SE of other users. Similarly, there also exists a fundamental EE trade-off among different users, i.e., the EE of one particular user will be generally decreased as EEs of other users increase. Therefore, to analyze the EE–SE trade-off, we shall first investigate the EE region of MAC.

Similar to the point-to-point AWGN channel, the SE and EE of user i can be defined as $\eta_i = R_i/B$ and $\epsilon_i = R_i/P_i$, respectively. Then, by substituting these definitions into the MAC capacity region, we can obtain the following expression

$$\sum_{i \in \mathcal{S}} \frac{\eta_i}{\epsilon_i} \geq 2^{\sum_{i \in \mathcal{S}} \eta_i} - 1, \forall \mathcal{S} \subset \{1, 2, \dots, K\}. \qquad (2.10)$$

Both SE and EE are involved in (2.10). Therefore, two fundamental and significant insights can be observed from it. The first is the EE trade-off among different users, which is similar to the SE trade-off in (2.9). The second insight is the SE–EE trade-off for each user in the MAC. Considering a particular user k, its SE–EE trade-off is a function of all other users' SE and EE, i.e., η_i and ϵ_i, $i \neq k$.

Assuming $K = 2$, the EE–SE relation of user 1 can be derived as

$$\epsilon_1(\epsilon_2, \eta_1, \eta_2) = \begin{cases} \dfrac{\eta_1}{2^{\eta_1} - 1}, & 0 \leq \epsilon_2 \leq \dfrac{\eta_2}{2^{\eta_1}(2^{\eta_2} - 1)} \\[2ex] \dfrac{\eta_1}{2^{\eta_1 + \eta_2} - (1 + \frac{\eta_2}{\epsilon_2})}, & \dfrac{\eta_2}{2^{\eta_1}(2^{\eta_2} - 1)} < \epsilon_2 \leq \dfrac{\eta_2}{2^{\eta_2} - 1}. \end{cases} \qquad (2.11)$$

With some simple mathematical analysis, we can derive that

- ϵ_1 decreases with η_1, indicating the EE–SE trade-off of a particular user.
- ϵ_1 non-increases with ϵ_2 and η_2, which means the EE of one user will be generally degraded if the other user increases its EE or SE.

The SE–EE trade-off and the EE region of the two-user MAC are illustrated in Fig. 2.1.

As we have discussed, the relation between EE and SE without considering circuit power can be expressed in closed form, for both point-to-point channel and MAC. In particular, the EE–SE curve is a cup shape curve for point-to-point transmission.

Whereas, considering circuit power, the relation between EE and SE is more complicated and cannot be expressed in closedform. It becomes a bell-shaped curve for point-to-point transmission, which will be discussed in detail later.

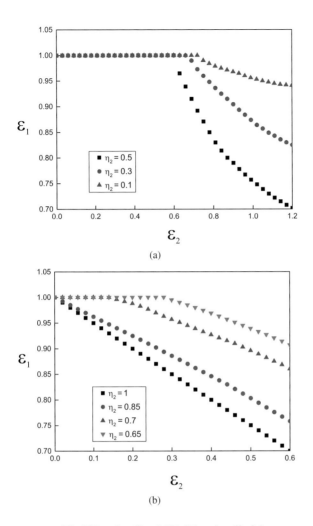

Figure 2.1 The EE trade-off and EE–SE trade-off of the two-user MAC, $\eta_1 = 1$. (a) The relationship between ϵ_1 and ϵ_2. (b) The relationship between ϵ_1 and η_2.

EE Region of TDMA and FDMA

EE regions in FDMA and TDMA systems are also attainable. In a FDMA system, we assume that the total bandwidth is B and the α_i proportion of the entire bandwidth is allocated for user i. Then the achievable rate region in FDMA system is given by [3]

$$
\bigcup_{(\alpha_1, ..., \alpha_M)} \left\{ (R_1, ..., R_M) \,\middle|\, R_i \le \alpha_i B \log_2 \left(1 + \frac{P_i}{\alpha_i B N_0} \right), \right.
$$

$$
\left. \sum_{i=1}^{M} \alpha_i = 1, \alpha_i \ge 0, \forall i = 1, 2, ... M \right\}. \tag{2.12}
$$

By substituting the definitions of SE and EE into (2.12), we can obtain the EE region for the FDMA system as

$$
\bigcup_{(\alpha_1,...,\alpha_M)} \left\{ (\epsilon_1,...,\epsilon_M) \,\middle|\, \epsilon_i \leq \frac{\eta_i}{\alpha_i \left(2^{\frac{\eta_i}{\alpha_i}} - 1 \right)}, \right.
$$

$$
\left. \sum_{i=1}^{M} \alpha_i = 1, \alpha_i \geq 0, \forall i = 1, 2, ...M \right\}.
$$

(2.13)

In a TDMA system, we assume that the total transmission time is T and $\alpha_i T$ is the time period allocated for each user. If we restrict the transmission power of each user in its own transmission period $\alpha_i T$ to be $P_{Ti} = P_i$, then the average achievable data rate in time period T can be expressed as

$$
\bigcup_{(\alpha_1,...,\alpha_M)} \left\{ (R_1,...,R_M) \,\middle|\, R_i \leq \alpha_i B \log_2 \left(1 + \frac{P_i}{BN_0} \right), \right.
$$

$$
\left. \sum_{i=1}^{M} \alpha_i = 1, \alpha_i \geq 0, \forall i = 1, 2, ...M \right\}.
$$

(2.14)

On the other hand, if we restrict the average transmission power of each user in the total transmission period T, that is $P_{Ti} = \frac{P_i}{\alpha_i}$, then the average achievable rate in T can be written as

$$
\bigcup_{(\alpha_1,...,\alpha_M)} \left\{ (R_1,...,R_M) \,\middle|\, R_i \leq \alpha_i B \log_2 \left(1 + \frac{P_i}{\alpha_i BN_0} \right), \right.
$$

$$
\left. \sum_{i=1}^{M} \alpha_i = 1, \alpha_i \geq 0, \forall i = 1, 2, ...M \right\}.
$$

(2.15)

For the first kind of power constraint, the definitions of SE and EE could be $\eta_i = \frac{R_i}{B}$ and $\epsilon_i = \frac{R_i}{\alpha_i P_i}$, respectively. For the second kind of power constraint, the definitions of SE and EE can be rewritten as $\eta_i = \frac{R_i}{B}$ and $\epsilon_i = \frac{R_i}{P_i}$, respectively.

By substituting the definitions of SE and EE into (2.14) and (2.15), we can obtain the EE expression in a TDMA system. However, under the two different constraints, the same expressions of the EE-region can be achieved, as

$$
\bigcup_{(\alpha_1,...,\alpha_M)} \left\{ (\epsilon_1,...,\epsilon_M) \,\middle|\, \epsilon_i \leq \frac{\eta_i}{\alpha_i \left(2^{\frac{\eta_i}{\alpha_i}} - 1 \right)}, \right.
$$

$$
\left. \sum_{i=1}^{M} \alpha_i = 1, \alpha_i \geq 0, \forall i = 1, 2, ...M \right\}.
$$

(2.16)

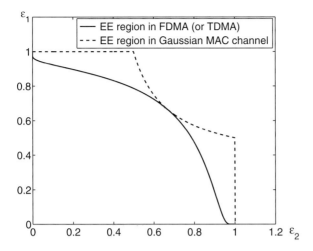

Figure 2.2 Achievable energy efficiency region in FDMA and TDMA systems, $\eta_1 = \eta_2 = 1$.

It is obvious that the EE-region of the TDMA system shown in (2.16) is the same as the EE-region of FDMA system shown in (2.13) assuming that the average power constraints in these systems are the same.

In Fig. 2.2, the achievable EE-regions in two-user FDMA and TDMA systems are shown and compared with the MAC, whose EE region is given in (2.10).

There is an intersection point of the two curves in Fig. 2.2. When $\alpha_i = \dfrac{\eta_i}{\eta_1 + \eta_2}, i = 1, 2$, the EE trade-off point in FDMA and TDMA is the same as that in MAC, that is, $\epsilon_1 = \epsilon_2 = \dfrac{\eta_1 + \eta_2}{2^{\eta_1 + \eta_2} - 1}$. Recalling that there is also a point that the capacity region of FDMA or TDMA is the same to the MAC when $\alpha_i = \dfrac{P_i}{P_1 + P_2}$. Interestingly, this point is exactly the same as the intersection point in Fig. 2.2.

2.3 Fundamental EE–SE Trade-Off

The EE–SE trade-off can be expressed in closed form without considering the fixed circuit power. However, if the circuit power cannot be ignored in practical networks, it is difficult to analyze the EE–SE relation in a closed form. In this section, we use an example of single-cell downlink OFDMA network to build a general EE–SE trade-off framework. We also demonstrate that EE is quasiconcave in SE and discuss the potential impact of channel power gain and circuit power consumption on the EE–SE trade-off.

Considering a single-cell downlink OFDMA networks with K active users and a total bandwidth B_t, which is equally divided into N subcarriers. Denote $\mathcal{K} = \{1, 2, \ldots, K\}$ as the set of K active users and $\mathcal{N} = \{1, 2, \ldots, N\}$ as the set of N subcarriers, each with a bandwidth of B. Let $p_{k,n}$ and $g_{k,n}$ be the transmit power and the channel power gain of user k ($k \in \mathcal{K}$) on subcarrier n ($n \in \mathcal{N}$), respectively. Then, for an AWGN channel with noise power density N_0, the channel capacity of user k on subcarrier n can be written as

$$r_{k,n} = B \log_2 \left(1 + \frac{p_{k,n} g_{k,n}}{N_0 B} \right). \tag{2.17}$$

The overall system throughput and the total transmit power can be expressed as

$$R = \sum_{k=1}^{K} \sum_{n=1}^{N} \rho_{k,n} r_{k,n},$$

$$P_t = \sum_{k=1}^{K} \sum_{n=1}^{N} \rho_{k,n} p_{k,n}, \tag{2.18}$$

respectively. In the above, $\rho_{k,n} \in \{1, 0\}$ is the subcarrier allocation indicator. We let $\rho_{k,n} = 1$ if subcarrier n is allocated to user k; $\rho_{k,n} = 1$ otherwise. The overall transmit power at the base station is constrained as $P_{\max}$, that is $P_t \le P_{\max}$. Moreover, as in many practical OFDMA systems, we assume that each subcarrier can only be used by at most one user to guarantee the orthogonality, i.e., $\sum_{k=1}^{K} \rho_{k,n} \le 1, \forall n \in \mathcal{N}$.

Let the set $\boldsymbol{\rho} = [\rho_{k,n}]_{K \times N}$ denote the feasible subcarrier assignment indicator matrix and the set $\boldsymbol{P} = [p_{k,n}]_{K \times N}$ denote the feasible power allocation matrix, which can be expressed as

$$\boldsymbol{\rho} \in \varrho \stackrel{\text{def}}{=} \left\{ [\rho_{k,n}]_{K \times N} \Big| \sum_{k=1}^{K} \rho_{k,n} \le 1, \forall n \in \mathcal{N}; \right.$$

$$\left. \rho_{k,n} = \{0, 1\}, \forall k \in \mathcal{K}, n \in \mathcal{N} \right\},$$

$$\boldsymbol{P} \in \mathcal{P} \stackrel{\text{def}}{=} \left\{ [p_{k,n}]_{K \times N} \Big| p_{k,n} \ge 0, \forall k \in \mathcal{K}, \forall n \in \mathcal{N}; \right. \tag{2.19}$$

$$\left. \sum_{k=1}^{K} \sum_{n=1}^{N} p_{k,n} \le P_{\max} \right\},$$

respectively.

To analyze the EE–SE trade-off, it is equivalent to maximize the EE for a given SE requirement, or data rate requirement. Let $\check{R}_k$ denote the data rate requirement of user k. As in practical LTE networks, we assume that there are two kinds of users: real-time users and non-real-time users. Let $\mathcal{K}_1 = \{1, 2, \ldots, K_0 - 1\}$ denote the set of $K_0 - 1$ ($K_0 \ge 1$) real-time users and $\mathcal{K}_2 = \{K_0, K_0+1, \ldots, K\}$ represent the set the remaining $K - K_0 + 1$ non-real-time users. The real-time users have a fixed data rate requirement, which is equal to $\check{R}_k, \forall k \in \mathcal{K}_1$, while the data rate requirement of the non-real-time users should be greater than $\check{R}_k, \forall k \in \mathcal{K}_2$.

We can now mathematically formulate the EE maximization problem as

$$
\max_{\rho \in \varrho, \, \boldsymbol{P} \in \mathcal{P}} \eta_{\text{EE}} \left(= \frac{\sum\limits_{k=1}^{K} \sum\limits_{n=1}^{N} \rho_{k,n} r_{k,n}}{\sum\limits_{k=1}^{K} \sum\limits_{n=1}^{N} \rho_{k,n} \left(\zeta p_{k,n} + \xi r_{k,n} \right) + P_s} \right), \tag{2.20}
$$

subject to

$$
\sum_{k=1}^{K} \sum_{n=1}^{N} \rho_{k,n} r_{k,n} \geq \check{R}, \tag{2.21}
$$

$$
\sum_{n=1}^{N} \rho_{k,n} r_{k,n} = \check{R}_k, \forall k \in \mathcal{K}_1, \tag{2.22}
$$

$$
\sum_{n=1}^{N} \rho_{k,n} r_{k,n} \geq \check{R}_k, \forall k \in \mathcal{K}_2. \tag{2.23}
$$

In the above, we can further assume that $\check{R} \geq \sum_{k=1}^{K} \check{R}_k$, indicating that the overall data rate requirement is no less than the summation of each user's data rate requirement.

2.3.1   EE–SE Relation

In the following discussion, we will study the fundamental EE–SE relation of the system in (2.21)–(2.23). By solving the problem in (2.20) for a given $\check{R}$, we can obtain the optimal EE as a function of SE. However it is impossible to express the EE function as a closed-form function due to the complicated optimization problem. Nevertheless, we can reveal some insightful properties of the EE function. In what follows, we first demonstrate that the EE is a quasiconcave function in SE. In addition, we also discuss how channel power gain and fixed circuit power consumption impact the EE–SE trade-off.

Assuming the number of subcarriers, N, is sufficiently large, the quasiconcavity of the EE function $\eta_{\text{EE}}(\boldsymbol{R})$ can be presented in the following theorem. Interested readers can refer to [5] for the detailed proof.

THEOREM 1   *For any achievable data rate vector, $\boldsymbol{R} = [R_k]_{K \times 1}$, achieved with a feasible subcarrier allocation and power allocation, the maximum EE, $\eta_{\text{EE}}^*(\boldsymbol{R}) = \max_{\rho \in \varrho, \, p_{k,n} \geq 0} \eta_{\text{EE}}(\boldsymbol{R})$, is strictly quasiconcave in $\boldsymbol{R}$ given sufficiently large number of subcarriers. Moreover, in the SE region $\left[\frac{\check{R}}{B_t}, \frac{\hat{R}}{B_t} \right]$, $\eta_{\text{EE}}^*(\eta_{\text{SE}})$ has the following monotonic properties.*

(a) It strictly decreases with η_{SE} and achieves its maximum at $\eta_{SE} = \frac{\check{R}}{B_t}$ if

$$\left. \frac{d\eta_{EE}^*\left(\eta_{SE}\right)}{d\eta_{SE}} \right|_{\eta_{SE} = \frac{\check{R}}{B_t}} \le 0,$$

(b) It strictly increases with η_{SE} and achieves its maximum at $\eta_{SE} = \frac{\hat{R}}{B_t}$ if

$$\left. \frac{d\eta_{EE}^*\left(\eta_{SE}\right)}{d\eta_{SE}} \right|_{\eta_{SE} = \frac{\check{R}}{B_t}} > 0 \text{ and } \left. \frac{d\eta_{EE}^*\left(\eta_{SE}\right)}{d\eta_{SE}} \right|_{\eta_{SE} = \frac{\hat{R}}{B_t}} \ge 0,$$

(c) It first strictly increases and then strictly decreases with η_{SE} and achieves its maximum at $\eta_{SE} = \frac{R_{EE,\,max}}{B_t}$ if

$$\left. \frac{d\eta_{EE}^*\left(\eta_{SE}\right)}{d\eta_{SE}} \right|_{\eta_{SE} = \frac{\check{R}}{B_t}} > 0 \text{ and } \left. \frac{d\eta_{EE}^*\left(\eta_{SE}\right)}{d\eta_{SE}} \right|_{\eta_{SE} = \frac{\hat{R}}{B_t}} < 0.$$

Here, $\hat{R}$ is the maximum throughput and $R_{EE,\,max}$ is the throughput that corresponds to the maximum EE, η_{EE}^{max}, under all constraints except the peak transmit power constraint in problem (2.20).

This theorem not only shows that the EE is a quasiconcave function of the overall SE but also presents an effective way to achieve the optimal EE–SE trade-off based on the quasiconcavity. According to [6, ch. 8], a unique global optimum always exists for any continuous and strictly quasiconcave function. Therefore, a unique and global optimal EE of the problem in (2.20) can always be achieved by this theorem.

The η_{EE}^*-versus-η_{SE} curve is referred to as the EE–SE trade-off curve in the sequel. In Fig. 2.3, we further illustrate the detailed EE–SE trade-off curves in the three possible cases in Theorem 1, as following.

- Case A (condition a): the EE decreases with SE for the entire feasible region where the optimal EE is achieved at the minimum achievable rate point.
- Case B (condition b): the EE is an increasing function of SE where the optimal EE is achieved at the maximum achievable rate point.
- Case C (condition c): the EE first increases and then decreases with SE in the feasible SE region where the maximum EE is achieved at the stationary point.

As discussed earlier, both EE and SE are related to the channel power gain and the fixed circuit power consumption. Therefore, it is important to investigate the impact of channel power gain and the circuit power consumption on the EE–SE trade-off. We present the following three properties.

Property 1. Given SE, the EE is a non-decreasing function of the channel power gain, $g_{k,n}$, and a strictly decreasing function of the circuit power consumption, P_c.

This property is rather intuitive since larger channel power gain leads to better channel capacity given a fixed transmit power, whereas larger circuit power consumption results in the degradation of EE. According to this property, it is better to schedule users with

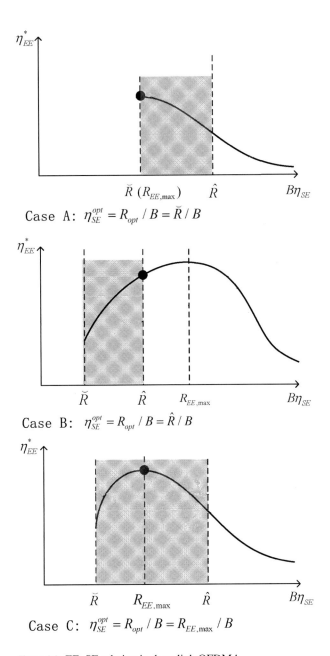

Figure 2.3 EE–SE relation in downlink OFDMA.

good channel quality, such as those users near the base station, for an improved EE–SE trade-off.

Property 2. The optimal SE, defined as $\eta_{SE}^{\text{opt}} \stackrel{\text{def}}{=} \dfrac{R_{\text{opt}}}{B_t}$, is a non-decreasing function of the static circuit power consumption, P_s. The maximum SE, defined as

$\eta_{\text{SE}}^{\max} \overset{\text{def}}{=} \dfrac{R_{\text{EE, max}}}{B_t}$, is a strictly increasing function of P_s. On the other hand, both $\eta_{\text{SE}}^{\text{opt}}$ and $\eta_{\text{SE}}^{\max}$ are independent of the dynamic circuit power consumption rate, ξ.

The independence of ξ on the maximal or optimal SE is interesting and a little bit unexpected. On some occasions if ξ or the static circuit power consumption, P_s, is large, a high SE may be still achievable despite the low EE performance. This leads to a relatively low and flat EE–SE curve, i.e., EE is insensitive to the change of SE. However, in this case, the locations of $\eta_{\text{SE}}^{\text{opt}}$ and $\eta_{\text{SE}}^{\max}$ are not impacted.

Property 3. Even for the case of very small circuit power consumption, i.e., $P_c \approx 0$, the optimal EE, $\eta_{\text{EE}}^{\text{opt}}$, is not necessarily achieved when the SE is minimized, i.e., $\eta_{\text{SE}} = \dfrac{\check{R}}{B_t}$.

The property 3 is a little bit counter-intuitive as the optimal EE is always achieved when the data rate is minimized in the single user point-to-point link. This property also indicates that even if the transmit power dominates the power consumption, the most energy-efficient communication scheme is not necessarily the least spectral-efficient. In this regard, it is possible for us to design both spectral-efficient and energy-efficient communication simultaneously.

2.3.2   Bounds on the EE–SE Curve

In the previous section, we demonstrated that EE is a quasiconcave function of SE, and introduced some insightful properties related to the EE–SE trade-off. In this section, we further analyze the upper and the lower bounds of the EE–SE trade-off curve.

From Theorem 1, the optimal EE–SE trade-off can be obtained by solving $\eta_{\text{EE}}^*(\eta_{\text{SE}})$ and $\dfrac{d\eta_{\text{EE}}^*(\eta_{\text{SE}})}{d\eta_{\text{SE}}}$. However, the exact and closed-form solution to (2.20) is rather difficult to achieve. In what follows, we apply the Lagrange dual decomposition (LDD) to approximately approach it. The LDD is an effective method to solve this kind of problems with a good accuracy and reasonable computational complexity.

The Lagrange dual problem of minimizing the total transmit power consumption for a given data rate requirement, $R \geq \check{R}$, can be expressed as

$$
\max_{\substack{\boldsymbol{\lambda}_1 \succeq \mathbf{0}, \boldsymbol{\lambda}_2 \succeq \mathbf{0}, \\ \lambda_3 \geq 0}} \min_{\boldsymbol{\rho} \in \varrho, \, \boldsymbol{P} \in \mathcal{P}} \left\{ \sum_{k=1}^{K} \sum_{n=1}^{N} \rho_{k,n} p_{k,n} + \sum_{n=1}^{N} \lambda_{1,n} \left(\sum_{k=1}^{K} \rho_{k,n} - \right) \right.
$$
$$
\left. + \sum_{k=1}^{K} \lambda_{2,k} \left(\check{R}_k - \sum_{n=1}^{N} \rho_{k,n} r_{k,n} \right) + \lambda_3 \left(R - \sum_{k=1}^{K_0-1} \check{R}_k - \sum_{k=K_0}^{K} \sum_{n=1}^{N} \rho_{k,n} r_{k,n} \right) \right\},
$$
$$
\tag{2.24}
$$

where $\boldsymbol{\lambda}_1 = [\lambda_{1,1}, \lambda_{1,2}, \ldots, \lambda_{1,N}]^T$, $\boldsymbol{\lambda}_2 = [\lambda_{2,1}, \lambda_{2,2}, \ldots, \lambda_{2,K}]^T$, and λ_3 are the Lagrange multipliers related to the corresponding constraints, and $\succeq$ denotes the component-wise inequality.

Consequently, the problem in (2.20) can be decomposed into two layers. The inner layer solves the subordinate problem for each subcarrier n, which can be formulated as

$$U_n = \min_{\rho \in \varrho, \, P \in \mathcal{P}} \sum_{k=1}^{K} \rho_{k,n} u_{k,n}, \, n \in \mathcal{N}, \tag{2.25}$$

subject to

$$
\begin{aligned}
u_{k,n} &= p_{k,n} + \lambda_{1,n} - \lambda_{2,k} r_{k,n}, \text{ if } k \in \mathcal{K}_1, \\
u_{k,n} &= p_{k,n} + \lambda_{1,n} - (\lambda_{2,k} + \lambda_3) r_{k,n}, \text{ if } k \in \mathcal{K}_2.
\end{aligned}
\tag{2.26}
$$

The outer layer is for the master problem and can be written as

$$\max_{\substack{\lambda_1 \succeq 0, \lambda_2 \succeq 0, \\ \lambda_3 \geq 0}} \left\{ \sum_{n=1}^{N} U_n - \sum_{n=1}^{N} \lambda_{1,n} + \sum_{k=1}^{K} \lambda_{2,k} \check{R}_k + \lambda_3 \left(R - \sum_{k=1}^{K_0-1} \check{R}_k \right) \right\}. \tag{2.27}$$

The problems in (2.25) and (2.27) can be iteratively solved. In the inner layer, if $\rho_{k,n} = 1$, the optimal transmit power can be expressed similarly in a water-filling way, as

$$
\begin{aligned}
p_{k,n} &= B \left[\frac{\lambda_{2,k}}{\ln 2} - \frac{N_0}{g_{k,n}} \right]^+, \text{ if } k \in \mathcal{K}_1, \\
p_{k,n} &= B \left[\frac{(\lambda_{2,k} + \lambda_3)}{\ln 2} - \frac{N_0}{g_{k,n}} \right]^+, \text{ if } k \in \mathcal{K}_2,
\end{aligned}
\tag{2.28}
$$

where $[x]^+ = \max(x, 0)$. The above equation should be solved for all $k \in \mathcal{K}$. After that, the optimal transmit power, $p_{k,n}$'s, should be substituted back into (2.26), and $\rho_{k,n}$'s should be set to 1 for the user with the minimum $u_{k,n}$ and 0 for all other users.

Moreover, the Lagrange multipliers can be updated by the subgradient method as

$$\lambda_{2,k}^{(i+1)} = \left[\lambda_{2,k}^{(i)} - s^{(i)} \left(\sum_{n=1}^{N} \rho_{k,n}^{(i)} r_{k,n}^{(i)} - \check{R}_k \right) \right]^+, \tag{2.29a}$$

$$\lambda_3^{(i+1)} = \left[\lambda_3^{(i)} - s^{(i)} \left(\sum_{k=K_0}^{K} \sum_{n=1}^{N} \rho_{k,n}^{(i)} r_{k,n}^{(i)} - \left(R - \sum_{k=1}^{K_0-1} \check{R}_k \right) \right) \right]^+, \tag{2.29b}$$

in which $s^{(i)}$ is a sufficiently small stepsize for the ith iteration. We shall note that the duality gap between the original problem and the LDD problem would not be zero due to the non-convex nature of the problem. However, the duality gap approaches zero when the number of subcarriers becomes large enough [7]. In practice, the duality gap is already very small for the case with $N = 64$ subcarriers.

Until now, the subcarrier allocation strategy has been derived based on the LDD method, as discussed earlier. In the next step, we shall consequently determine transmit power allocation, and derive upper and lower bounds, respectively.

Upper-Bound Power Allocation Strategy

For a given subcarrier allocation result, $\rho = [\rho_{k,n}]_{K,N}$, from the LDD method, we design a two-stage power allocation strategy. In the first stage, transmit power is allocated by the water-filling method for each user to fulfill its own data rate requirement, $\check{R}_k$. In the second stage, the extra power can be allocated among the subcarriers of all non-real-time users also by the water-filling method until the data rate reaches R.

It is very straightforward that this power allocation strategy achieves an upper bound of the transmit power, which also corresponds to the upper bound of the EE–SE trade-off curve.

Lower-Bound Power Allocation Strategy

A lower bound on the minimum transmit power can be achieved by relaxing the binary variables, $\rho_{k,n}$'s, into continuous real variables within $[0,1]$, which can be also interpreted as time sharing of the subcarrier. With this manipulation, the data rate expression can be rewritten as

$$ r_{k,n} = B \log_2 \left(1 + \frac{p_{k,n} g_{k,n}}{\rho_{k,n} B N_0} \right). \tag{2.30} $$

Then, the original problem in (2.20) can be transformed into a convex optimization problem and then be solved by the standard methods [8]. Since the binary constraint on subcarrier allocation has been relaxed, the result from this method can serve as a lower bound of the total transmit power, which also corresponds to the lower bound of the EE–SE trade-off curve.

An example of EE–SE trade-off curve is depicted in Fig. 2.4. In the example, there are 72 OFDM subcarriers and each subcarrier has a bandwidth of 15 kHz. The frequency-selective Rayleigh fading is according to the ITU Pedestrian-B model and has an identical average channel-gain-to-noise ratio of 20 dB. The noise power density is -174 dBm/Hz. There are two real-time users and four non-real-time users. It is assumed that the data rate requirement for each real-time user is 200 kbps and the minimum data rate requirement for each non-real-time user is 50 kbps. The power amplifier efficiency is set to be 38%. From the figure, it is clear that the EE–SE trade-off has a bell-shaped curve, which is also quasiconcave. Figure 2.4 also shows that the maximum EE point decreases while the maximum SE point increases with the static circuit power consumption. However, the latter is almost independent to the dynamic circuit power factor. These results demonstrate the effectiveness of the analysis in this section.

We further plot the optimal EE with different SE in Fig. 2.5. The results therein indicate that the optimal EE design also leads to the same throughput when the minimum data rate requirement of each user is no greater than a given threshold. However, when the minimum data rate requirement is greater than the threshold, it is better to operate at the minimum data rate point to maximize the system EE.

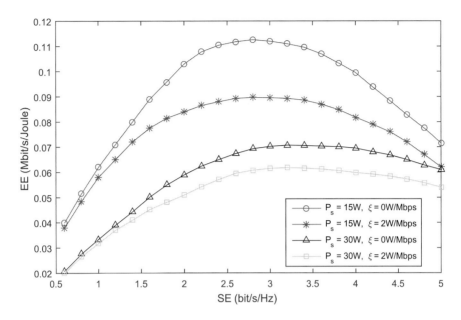

Figure 2.4 The EE–SE trade-off.

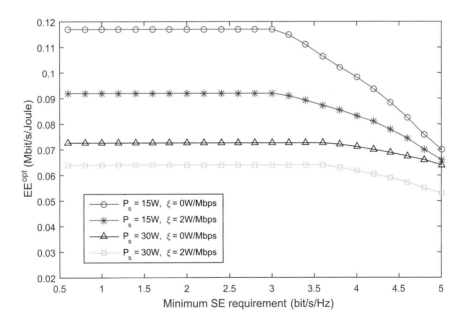

Figure 2.5 The optimal EE versus SE.

2.3.3 Further Discussion

In this section, we have discussed the EE–SE trade-off with the consideration of circuit power consumption. The fundamental results introduced here can be applied to many other different network models and scenarios. In [9], the EE–SE trade-off in downlink OFDMA networks has been further investigated by taking into account the user fairness. It shows that a certain EE degradation should be compromised as a cost to ensure users' rate fairness. The EE–SE trade-off has also been extended into the amplify-and-forward-based relay network, and the result shows that the EE–SE trade-off curve is also quasiconcave in this scenario [10]. In [11, 12], the EE–SE trade-off in type-I ARQ system and cognitive radio has been studied, respectively. The EE–SE study in homogeneous cellular networks with random distributed base stations shows that, with respect to the outage constraint, the EE–SE trade-off only occurs under a given network situation [13].

In [14, 15], an alternative multi-objective optimization approach has been applied to analyze the EE–SE trade-off. The quasiconcave EE–SE relation is also revealed in these works. Moreover, a practical theoretical framework for analyzing the EE–SE trade-off in single-cell cellular networks to achieve tractable results has been developed in [16]. Leveraging the stochastic geometry method, the framework has also been extended to the multicell scenario with the presence of intercell interference. The EE–SE trade-off and the corresponding upper and lower bounds have also been investigated for video transmission over mobile ad hoc networks in [17].

In addition to the EE–SE trade-off in practical networks, there are also three other fundamental trade-offs regarding the green radio design [18].

- Deployment efficiency (DE)–EE trade-off: The EE can be increased by shrinking the cell radius. However, higher density of base stations would certainly lead to increased deployment complexity, or decreased DE. Moreover, more deployment of base stations also results in larger circuit power consumption. With this regard, the energy-efficient design should also be in coordination with DE.
- Bandwidth-power trade-off: In traditional SE design, the relation between bandwidth and power is monotonic. However, considering EE design when circuit power consumption scales with the transmission bandwidth, their relation is non-monotonic. To this end, the optimal bandwidth and power trade-off should be revisited in the EE oriented design.
- Delay–EE trade-off: Without considering fixed-circuit power, the relation between per-bit power and packet delay is monotonically decreasing. However, as the packet delay increases, more circuit power consumption would be introduced. With this regard, the delay–EE relation would be no longer monotonic, and there would exist an optimal delay–EE trade-off.

2.4   EE Design in Orthogonal Systems

In the previous section, we introduced the EE–SE relation of an OFDMA network. In this section, we take an uplink OFDMA network as an example to show how to maximize the overall EE, summation of EE, and minimal EE of orthogonal systems. Through this section, we try to illustrate the problem formulation, mathematical analysis, and algorithm development for practical energy-efficient wireless resource allocation. Note that the detailed mathematical optimization theory is illustrated in Appendix 2.1.

Similar to Section 2.3, we also consider a single-cell uplink OFDMA system with K users and N subcarriers, each of which has a bandwidth of B Hz. In OFDMA systems, each user can transmit on several subcarriers, whereas each subcarrier can only be occupied by one user to ensure orthogonality.

The overall transmit rate of user k can be written as

$$R_k = \sum_{n=1}^{N} R_{k,n} = \sum_{n=1}^{N} \rho_{k,n} B \log_2 \left(1 + \frac{p_{k,n} h_{k,n}}{\rho_{k,n} B N_0} \right), \rho_{k,n} \in \{0,1\}. \qquad (2.31)$$

In the above, $p_{k,n}$ denotes the transmit power and $\rho_{k,n}$ stands for the binary channel allocation indicator.

In the OFDMA network, the overall power consumption of each user can be expressed as

$$P_k = \sum_{n=1}^{N} \left(\zeta p_{k,n} + \rho_{k,n} P_e \right) + P_{\text{fix}}. \qquad (2.32)$$

Here, we use P_e to denote the radio frequency (RF) circuit power consumption per used subcarrier and P_{fix} to denote the fixed power consumption irrespective of the number of used subcarriers. Note that the overall circuit power $P_c \stackrel{\text{def}}{=} \sum_{n=1}^{N} \rho_{k,n} P_e + P_{\text{fix}}$ is consistent with the definition in (2.2).

The EE for each individual user, say, user k, can be expressed as

$$\eta_k = \frac{R_k}{P_k}. \qquad (2.33)$$

Instead of the GEE in Section 2.3, we aim at maximizing the WSEE or WMEE in this section. Since the WSEE or WMEE is related to the EE of each individual user, this objective function has the merit of providing better insight on the EE trade-off among users. Users in practical cellular networks may have different priorities as well as different status of battery levels. Through maximizing the WSEE or WMEE, one can allocate more wireless resources to those users with higher priorities or lower battery levels by setting different weight values to different users. In this way, higher EEs and better quality of experience (QoE) could be achieved by those users. Moreover, we can also provide fairness among users by maximizing individual EE rather than maximizing the overall system EE or GEE.

2.4.1 Weighted Summation EE Maximization

Based on the discussion in the previous section, the maximization of WSEE can be mathematically defined as

$$\eta_{\text{ws}}^{\max} = \max_{(\boldsymbol{\rho}, \mathbf{p}) \in \Theta} \left\{ \sum_{k \in \mathcal{K}} \epsilon_k \eta_k \right\}, \qquad (2.34)$$

subject to

$$\rho_{k,n} \in \{0,1\}, \forall k, n, \tag{2.35}$$

$$\sum_{k=1}^{K} \rho_{k,n} \leqslant 1, \forall n, \tag{2.36}$$

$$\sum_{n=1}^{N} p_{k,n} \leqslant P_{\max}, \forall k, \tag{2.37}$$

$$p_{k,n} \geqslant 0, \forall k, n, \tag{2.38}$$

$$R_k \geqslant R_{\min}, \forall k. \tag{2.39}$$

In the above, ϵ_k is the weight for user k, Θ denotes the feasible subcarrier and power allocation set, (2.35) is the constraint of binary subcarrier allocation, (2.36) limits that each subcarrier can only be allocated to at most one user, (2.37) and (2.38) are the transmit power constraints of each user, and (2.39) is the QoS requirement.

We shall note that the WSEE maximization problem might be infeasible due to the potential conflicts of user QoS requirement and the maximum transmit power limitation. In this way, some users should be dropped through the admission control strategy. Here, admission control is not considered, and we assume that the problem is always feasible and try to find some effective algorithms to approach it.

We notice that the WSEE optimization problem aims to maximize the summation of several fractional functions, and thus is a sum-of-ratios optimization problem, which can be solved by the method introduced in Appendix 2.1. However, due to the binary subcarrier allocation indicator, $\rho_{k,n} \in \{0,1\}, \forall k, n$, the problem cannot be directly solved. To make it more tractable, we first relax $\rho_{k,n}$ into continuous variables within [0,1]. After that, the problem can be rewritten as

$$\tilde{\eta}_{\mathrm{ws}}^{\max} = \max_{(\tilde{\boldsymbol{\rho}}, \mathbf{p}) \in \tilde{\Theta}} \left\{ \sum_{k=1}^{K} \epsilon_k \tilde{\eta}_k \right\} = \max_{(\tilde{\boldsymbol{\rho}}, \mathbf{p}) \in \tilde{\Theta}} \left\{ \sum_{k=1}^{K} \epsilon_k \frac{\tilde{R}_k}{\tilde{P}_k} \right\}, \tag{2.40}$$

where $\tilde{R}_k$ and $\tilde{P}_k$ are the data rate and power consumption after the binary subcarrier allocation $\rho_{k,n}$ is replaced with $\tilde{\rho}_{k,n}$, respectively, and $\tilde{\Theta}$ is the feasible subcarrier allocation and power control region with constraint (2.35) being replaced by

$$0 \leqslant \tilde{\rho}_{k,n} \leqslant 1, \forall k, n. \tag{2.41}$$

We shall note that since the original constraint (2.35) has been relaxed, the solution to (2.40) serves as an upper bound of the original problem in (2.34), that is $\tilde{\eta}_{\mathrm{ws}}^{\max} \geq \eta_{\mathrm{ws}}^{\max}$, since $\Theta \subset \tilde{\Theta}$.

To tackle the problem, we first present the following important property.

PROPOSITION 1　*The generalized EE, $\tilde{\eta}_k (\tilde{\boldsymbol{\rho}}, \mathbf{p}) = \dfrac{\tilde{R}_k}{\tilde{P}_k}, \forall k$, is jointly quasiconcave in variables $\tilde{\rho}_{k,n}$ and $p_{k,n}, \forall k, n$.*

Proof: We will first show that $\tilde{R}_{k,n}$ is a concave function over $\tilde{\rho}_{k,n}$ and $p_{k,n}$. Define $x \triangleq B\tilde{\rho}_{k,n}, y \triangleq \frac{p_{k,n}h_{k,n}}{N_0}$, and $f(x,y) = -x\log_2\left(1 + \frac{y}{x}\right)$, then $\tilde{R}_{k,n} = -f(x,y)$. The Hessian of $f(x,y)$ is

$$
\mathbf{H} = \begin{bmatrix} \dfrac{y^2/x}{(x+y)^2} & -\dfrac{y}{(x+y)^2} \\[2ex] -\dfrac{y}{(x+y)^2} & \dfrac{x}{(x+y)^2} \end{bmatrix}, \tag{2.42}
$$

which is positive semi-defined, since the eigenvalues of $\mathbf{H}$ are

$$
\lambda_1 = 0,
$$
$$
\lambda_2 = \frac{x^2 + y^2}{x^3 + 2x^2y + xy^2} \geq 0.
$$

Therefore $\tilde{R}_{k,n}$ is concave. Clearly, $\tilde{R}_k$ is concave since it is a linear combination of $\tilde{R}_{k,n}$. We further define $\bar{\eta}_k(\tilde{\rho}, \mathbf{p}) = -\tilde{\eta}_k(\tilde{\rho}, \mathbf{p}) = -\dfrac{\tilde{R}_k}{\tilde{P}_k}$ and its sublevel sets as

$$
\tau_\alpha = \{\tilde{\rho}_{k,n} \geqslant 0, p_{k,n} \geqslant 0, \forall n | \bar{\eta}_k(\tilde{\rho}, \mathbf{p}) \leq \alpha\}, \tag{2.43}
$$

which equals

$$
\tau_\alpha = \left\{\tilde{\rho}_{k,n} \geqslant 0, p_{k,n} \geqslant 0, \forall n | -\alpha\tilde{P}_k - \tilde{R}_k \leqslant 0\right\}. \tag{2.44}
$$

We see that τ_α is convex due to the convexity of $-\alpha\tilde{P}_k - \tilde{R}_k$, which leads to the quasiconcavity of $\tilde{\eta}_k(\tilde{\rho}, \mathbf{p})$. This ends the proof.   □

According to this property, we can equivalently transform the problem into

$$
\max_{(\tilde{\rho}, \mathbf{p}) \in \tilde{\Theta}, \psi} \left\{ \sum_{k=1}^{K} \psi_k \right\}, \tag{2.45}
$$

subject to

$$
\psi_k \leq \epsilon_k \frac{\tilde{R}_k}{\tilde{P}_k}, \forall k. \tag{2.46}
$$

Then, based on the sum-of-ratios optimization method (please see the details in Appendix 2.1), we have the following theorem.

THEOREM 2   *If $(\tilde{\rho}^*, \mathbf{p}^*, \mathbf{\Psi}^*)$ is the optimal solution to (2.45), then there exists $\kappa^* = (\kappa_1^*, \kappa_2^*, \ldots, \kappa_K^*)$, such that $(\tilde{\rho}^*, \mathbf{p}^*)$ is the optimal solution to the following problem for $\kappa = \kappa^*$ and $\mathbf{\Psi} = \mathbf{\Psi}^*$*

$$
\max_{(\tilde{\rho}, \mathbf{p}) \in \tilde{\Theta}} \left\{ \sum_{k=1}^{K} \kappa_k \left(\epsilon_k \tilde{R}_k - \psi_k \tilde{P}_k \right) \right\}. \tag{2.47}
$$

And $(\widetilde{\boldsymbol{\rho}}^*, \mathbf{p}^*)$ *also satisfies the following system of equations for* $\boldsymbol{\kappa} = \boldsymbol{\kappa}^*$ *and* $\boldsymbol{\Psi} = \boldsymbol{\Psi}^*$:

$$\kappa_k = \frac{1}{\tilde{P}_k}, \forall k, \tag{2.48}$$

$$\epsilon_k \tilde{R}_k - \psi_k \tilde{P}_k = 0, \forall k. \tag{2.49}$$

According to this theorem, to solve the problem in (2.40) is equivalent to solving the problem in (2.47). Moreover, the latter can be iteratively solved by the following two nested steps: the inner step solves $(\widetilde{\boldsymbol{\rho}}^*, \mathbf{p}^*)$ for given $(\boldsymbol{\kappa}, \boldsymbol{\Psi})$, and the outer step updates the parameter $(\boldsymbol{\kappa}, \boldsymbol{\Psi})$ satisfying (2.48) and (2.49).

We now discuss the optimality of the algorithm. As shown in [24], the sum-of-ratios algorithm achieves at least a KKT point of the original problem, which means at least a local optimal solution can be obtained. In addition, as we have proved that the generalized EE is quasiconcave, the local optimal solution is also the global optimal one.

Although the global optimum of the problem in (2.40) can be obtained, the proposed algorithm is not exactly the optimal one due to the continuous relaxation of the binary channel allocation indicators. That is, the resulting optimal solution does not necessarily guarantee that $\rho_{k,n}$'s are binary variables. In what follows, we shall develop a sub optimal algorithm to approach the original problem based on the LDD method and take into account the binary variables as well.

The suboptimal solution is based on the resulting Lagrange parameters, κ_k^* and ψ_k^*, $\forall k$, in the upper-bound algorithm. Once κ_k^* and ψ_k^*, $\forall k$ are obtained, the objective function for the suboptimal problem can be redefined as

$$\max_{(\boldsymbol{\rho}, \mathbf{p}) \in \Theta} \sum_{k=1}^{K} \kappa_k^* \left(\epsilon_k R_k - \psi_k^* P_k \right). \tag{2.50}$$

In the above, $\boldsymbol{\rho}$ is now a binary variable.

We can rewrite the Lagrangian function of (2.50) as

$$
\begin{aligned}
L(\boldsymbol{\rho}, \mathbf{p}, \boldsymbol{\lambda}) = & \sum_{k=1}^{K} \kappa_k^* \left(\epsilon_k R_k - \psi_k^* P_k \right) + \sum_{n=1}^{N} \lambda_n^{(1)} \left(1 - \sum_{k=1}^{K} \rho_{k,n} \right) \\
& + \sum_{k=1}^{K} \lambda_k^{(2)} \left(P_{\max} - \sum_{n=1}^{N} p_{k,n} \right) + \sum_{k=1}^{K} \lambda_k^{(3)} \left(R_k - R_{\min} \right),
\end{aligned}
\tag{2.51}
$$

where $\lambda_n^{(1)}$, $\lambda_k^{(2)}$, and $\lambda_k^{(3)}$ are the Lagrange multipliers related to the corresponding constraints. Furthermore, the Lagrange dual problem can be expressed as

$$\min_{\boldsymbol{\lambda}} \max_{\boldsymbol{\rho}, \mathbf{p}} L(\boldsymbol{\rho}, \mathbf{p}, \boldsymbol{\lambda}), \tag{2.52}$$

subject to $\qquad \lambda_n^{(1)} \geqslant 0, \lambda_k^{(2)} \geqslant 0, \lambda_k^{(3)} \geqslant 0, \forall k, n. \tag{2.53}$

Since the problem is now non-convex because of the binary $\rho_{k,n}$, the LDD method and the KKT condition cannot be applied to find the optimal solution. In the following, we decouple the problem into two sub-problems: the channel assignment sub-problem and the power allocation sub-problem.

In the channel assignment sub-problem, we again relax the variables, ρ. In this way, we can utilize the KKT condition to find the optimal Lagrange multipliers, $\lambda_n^{(1)*}$, $\lambda_k^{(2)*}, \lambda_k^{(3)*}, \forall k, n$, and the optimal results, $\tilde{\rho}_{k,n}^*, p_{k,n}^*, \forall k, n$. By substituting these results into (2.52), the channel assignment problem can be formulated as

$$\max_{\rho} \sum_{k=1}^{K} \sum_{n=1}^{N} \rho_{k,n} u_{k,n} + v, \tag{2.54}$$

subject to

$$\rho_{k,n} \in \{0, 1\}, \forall k, n, \tag{2.55}$$

$$\sum_{k=1}^{K} \rho_{k,n} \leq 1, \forall n, \tag{2.56}$$

where

$$u_{k,n} = \left(\kappa_k^* \epsilon_k + \lambda_k^{(3)*}\right) R_{k,n}^* - \lambda_n^{(1)*} - \psi_k^* P_e, \forall k, n,$$

and

$$v = - \sum_{k=1}^{K} \sum_{n=1}^{N} \psi_k \left(\zeta p_{k,n}^* + P_{\text{fix}}\right) + \sum_{n=1}^{N} \lambda_n^{(1)*} + \sum_{k=1}^{K} \lambda_k^{(2)*} \left(P_{\text{max}} - \sum_{n=1}^{N} p_{k,n}^*\right)$$
$$- \sum_{k=1}^{K} \lambda_k^{(3)*} R_{\text{min}},$$

is a constant irrespective to $\rho_{k,n}$. Apparently, this problem is an assignment problem whose optimal solution can be obtained by the classical Hungarian algorithm with a computational complexity of $O(N^3)$ [35].

Once the channel assignment is obtained, each user's subcarrier set can be denoted as $\mathcal{S}_k$. Next, the power allocation problem can be solved for each user with the following problem

$$\max_{\mathbf{p}} \frac{\sum_{n \in \mathcal{S}_k} R_{k,n}}{\sum_{n \in \mathcal{S}_k} \left(\zeta p_{k,n} + P_e\right) + P_{\text{fix}}}, \forall k, \tag{2.57}$$

subject to

$$\sum_{n \in \mathcal{S}_k} p_{k,n} \leq P_{\text{max}}, \tag{2.58}$$

$$p_{k,n} \geq 0, n \in \mathcal{S}_k, \tag{2.59}$$

$$\sum_{n \in \mathcal{S}_k} R_{k,n} \geq R_{\text{min}}. \tag{2.60}$$

Note that this problem is a standard convex–concave fractional programming and therefore the Dinkelbach algorithm can be used to solve it optimally. The detailed theory and approaches are discussed in Appendix 2.1.

Although both the Hungarian algorithm and the Dinkelbach algorithm optimally solve the channel assignment and power control problem, respectively, the proposed solution to the problem in (2.50) is not necessarily optimal due to the problem decoupling as well as its non-convexity. However, at least a suboptimal solution can be obtained since the result corresponds to a KKT point of the problem. Moreover, the main merit of the proposed algorithm is its low computational complexity, which can be implemented in practical networks.

2.4.2   Maximum-Minimal EE Maximization

We now consider the WMEE maximization problem, which aims at maximizing the minimum EE. The mathematical optimization problem can be formulated as

$$\eta_{\text{mm}}^{\text{max}} = \max_{(\boldsymbol{\rho}, \mathbf{p}) \in \boldsymbol{\Theta}} \min_{k \in \mathcal{K}} \{\eta_k\}. \tag{2.61}$$

Again, binary subcarrier allocation indicator is involved. Thus, we need to relax the above problem into a continuous one, as

$$\tilde{\eta}_{\text{mm}}^{\text{max}} = \max_{(\tilde{\boldsymbol{\rho}}, \mathbf{p}) \in \tilde{\boldsymbol{\Theta}}} \min_{k \in \mathcal{K}} \left\{ \frac{\tilde{R}_k}{\tilde{P}_k} \right\}. \tag{2.62}$$

This problem is a generalized fraction problem (GFP) as described in Appendix 2.1. We can use the generalized Dinkelbach algorithm to solve it. However, the "max-min" operation renders it difficult to solve. In the following, we shall introduce an alternative algorithm to solve the problem.

First, according to the fact that a quasiconvex function attains its maximum on the vertex of a convex polyhedron [32], we have the following proposition.

PROPOSITION 2   *The WMEE problem is equivalent to*

$$\min_{k \in \mathcal{K}} \left\{ \frac{\tilde{R}_k}{\tilde{P}_k} \right\} = \min_{\mathbf{y} \in \mathcal{Y}} \left\{ \tilde{\eta}_{\text{mm}}^{\text{sum}} (\mathbf{y}, \tilde{\boldsymbol{\rho}}, \mathbf{p}) = \frac{\mathbf{y}\mathbf{R}^{\text{T}}}{\mathbf{y}\mathbf{P}^{\text{T}}} \right\}, \tag{2.63}$$

where $\mathcal{Y} \triangleq \left\{ (y_1, \ldots, y_K) \,|\, y_k \geq 0, \forall k, \sum_{k=1}^{K} y_k = 1 \right\}$, $\mathbf{P} = \left\{ \tilde{P}_1, \tilde{P}_2, \cdots \tilde{P}_K \right\}$, *and* $\mathbf{R} = \left\{ \tilde{R}_1, \tilde{R}_2, \cdots \tilde{R}_K \right\}$.

Furthermore, we can prove the quasiconcavity of $\tilde{\eta}_{\text{mm}}^{\text{sum}}$, as presented in the following proposition.

PROPOSITION 3 $\tilde{\eta}_{\text{mm}}^{\text{sum}} (\mathbf{y}, \tilde{\boldsymbol{\rho}}, \mathbf{p})$ *is quasiconcave over* $(\tilde{\boldsymbol{\rho}}, \mathbf{p})$ *for a given* $\mathbf{y} \in \mathcal{Y}$, *and quasilinear over* $\mathbf{y}$, *for a given* $(\tilde{\boldsymbol{\rho}}, \mathbf{p}) \in \tilde{\boldsymbol{\Theta}}$.

Proof: We can first prove that $\mathbf{y}\mathbf{R}^{\text{T}}$ is concave. Define $\bar{\eta}_{\text{mm}}^{\text{sum}} (\tilde{\boldsymbol{\rho}}, \mathbf{p}) = -\tilde{\eta}_{\text{mm}}^{\text{sum}} (\tilde{\boldsymbol{\rho}}, \mathbf{p}) = -\frac{\mathbf{y}\mathbf{R}^{\text{T}}}{\mathbf{y}\mathbf{P}^{\text{T}}}$, then its sublevel set can be expressed as

$$\tau_{\alpha} = \left\{ \tilde{\rho}_{k,n} \geq 0, p_{k,n} \geq 0, \forall k, n \,|\, \bar{\eta}_{\text{mm}}^{\text{sum}} (\tilde{\boldsymbol{\rho}}, \mathbf{p}) \leq \alpha \right\}, \tag{2.64}$$

which is equivalent to

$$\tau_\alpha = \left\{ \tilde{\rho}_{k,n} \geqslant 0, p_{k,n} \geqslant 0, \forall k, n | -\alpha \mathbf{y} \mathbf{P}^{\mathrm{T}} - \mathbf{y} \mathbf{R}^{\mathrm{T}} \leq 0 \right\}. \tag{2.65}$$

Since $-\alpha \mathbf{y} \mathbf{P}^{\mathrm{T}} - \mathbf{y} \mathbf{R}^{\mathrm{T}}$ is convex over $(\tilde{\rho}, p)$, τ_α is convex, which leads to the quasiconcavity of $\tilde{\eta}_{\mathrm{mm}}^{\mathrm{sum}}(\tilde{\rho}, \mathbf{p})$. Similarly, define the sublevel sets of $\tilde{\eta}_{\mathrm{mm}}^{\mathrm{sum}}(\mathbf{y}) = \dfrac{\mathbf{y} \mathbf{R}^{\mathrm{T}}}{\mathbf{y} \mathbf{P}^{\mathrm{T}}}$ for given $(\tilde{\rho}, p)$, as

$$S_\alpha = \left\{ y_k \geqslant 0, \forall k | \tilde{\eta}_y(\mathbf{y}) \leq \alpha \right\}, \tag{2.66}$$

which is equivalent to

$$S_\alpha = \left\{ y_k \geqslant 0, \forall k | \mathbf{y} \mathbf{R}^{\mathrm{T}} - \alpha \mathbf{y} \mathbf{P}^{\mathrm{T}} \leq 0 \right\}. \tag{2.67}$$

Since $\mathbf{y} \mathbf{R}^{\mathrm{T}} - \alpha \mathbf{y} \mathbf{P}^{\mathrm{T}}$ is both convex and concave over $y_k, \forall k$, it is quasilinear. $\qquad\square$

Since $\tilde{\eta}_{\mathrm{mm}}^{\mathrm{sum}}$ is quasiconcave, we can further apply the Sion's min-max theorem [33] to finally convert the problem into a better tractable one, as

$$\max_{(\tilde{\rho}, \mathbf{p}) \in \tilde{\Theta}} \min_{k \in \mathcal{K}} \left\{ \frac{\tilde{R}_k}{\tilde{P}_k} \right\} = \max_{(\tilde{\rho}, \mathbf{p}) \in \tilde{\Theta}} \min_{\mathbf{y} \in \mathcal{Y}} \left\{ \frac{\mathbf{y} \mathbf{R}^{\mathrm{T}}}{\mathbf{y} \mathbf{P}^{\mathrm{T}}} \right\} = \min_{\mathbf{y} \in \mathcal{Y}} \max_{(\tilde{\rho}, \mathbf{p}) \in \tilde{\Theta}} \left\{ \frac{\mathbf{y} \mathbf{R}^{\mathrm{T}}}{\mathbf{y} \mathbf{P}^{\mathrm{T}}} \right\}. \tag{2.68}$$

Now, we can solve the problem in (2.68) in two steps: the inner layer finds the optimal $\tilde{\rho}, \mathbf{p}$ for a given $\mathbf{y}$, and the outer layer solves the optimal $\mathbf{y}^*$. The detailed algorithm is similar to Algorithm 6 in Appendix 2.1.

Interestingly, the GEE, defined as the ratio of the overall achievable data rate and the overall power consumption, serves as an upper bound for the WMEE. This can be simply proved by the setting $y_k = \dfrac{1}{K}, \forall k$ in (2.63), as

$$\tilde{\eta}_{\mathrm{mm}}^{\mathrm{spup}} = \max_{(\tilde{\rho}, \mathbf{p}) \in \tilde{\Theta}} \left\{ \frac{\mathbf{y} \mathbf{R}^{\mathrm{T}}}{\mathbf{y} \mathbf{P}^{\mathrm{T}}} \right\} = \max_{(\tilde{\rho}, \mathbf{p}) \in \tilde{\Theta}} \left\{ \frac{\sum_{k=1}^{K} \tilde{R}_k}{\sum_{k=1}^{K} \tilde{P}_k} \right\} \geq \tilde{\eta}_{\mathrm{mm}}^{\mathrm{max}}. \tag{2.69}$$

The GEE can be achieved by the Dinkelbach algorithm (please see the details in Appendix 2.1). We denote

$$B(\theta) = \sum_{k=1}^{K} \tilde{R}_k - \theta \sum_{k=1}^{K} \tilde{P}_k. \tag{2.70}$$

Then $\tilde{\eta}_{\mathrm{mm}}^{\mathrm{spup}}$ is achieved if and only if $\max\limits_{(\tilde{\rho}, p) \in \tilde{\Theta}} B\left(\tilde{\eta}_{\mathrm{mm}}^{\mathrm{spup}}\right) = 0$.

Again, the above algorithm only achieves an upper bound to the WMEE maximization problem due to the relaxation of binary variables. Therefore, in the following, we shall develop a suboptimal algorithm to the original problem in (2.61) while considering the binary subcarrier allocation constraint. Based on the optimal EE achieved by the upper-bound algorithm, a suboptimal heuristic problem can be formulated, as

$$\max_{(\boldsymbol{\rho},\mathbf{p})\in\Theta} \min_{k\in\mathcal{K}} \left\{ R_k - \tilde{\eta}_{\text{mm}}^{\text{max}}(k) P_k \right\},$$

where $\tilde{\eta}_{\text{mm}}^{\text{max}}(k)$ is the k-th element in the optimal EE achieved by the upper-bound algorithm. The idea of the suboptimal heuristic algorithm is to achieve the EE as close as to the optimal EE for each individual user. We can rewrite the problem in a parametric optimization format, as

$$\max_{(\boldsymbol{\rho},\mathbf{p})\in\Theta} \tau, \tag{2.71}$$

subject to

$$R_k - \tilde{\eta}_{\text{mm}}^{\text{max}}(k) P_k \geq \tau, \forall k. \tag{2.72}$$

Then, the Lagrange multiplier method can be applied to solve it. The Lagrange function can be written as

$$\begin{aligned}
L(\boldsymbol{\rho},\mathbf{p},\boldsymbol{\lambda}) =& \tau + \sum_{n=1}^{N} \lambda_n^{(1)} \left(1 - \sum_{k=1}^{K} \rho_{k,n} \right) \\
&+ \sum_{k=1}^{K} \lambda_k^{(2)} \left(P_{\text{max}} - \sum_{n=1}^{N} p_{k,n} \right) \\
&+ \sum_{k=1}^{K} \lambda_k^{(3)} (R_k - R_{\text{min}}) \\
&+ \sum_{k=1}^{K} \lambda_k^{(4)} \left[\left(R_k - \tilde{\eta}_{\text{mm}}^{\text{max}}(k) P_k \right) - \tau \right],
\end{aligned} \tag{2.73}$$

where $\lambda_n^{(1)}$, $\lambda_k^{(2)}$, $\lambda_k^{(3)}$ and $\lambda_k^{(4)}$ are Lagrange multipliers related to the corresponding constraints in (2.35)–(2.39). Furthermore, the Lagrange dual decomposition is

$$\min_{\boldsymbol{\lambda}} \max_{\boldsymbol{\rho},\mathbf{p}} L(\boldsymbol{\rho},\mathbf{p},\boldsymbol{\lambda}), \tag{2.74}$$

subject to

$$\lambda_n^{(1)} \geqslant 0, \lambda_k^{(2)} \geqslant 0, \lambda_k^{(3)} \geqslant 0, \lambda_k^{(4)} \geqslant 0, \forall k, n. \tag{2.75}$$

This problem can be easily solved by a similar method to the WSEE problem.

2.4.3   Numerical Results

We now present simulation results to evaluate the performance of the algorithms developed in this section. A simple two-user scenario is considered. The base station has a radius of 500 m. There are 2–8 subcarriers in the system. The channel of each user follows independent and identically distributed (i.i.d.) Rayleigh fading model. Other parameters are summarized in Table 2.2.

Table 2.2 Simulation parameters

Parameter	Value
Cell radius	500 m
subcarriers bandwidth, B	$\{1,2,3,4,5\} * 0.25$ MHz
Noise spectral density, N_0	-174 dBm/Hz
Path loss model	$128.1+37.6\log_{10}(d[\text{km}])$ dB
Shadowing standard deviation	10 dB
$R_{\min}$	100 kbps
Number of users, K	2
Number of subcarriers, N	2–8
P_{fix}	1 W–3 W
P_{e}	0.42 W
Power efficiency	50%

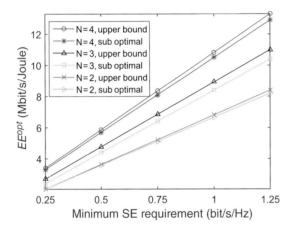

Figure 2.6 Weighted-sum EE versus subcarrier bandwidth. $P_{\max} = 26.7$ dBm, $P_{\text{fix}} = 2.13$ W.

In Fig. 2.6, we show the performance of WSEE algorithms where the weights are set equal for all users. Particularly, the upper-bound algorithm and the suboptimal algorithm are compared. As we can observe from the figure, the performance gap between the upper-bound algorithm and the suboptimal algorithm is very small, which verifies the effectiveness of our proposed algorithm. In addition, the WSEE increases as the number of subcarriers increases. This is because that more subcarriers will bring about more communication freedom for EE improvement.

In Fig. 2.7, the algorithms for maximizing the WMEE are compared, i.e., the upper-bound algorithm, the algorithm to maximize the GEE, and the suboptimal algorithm. From the figure, the GEE achieves the highest EE performance, as has been discussed before. However, this upper bound is very loose. On the other hand, the upper-bound algorithm can achieve a tight upper bound as we can observe that the WMEE gap between the upper-bound algorithm and suboptimal algorithm is very small. This result

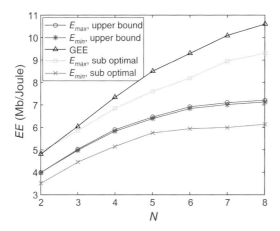

Figure 2.7 The minimum EE versus number of subcarriers. $P_{\max} = 26.7$ dBm, $P_{\text{fix}} = 2.13$ W, $B = 1.25$ MHz.

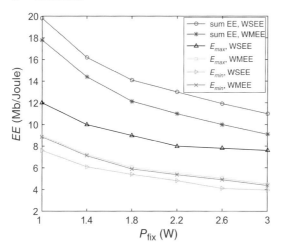

Figure 2.8 EE comparison of different algorithms. $N = 4$, $P_{\max} = 26.7$ dBm, $B = 1.25$ MHz.

verifies the effectiveness of our proposed algorithms for WMEE maximization. Again, from the figure, as the subcarrier number, N, increases, the EE performance increases.

We further compare the sum EE, the maximum EE, and the minimum EE of the two users with WSEE and WMEE maximization algorithms in Fig. 2.8. Both upper bounds for WSEE and WMEE maximization algorithms are plotted. From the figure, the WSEE algorithm has a better sum EE than the WMEE algorithm. However, from this result, WSEE algorithm doesn't necessarily guarantee user fairness. On the other hand, although the sum EE in the WMEE algorithm is smaller than that in the WSEE algorithm, the two users achieve almost the same EE as can be observed from the figure. In this regard, the WMEE maximization can achieve fair EE performance among users.

Our proposal in this section indeed provides an insightful reference for network operator to choose which algorithm to be used in practice.

2.5   EE Design in Non-Orthogonal Systems

In the previous section, we illustrated how to use the optimization theory to maximize various EEs for orthogonal systems, such as OFDMA networks. In this section, we will further analyze the EE trade-off problem in non-orthogonal systems, where non-convex optimization theory will be applied.

As shown in Fig. 2.9, we consider a network with K pairs of users communicating with each other, denoted as $\mathcal{K} = \{1, 2, \ldots, K\}$. Let $\{\mathrm{Tx}_1, \mathrm{Tx}_2, \ldots, \mathrm{Tx}_K\}$ and $\{\mathrm{Rx}_1, \mathrm{Rx}_2, \ldots, \mathrm{Rx}_K\}$ denote the transmitter and the receiver, respectively, and B denote the overall bandwidth of the considered system. Moreover, let $h_{m,k}$ denote the channel power gain between Tx_m and Rx_k.

The capacity region of interference channel is still an open challenge and the best known achievable region is the Han–Kobayashi region [36], which is very difficult to achieve. Here, we simply treat interference as noise, which serves as a lower bound of the data rate for interference networks.

The data rate of user pair k in the interference network can be expressed as

$$R_k = B\log_2\left(1 + \frac{p_k h_{k,k}}{\sum\limits_{m=1, m\neq k}^{K} p_m h_{m,k} + BN_0}\right), \forall k, \tag{2.76}$$

where p_k is the transmit power of the transmitter k and N_0 represents the variance of the noise spectral density. Denote $\mathbf{p} = \{p_k\}$. The EE of each user pair can be defined as

$$\eta_k = \frac{R_k}{p_k + P_c}, \forall k. \tag{2.77}$$

where p_k and P_c are the RF power consumption and the fixed circuit power consumption respectively. Here, for notational simplicity, we have omitted the power amplifier efficiency, which would not lose the generality of the following analysis.

Instead of maximizing a single-objective EE function, the emphasis of this section is to maximize the EE of each user from the perspective of multi-objective optimization

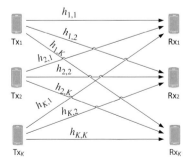

Figure 2.9 The model of non-orthogonal systems.

theory, as detailed in Appendix 2.1. Therefore, the optimization problem can be mathematically formulated as

$$\eta_{\max} = \max_{\mathbf{p} \in \mathcal{P}} \{\eta_1, \eta_2, \ldots, \eta_K\}, \tag{2.78}$$

where $\mathcal{P}$ is the feasible power allocation strategy set satisfying

$$0 \leq p_k \leq P_{\max}, \forall k, \tag{2.79}$$

$$R_k \geq R_{\min}, \forall k. \tag{2.80}$$

In the above, (2.79) limits the maximum transmit power for each user and (2.80) guarantees the minimum data rate for each user.

2.5.1   Utopia EE

We use the weighted Tchebycheff method to solve (2.78) (please see the details in Appendix 2.1). We shall first introduce the utopia EE for each individual user, as defined in the following.

Definition 1. The utopia EE for user k, η_k^{u}, is defined as the maximal EE this user could achieve, i.e., $\eta_k^{\mathrm{u}} = \max_{\eta \in \mathcal{F}} \{\eta_k\}$, where $\mathcal{F}$ denotes the set of all Pareto-optimal EEs.

According to the definition, the utopia EE can be solved by the following problem

$$\eta_k^{\mathrm{u}} = \max \eta_k, \tag{2.81}$$

subject to (2.79) and (2.80).

The above problem is a single-ratio fractional program, therefore Dinkelbach-like algorithm can be utilized to solve it. Let us define $\mathcal{U}(\varphi) = \max_{\mathbf{p} \in \mathcal{P}} \{R_k - \varphi P_k\}$, then η_k^{u} can be achieved if and only if $\mathcal{U}(\eta_k^{\mathrm{u}}) = 0$.

Unfortunately, it is still difficult to obtain the optimum value since the objective function is non-concave due to the interference in the denominator of R_k. However, this problem has a difference-of-convex (d.c.) structure, which can be solved by the method elaborated as follows.

The objective function can be rewritten as

$$f(\mathbf{p}) = R_k - \varphi P_k \overset{\Delta}{=} f_{\mathrm{cave1}}(\mathbf{p}) - f_{\mathrm{cave2}}(\mathbf{p}), \tag{2.82}$$

where

$$f_{\mathrm{cave1}}(\mathbf{p}) = B\log_2\left(\sum_{m=1}^{K} p_m h_{m,k} + BN_0\right),$$

and

$$f_{\mathrm{cave2}}(\mathbf{p}) = B\log_2\left(\sum_{m=1, m \neq k}^{K} p_m h_{m,k} + BN_0\right) + \varphi(p_k + P_{\mathrm{c}}).$$

We can observe that both $f_{\text{cave1}}(\mathbf{p})$ and $f_{\text{cave2}}(\mathbf{p})$ are strictly concave on $\mathbf{p}$. Therefore, the original problem can be expressed as the following d.c. problem

$$\max_{\mathbf{p}\in\mathcal{P}} \{f_{\text{cave1}}(\mathbf{p}) - f_{\text{cave2}}(\mathbf{p})\}. \tag{2.83}$$

There are many algorithms that can effectively solve the d.c. problem, such as the d.c. algorithm (DCA) [37]. In this problem, the function $f_{\text{cave2}}(\mathbf{p})$ is differentiable. Therefore, the concave-convex procedure (CCCP) method can be utilized to solve the problem in (2.83) [38]. Therefore, we can develop the following theorem to solve the above problem by following the majorization-maximization approximation.

THEOREM 3 *The problem in (2.83) can be iteratively solved by the following concave programming*

$$\mathbf{p}^{(j+1)} = \arg\max_{\mathbf{p}\in\mathcal{P}} \left\{ f_{\text{cave1}}(\mathbf{p}) - \mathbf{p}^{\mathrm{T}} * \nabla f_{\text{cave2}}\left(\mathbf{p}^{(j)}\right) \right\}, \tag{2.84}$$

where $\mathbf{p}^{\mathrm{T}}$ *denotes the transpose of* $\mathbf{p}$ *and* $\nabla f_{\text{cave2}}\left(\mathbf{p}^{(j)}\right) \triangleq \left[\nabla_1^{(j)}, \nabla_2^{(j)}, \ldots, \nabla_K^{(j)}\right]$ *denotes the gradient of* $f_{\text{cave2}}(\mathbf{p})$ *at* $\mathbf{p}^{(j)}$, *where*

$$\nabla_i^{(j)} = \begin{cases} \dfrac{Bh_{i,k}}{\left(\displaystyle\sum_{m=1,m\neq k}^{K} p_m^{(j)} h_{m,k} + BN_0\right)\ln 2}, & \forall i \neq k, \\[4ex] \varphi, & i = k. \end{cases}$$

In the above theorem, we can further prove that the first part in (2.84), $f_{\text{cave1}}(\mathbf{p})$, is concave, whereas the second part, $-\mathbf{p}^{\mathrm{T}} * \nabla f_{\text{cave2}}\left(\mathbf{p}^{(j)}\right)$, is linear. Therefore, it is a standard concave optimization problem and classic convex optimization methods can be used to solve it. The detailed procedures of applying the CCCP approach to solve our problem can be summarized in Algorithm 1.

Algorithm 1 CCCP for the d.c. problem

1: **Initialize**
2:    Set $\epsilon > 0$, $j = 0$, and $\forall \mathbf{p}^{(0)} \in \mathcal{P}$.
3: **Do**
4:    $\mathbf{p}^{(j+1)} = \arg\max_{\mathbf{p}\in\mathcal{P}} \left\{ f_{\text{cave1}}(\mathbf{p}) - \mathbf{p}^{\mathrm{T}} * \nabla f_{\text{cave2}}\left(\mathbf{p}^{(j)}\right) \right\}.$
5:    $j = j + 1$.
6: **Until** $\left\| \mathbf{p}^{(j)} - \mathbf{p}^{(j-1)} \right\| < \epsilon$.

Furthermore, the detailed procedures of our algorithm to find the Utopia EE for user k are summarized in Algorithm 2.

Algorithm 2 The algorithm to find the Utopia EE for user k.

1: **Initialize**

2:    Set $\epsilon > 0$, $j = 0$, and $\forall \mathbf{p}^{(0)} \in \mathcal{P}$.

3: **Do**

4:    $\varphi_j = \eta_k^{(j)}$.

5:    Calculate $\mathbf{p}^{(j+1)} = \arg \max_{\mathbf{p} \in \mathcal{P}} \left\{ R_k - \varphi_j P_k \right\}$ by CCCP.

6:    $j = j + 1$.

7: **Until** $\left| \max_{\mathbf{p} \in \mathcal{P}} \left\{ R_k - \varphi_{j-1} P_k \right\} \right| < \epsilon$.

We now show that the above algorithm can converge. First, according to the CCCP, the objective function in (2.84) is non-decreasing on the generated sequence $\left\{ \mathbf{p}^{(j)} \right\}$, which will eventually converge to the stationary point $\mathbf{p}^{(\infty)}$ when $\mathbf{p}^{(j+1)} = \mathbf{p}^{(j)}$. From [38], the stationary point also satisfies the Karush–Kuhn–Tucker (KKT) conditions of the problem in (2.83). Therefore, by jointly utilizing the Dinkelbach method and the CCCP, at least a local optimal solution can be obtained. In fact, it is quite challenging to prove the global optimality. However, from [37, 39], the global optimum can be always achieved if the starting point is appropriately chosen.

2.5.2   Pareto-Optimal EE

After the utopia EE for each user is attained, we can apply the weighted Tchebycheff method to convert the multi-objective optimization problem into a single-objective function one, as

$$
\begin{aligned}
\eta^{\text{wt}} &= \min_{\mathbf{p} \in \mathcal{P}} \max_{k \in \mathcal{K}} \left\{ \phi_k \left(\eta_k^{\text{u}} - \frac{R_k}{P_k} \right) \right\} \\
&\triangleq \min_{\mathbf{p} \in \mathcal{P}} \max_{k \in \mathcal{K}} \left\{ \frac{\phi_k \left(\eta_k^{\text{u}} P_k - R_k \right)}{P_k} \right\},
\end{aligned}
\tag{2.85}
$$

where ϕ_k is an arbitrary positive weight for user k. The details of the weighted-Tchebycheff method are discussed in Appendix 2.1.

The utopia EE and the Pareto-optimal EE in a two-user case is illustrated in Fig. 2.10. In the figure, the shadowed area depicts all achievable EE for the two users in the problem in (2.78) and its boundary is the Pareto-optimal EE, as illustrated by the solid curve. Furthermore, η_1^{u} and η_2^{u} are the utopia EE for each user, which is exactly the maximum EE that the user can achieve. This figure also clearly illustrates how the weighted Tchebycheff method achieves the Pareto-optimal EE by minimizing the distance between the utopia EE point and achieved EE point, i.e., η_1^* and η_2^*. Moreover, by varying the weight $\phi_k, k = 1, 2$, all Pareto-optimal EEs on the boundary can be achieved.

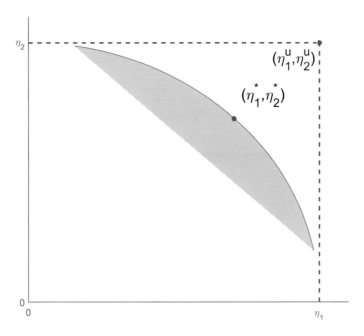

Figure 2.10 The weighted Tchebycheff method for a two-user case.

We now introduce an algorithm to solve the problem in (2.85) for given ϕ_k and η_k^{u}. The objective function in the weighted Tchebycheff method can be rewritten as

$$\eta^{\mathrm{wt}} = \min_{\mathbf{p}\in\mathcal{P}} \max_{k\in\mathcal{K}} \left\{ \frac{\phi_k \left(\eta_k^{\mathrm{u}} P_k - R_k \right)}{P_k} \right\}. \tag{2.86}$$

The above problem is also a fractional optimization and can be solved by the Dinkelbach method [19], which is discussed in Appendix 2.1 in detail. We have the following theorem, whose proof is similar to that in [26].

THEOREM 4 *Define*

$$V\left(\alpha\right) = \min_{\mathbf{p}\in\mathcal{P}} \max_{k\in\mathcal{K}} \left\{ \phi_k \left(\eta_k^{\mathrm{u}} P_k - R_k \right) - \alpha P_k \right\}, \tag{2.87}$$

then η^{wt} is achieved if and only if $V\left(\eta^{\mathrm{wt}}\right) = 0$.

Unfortunately, the problem in (2.87) is also non-convex due to the involved interference in R_k. In the following, we develop a suboptimal algorithm to achieve the local optimum point by jointly utilizing the Lagrange method and the CCCP method. First, define $l = \max_{k\in\mathcal{K}} \left\{ \phi_k \left(\eta_k^{\mathrm{u}} P_k - R_k \right) - \alpha P_k \right\}$. Then, based on the parametric method, the problem in (2.87) can be equivalently transformed into

$$\min_{\mathbf{p}\in\mathcal{P}} l, \tag{2.88}$$

subject to

$$l \geq \phi_k \left(\eta_k^{\mathrm{u}} P_k - R_k \right) - \alpha P_k, \forall k. \tag{2.89}$$

According to the LDD method, we can express the Lagrangian function of the above problem as

$$L \left(\mathbf{p}, l, \boldsymbol{\lambda}, \boldsymbol{\mu}, \boldsymbol{\nu} \right) = l + \sum_{k=1}^{K} \lambda_k \left\{ \phi_k \left(\eta_k^{\mathrm{u}} P_k - R_k \right) - \alpha P_k - l \right\}$$

$$+ \sum_{k=1}^{K} \mu_k \left(p_k - P_{\max} \right)$$

$$+ \sum_{k=1}^{K} \nu_k \left\{ \left(2^{\frac{R_{\min}}{B}} - 1 \right) \left(\sum_{m=1, m \neq k}^{K} p_m h_{m,k} + BN_0 \right) - p_k h_{k,k} \right\},$$

where $\boldsymbol{\lambda}, \boldsymbol{\mu}, \boldsymbol{\nu}$ are Lagrange multiplier vectors corresponding to the respective constraints. Furthermore, the Lagrangian dual problem can be written as

$$\max_{\boldsymbol{\lambda} \geq 0, \boldsymbol{\mu} \geq 0, \boldsymbol{\nu} \geq 0} \min_{\mathbf{p} \geq 0, l} L \left(\mathbf{p}, l, \boldsymbol{\lambda}, \boldsymbol{\mu}, \boldsymbol{\nu} \right). \tag{2.90}$$

Then, the Algorithm 3 is presented to solve the above problem, which contains the inner loop and the outer loop.

Inner Loop

The inner loop aims to solve $\min_{\mathbf{p} \geq 0, l} L \left(\mathbf{p}, l \right)$ for a given $\left(\boldsymbol{\lambda}, \boldsymbol{\mu}, \boldsymbol{\nu} \right)$. It is also non-convex but has a d.c. structure. Therefore, we can decompose $L \left(\mathbf{p}, l \right)$ into a subtraction of a concave function and a linear function, as

$$L \left(\mathbf{p}, l \right) = L_{\mathrm{vex1}} \left(\mathbf{p}, l \right) - L_{\mathrm{vex2}} \left(\mathbf{p} \right), \tag{2.91}$$

where

$$L_{\mathrm{vex1}} \left(\mathbf{p}, l \right) = l - \sum_{k=1}^{K} \lambda_k \left\{ \phi_k B \log_2 \left(\sum_{m=1}^{K} p_m h_{m,k} + BN_0 \right) \right\}$$

$$+ \sum_{k=1}^{K} \lambda_k \left\{ \phi_k \eta_k^{\mathrm{u}} P_k - \alpha P_k - l \right\} + \sum_{k=1}^{K} \mu_k \left(p_k - P_{\max} \right)$$

$$+ \sum_{k=1}^{K} \nu_k \left\{ \left(2^{\frac{R_{\min}}{B}} - 1 \right) \left(\sum_{m=1, m \neq k}^{K} p_m h_{m,k} + BN_0 \right) - p_k h_{k,k} \right\},$$

and

$$L_{\mathrm{vex2}} \left(\mathbf{p} \right) = - \sum_{k=1}^{K} \lambda_k \left\{ \phi_k B \log_2 \left(\sum_{m=1, m \neq k}^{K} p_m h_{m,k} + BN_0 \right) \right\}.$$

Now, $\min\limits_{\mathbf{p}\geq 0,l} L(\mathbf{p},l)$ has been expressed as the minimization of a d.c. function in a convex set, as

$$\min_{\mathbf{p}\in\mathcal{P},l}\{L_{\text{vex1}}(\mathbf{p},l) - L_{\text{vex2}}(\mathbf{p})\}. \qquad (2.92)$$

The CCCP method can be used to effectively solve it. Similar to Theorem 3, we have the following theorem.

THEOREM 5   *The problem in* (2.92) *can be iteratively solved by the following sequential convex programming*

$$\mathbf{p}^{(j+1)} = \arg\min_{\mathbf{p}\in\mathcal{P}}\left\{L_{\text{vex1}}(\mathbf{p},l) - \mathbf{p}^{\mathrm{T}} * \nabla L_{\text{vex2}}\left(\mathbf{p}^{(j)}\right)\right\}, \qquad (2.93)$$

where $L_{\text{vex2}}\left(\mathbf{p}^{(j)}\right) \triangleq \left[\nabla_1^{(j)}, \nabla_2^{(j)}, \ldots, \nabla_K^{(j)}\right]$ *denotes the gradient of* $L_{\text{vex2}}(\mathbf{p})$ *at* $\mathbf{p}^{(j)}$, *where*

$$\nabla_i^{(j)} = -\sum_{k=1,k\neq i}^{K}\lambda_k\phi_k B\left\{\frac{h_{i,k}}{\left(\sum\limits_{m=1,m\neq k}^{K} p_m^{(j)} h_{m,k} + BN_0\right)\ln 2}\right\}.$$

We can also prove that the first part of (2.93), $L_{\text{vex1}}(\mathbf{p},l)$, is convex, and the second part of (2.93), $-\mathbf{p}^{\mathrm{T}} * \nabla L_{\text{vex2}}\left(\mathbf{p}^{(j)}\right)$, is linear. Therefore, (2.93) is a standard convex optimization problem and can be easily solved.

Outer Loop
Since the Lagrangian dual function is differentiable, we utilize the subgradient method to solve the outer loop, i.e., finding the optimal $(\boldsymbol{\lambda}, \boldsymbol{\mu}, \boldsymbol{\nu})$ for given $(\mathbf{p}, l)$. The subgradient update equations can be written as

$$\begin{aligned}
\lambda_k &= \left[\lambda_k + \varsigma_1\left\{\phi_k\left(\eta_k^{\mathrm{u}} P_k - R_k\right) - \alpha P_k - l\right\}\right]^+, \forall k, \\
\mu_k &= \left[\mu_k + \varsigma_2\left(p_k - P_{\max}\right)\right]^+, \forall k, \qquad (2.94) \\
\nu_k &= \left[\nu_k + \varsigma_3\left(\vartheta - p_k h_{k,k}\right)\right]^+, \forall k,
\end{aligned}$$

where $\vartheta = \left(2^{\frac{R_{\min}}{B}} - 1\right)\left(\sum_{m=1,m\neq k}^{K} p_m h_{m,k} + BN_0\right)$, ς_1, ς_2, and ς_3 are all positive stepsizes.

We now summarize the proposed algorithm to achieve the Pareto-optimal EE, which contains three nested steps. The first step leverages the Dinkelbach method to update α in (2.87) until $V(\alpha) = 0$. The second step is a standard subgradient method to update the Lagrangian multipliers in (2.94). The third step utilizes the CCCP method to solve (2.93). The detailed procedures of the three-step algorithm can be found in Algorithm 3.

Again, since all three steps can converge, the convergence of the proposed algorithm can be guaranteed. However, due to the non-convexity of the optimization problem, the global optimality cannot be ensured although it converges. However, our algorithm can at least achieve a local optimal point that satisfies the KKT condition. Nevertheless,

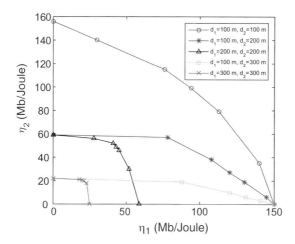

Figure 2.11 The EE trade-off with different communication distances. © 2016 IEEE. Reprinted, with permission, from Yu, G., 2016, "Energy Efficiency Tradeoff in Interference Channels," IEEE Access, Vol. 14, No. 6, pp. 3207–3218.

if carefully choosing the start point in the CCCP method, global optimum is often attainable.

Algorithm 3 The algorithm to find the Pareto-optimal EE.

1: **Initialize**

2:     Set $\epsilon_1, \epsilon_2 > 0$, $j = 0$, and $\forall \mathbf{p}^{(0)} \in \mathcal{P}$.

3:     Calculate $\eta_k^{\mathrm{u}}, \forall k$.

4: **Do**

5:     Calculate $\alpha_j = \max\limits_{k \in \mathcal{K}} \left\{ \dfrac{\phi_k \left(\eta_k^{\mathrm{u}} P_k^{(j)} - R_k^{(j)} \right)}{P_k^{(j)}} \right\}$.

6:     Initialize $\boldsymbol{\lambda}, \boldsymbol{\mu}, \boldsymbol{\nu}, i = 0, l^{(0)} = 0$.

7:     **Do**

8:        Using the CCCP to find $(\mathbf{p}^*, l^*) = \arg\min\limits_{\mathbf{p} \in \mathcal{P}, l} \{ L_{\mathrm{vex1}}(\mathbf{p}, l) - L_{\mathrm{vex2}}(\mathbf{p}) \}$.

9:        Update $\boldsymbol{\lambda}, \boldsymbol{\mu}, \boldsymbol{\nu}, i = i + 1, \mathbf{p}^{(i)} = \mathbf{p}^*, l^{(i)} = l^*$.

10:    **Until** $\left\| \left(\mathbf{p}^{(i)}, l^{(i)} \right) - \left(\mathbf{p}^{(i-1)}, l^{(i-1)} \right) \right\| < \epsilon_1$.

11:    $j = j + 1, \mathbf{p}^{(j)} = \mathbf{p}^*$.

12: **Until** $\left| \min\limits_{\mathbf{p} \in \mathcal{P}} \max\limits_{k \in \mathcal{K}} \left\{ \phi_k \left(\eta_k^{\mathrm{u}} P_k - R_k \right) - \alpha_{j-1} P_k \right\} \right| < \epsilon_2$.

2.5.3   Numerical Results

A system with two pairs of users is considered. Let d_i denote the distance between the i-th user pair, which varies from 100–300 m. We assume that the channel models

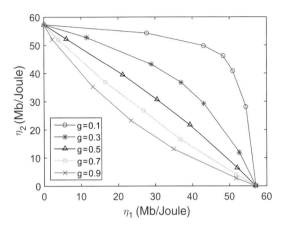

Figure 2.12 The EE trade-off in the two-user case with different g. © 2016 IEEE. Reprinted, with permission, from Yu, G., 2016, "Energy Efficiency Tradeoff in Interference Channels," IEEE Access, Vol. 14, No. 6, pp. 3207–3218.

between users are according to i.i.d Rayleigh fading and the path loss exponent is 4. Furthermore, we assume that each user has the same maximum transmit power, $P_{\max}$, and the fixed circuit power consumption is assumed to be 24 dBm. The minimum data rate requirement, $R_{\min}$, is not considered in the simulation. Moreover, for simplicity, we further assume that the average interference power gain to all other users are identical for each user, that is, $\bar{h}_{k,m} = g\bar{h}_{k,k}, \forall m \neq k$, where g is the interference-to-signal ratio.

Figure 2.11 plots the Pareto-optimal EE for the two-user case with different communication distances, where d_1 and d_2 denote the communication distances between user pairs. The EE trade-off can be easily observed from the figure. Moreover, the EE decreases with the communication distance. We further compare the EE trade-off curves with different interference-to-signal ratios, g, in Fig. 2.12. Here, both users have the same communication distance of 200 m. Also, the EE trade-off between the two users can be easily observed from the figure. Both users' EE decreases with g since large interference will certainly degrade both the spectral and energy efficiency.

Appendix 2.1 Optimization Theory for EE Design

This section discusses the mathematical optimization theory for EE design. Different from SE optimization problems, EE optimization problems generally involve fractional objective functions, which are known as fractional programming and are challenging to solve in general. New mathematical optimization theory and method are required to deal with it. The fractional programming can be effectively solved by the Dinkelbach approach, by converting the fractional objective function into a subtractive form. In this section, we will introduce the basic methodology of fractional programming and its extensions, such as sum-of-ratios optimization and generalized fractional programming

(GFP). We will also introduce the multi-objective optimization theory and attempt to deal with the EE trade-off problem.

A2.1.1   Fractional Programming and the Dinkelbach Algorithm

Fractional programming solves the mathematical optimization problems with fractional objective functions, which are in general nonlinear and non-convex. The ratio function to be optimized often reflects some kind of efficiency, defined as the ratio of utility and cost. In the EE optimization problem, the utility corresponds to the achievable data rate while the cost corresponds to the consumed power.

Definition 2 (Fractional programming problem). Let us define $f, g, h_j, j = 1, \ldots, M$ as some real-valued functions on a set $\mathcal{S}_0 \in \mathbb{R}^n$. The nonlinear programming

$$\max_x \frac{f(x)}{g(x)}, \tag{A2.1}$$

subject to

$$h_j(x) \leq 0, j = 1, \ldots, M,$$
$$g(x) > 0,$$

is a fractional programming. Specifically, when f is nonnegative and concave, g is positive and convex, and $\mathcal{S}_0$ is a convex set, the problem is a concave fractional programming.

The Dinkelbach algorithm is effective to solve fractional programming problems. Define an auxiliary function

$$F(\lambda) = \max_x \{f(x) - \lambda g(x)\}. \tag{A2.2}$$

Then, the following theorem presents the solution to the problem in (A2.1).

THEOREM 6 x^* *is the optimal solution to the fractional programming in* (A2.1) *if and only if* x^* *also solves the problem in* (A2.2) *with the same constraints and when* $F(\lambda) = 0$.

According to Theorem 6, the fractional objective function now can be transformed into a subtractive function and can be solved by standard methods. Particularly, if f is concave and g is convex, the problem in (A2.2) can be solved by classical convex optimization methods, such as interior method, subgradient method, etc. To solve (A2.1), an iterative updating is further required to find the root of $F(\lambda) = 0$. The Dinkelbach algorithm [19] is an effective way to accomplish this, as presented in Algorithm 4.

Algorithm 4 Dinkelbach algorithm to solve single-ratio fractional programming

---
1: Initialize $x_0 \in \mathcal{S}_0$. Set $\varepsilon > 0$;
2: Compute $\lambda_1 = f(x_0)/g(x_0)$. Set $k = 1$;
3: **while** $|F(\lambda_k)| \geq \varepsilon$ **do**
4:      $x^* = \arg\max_{x \in \mathcal{S}_0} \{ f(x) - \lambda_k g(x) \}$
5:      $k = k + 1$;
6:      $\lambda_k = f(x^*)/g(x^*)$;
7: end **while**

The Dinkelbach algorithm can deal with all fractional programming problems, no matter whether the problem is concave. However, for concave fractional programming, a simple approach can be used to solve it.

Let us denote $t = \dfrac{1}{g(x)}$, then the concave fractional program can be equivalently transformed into the following parameter-free concave problem

$$\max_{\{x,t\}} tf(x), \tag{A2.3}$$

subject to

$$h_j(x) \leq 0, j = 1, \ldots, M,$$
$$tg(x) \leq 1,$$
$$t \geq 0.$$

Now the transformed problem is a concave one and can be easily solved by some classical methods. The fractional programming has been applied to design the energy-efficient resource allocation for green radio networks in [20–23].

A2.1.2 Sum-of-Ratios Optimization

The fractional programming aims to solve a single ratio problem, e.g., the GEE metric in (2.3). Whereas for the other EE metrics, such as WSEE and WMEE, the objective function involves multiple ratios, which can not be directly solved by the single-ratio Dinkelbach algorithm. In the following exploration, we will introduce two alternative methods to solve multi-ratio fractional programming, namely, sum-of-ratios optimization [24] and generalized fractional programming [25, 26].

Regarding the WSEE metric defined in (2.4), the objective function is the summation of several fractional functions, which is known as the sum-of-ratios programming. The sum-of-ratios optimization has been employed to solve various WSEE maximization problems in [27–30].

Similar to the definition of fractional programming, the sum-of-ratios programming is defined as follows.

Definition 3 (Sum-of-ratios programming). Define $f_i, g_i, i = 1, \ldots, N, h_j, j = 1, \ldots, M$ as some real-valued functions on a set $\mathcal{S}_0 \in \mathbb{R}^n$. The sum-of-ratios programming has the following format

$$\max_{x} \sum_{i=1}^{N} \frac{f_i(x)}{g_i(x)}, \tag{A2.4}$$

subject to

$$h_j(x) \leq 0, j = 1, \ldots, M,$$
$$g_i(x) > 0, i = 1, \ldots, N.$$

The sum-of-ratios programming can be solved by utilizing parametric algorithm and by converting the fractional format into subtractive format as well. It is easy to equivalently convert the above problem into the following one

$$\max_{x, \beta} \sum_{i=1}^{N} \beta_i, \tag{A2.5}$$

subject to

$$f_i(x) \geq \beta_i g_i(x), i = 1, \ldots, N,$$
$$h_j(x) \leq 0, j = 1, \ldots, M,$$
$$g_i(x) > 0, i = 1, \ldots, N.$$

Then, we introduce the following theorem to further transform the problem into a better tractable one [24].

THEOREM 7   *If (x^*, β^*) is the solution to the problem (A2.5), then there exist $u_i, i = 1, \ldots, N$, such that x^* is a solution to the following problem for $u = u^*$ and $\beta = \beta^*$*

$$\max_{x} \sum_{i=1}^{N} u_i (f_i(x) - \beta_i g_i(x)), \tag{A2.6}$$

subject to

$$h_j(x) \leq 0, j = 1, \ldots, M.$$

And x^* also satisfies the following system of equations for $u = u^*$ and $\beta = \beta^*$

$$u_i = \frac{1}{g_i(x)}, i = 1, \ldots, N, \tag{A2.7}$$
$$f_i(x) - \beta_i g_i(x) = 0, i = 1, \ldots, N. \tag{A2.8}$$

According to the above theorem, the sum-of-ratios optimization problem can be iteratively solved by two nested loops. The inner loop solves the problem in (A2.6) and the outer loop finds the optimal u^* and β^* satisfying (A2.7) and (A2.8). If $f_i, \forall i$ are concave and $g_i, \forall i$ are convex, the inner problem is concave and its optimum can be obtained. In this case, the duality gap between the equivalent problem (A2.6) and the original problem (A2.5) is zero, which means that the problem can be optimally solved. Moreover, the outer loop can be solved by some numerical analysis approaches like Newton's method.

A2.1.3 Generalized Fractional Programming

Generalized fractional programming (GFP) is another extension of the single-ratio fractional programming into the multi-ratio scenario, which aims to maximize the minimum of several ratio functions.

Definition 4 (Generalized fractional programming). Define $f_i, g_i, i = 1, \ldots, N, h_j$, $j = 1, \ldots, M$, as some real-valued functions on a set $\mathcal{S}_0 \in \mathbb{R}^n$. The GFP has the following format

$$\max_x \min_i \frac{f_i(x)}{g_i(x)}, \tag{A2.9}$$

subject to

$$h_j(x) \leq 0, j = 1, \ldots, M,$$

$$g_i(x) > 0, i = 1, \ldots, N.$$

Also, we can use the parametric algorithm to convert the fractional problem into a subtractive form, as

$$F(\mu) = \max_x \min_i \{f_i(x) - \mu g_i(x)\}. \tag{A2.10}$$

Then, similar to the single-ratio fractional programming, problem (A2.9) and problem (A2.10) have the same optimal solutions if $F(\mu) = 0$.

Accordingly, a Dinkelbach-type algorithm can be introduced to solve the above problem, as summarized in the following.

Algorithm 5 Dinkelbach-type algorithm to solve GFP

1: Initialize $x_0 \in \mathcal{S}_0$. Set $\varepsilon > 0$;
2: Compute $\mu_1 = \min_i f_i(x_0)/g_i(x_0)$. Set $k = 1$;
3: **while** $|F(\mu_k)| \geq \varepsilon$ **do**
4:     $x^* = \arg\max_{x \in \mathcal{S}_0} \min_i \{f_i(x) - \mu_k g_i(x)\}$
5:     $k = k + 1$;
6:     $\mu_k = \min_i f_i(x^*)/g_i(x^*)$;
7: **end while**

This algorithm is very similar to the single-ratio Dinkelbach algorithm except that a *min* operation is involved. It is useful to apply if the step (4) can be easily solved. However, the main drawback of the above algorithm is its slow convergence rate, which is superlinear as indicated in [26].

A fast Dinkelbach-type algorithm was developed in [31], which applies the Newton-like algorithm to update the ratio μ. Assuming x^* to be the optimal solution, the parametric subtractive problem can be reformulated as

$$F(\mu) = \max_x \min_i \left\{ \frac{f_i(x) - \mu g_i(x)}{g_i(x^*)} \right\}. \tag{A2.11}$$

In practice, the optimal x^* is impossible to be known a priori. Therefore, the previous iteration point, x_{k-1}, could be used as an approximation of x^*. Using (A2.11) instead of step (4) in the Dinkelbach-type algorithm, the convergence rate becomes quadratic.

The preceding algorithms can effectively solve the GFP. However, the *min* operation in step (4) renders it difficult to solve in some cases. In the following, we introduce a new dual problem to solve the GFP.

We assume that $f_i, \forall i$ are concave on $\mathcal{S}_0$ and $g_i, \forall i$, are positive and convex on $\mathcal{S}_0$. Moreover, either $f_i, \forall i$ are nonnegative or $g_i, \forall i$ are affine on $\mathcal{S}_0$.

Let $f(x) = [f_1(x), f_2(x), \dots, f_N(x)]$ and $g(x) = [g_1(x), g_2(x), \cdots, g_N(x)]$. Then, according to the fact that a quasiconvex function attains its maximum on the vertex of a convex polyhedron [32], we have

$$\min_i \frac{f_i(x)}{g_i(x)} = \min_{y \in Y} \frac{yf(x)}{yg(x)},$$

where $Y \triangleq \{(y_1, ..., y_N) | y_n \geq 0, \forall n, \sum_{n=1}^{N} y_n = 1\}$.

Moreover, by applying Sion's mini-max theorem [33], the problem can be eventually transformed into

$$\max_x \min_i \frac{f_i(x)}{g_i(x)} = \max_x \min_{y \in Y} \frac{yf(x)}{yg(x)} = \min_{y \in Y} \max_x \frac{yf(x)}{yg(x)}.$$

Now the problem can be solved in two steps: the inner layer solves the optimal x^* for a given y and the outer layer finds the optimal y^*. Let us define $c(y) = \max_x \frac{yf(x)}{yg(x)}$. Obviously, the inner problem is a single-ratio fractional problem and can be easily solved by the Dinkelbach algorithm.

To solve the outer layer, we define $F(y, c(y)) = \max_x y(f(x) - c(y)g(x))$. Then, the optimal y^* can be achieved if $\min_y F(y, c(y)) = 0$. Based on the above analysis, we can introduce an iterative algorithm to solve the dual problem, as detailed in Algorithm 6.

Algorithm 6 Dual algorithm to solve the GFP

1: Initialize $y_0 \in Y, \varepsilon, \varphi$, and $k = 0$.

2: $x^* = \arg \max_{x \in \mathcal{S}_0} \{y_k f(x) - \varphi y_k g(x)\}$

3: If $\left| \max_{x \in \mathcal{S}_0} \{y_k f(x) - \varphi y_k g(x)\} \right| \leq \varepsilon$, then

4:     Goto step (9)

5: Else

6:     $\varphi = \dfrac{y_k f(x)}{y_k g(x)}$

7:     Goto step (2)

8: End if

9: $y^* = \arg \min_{y \in Y} \max_{x \in \mathcal{S}_0} \{yf(x) - \varphi yg(x)\}$

10: If $\left| \max_{x \in \mathcal{S}_0} \{y^* f(x) - \varphi y^* g(x)\} \right| \leq \varepsilon$, then

11:     Exit

12: Else

13:     $k = k + 1, y_k = y^*$, go to step (2)

14: End if

A2.1.4 Multi-Objective Optimization and Weighted Tchebycheff Method

In the previous subsection, we have introduced the fractional programming and its extensions to solve EE-oriented resource optimization problems. The optimization solutions for both single-ratio and multi-ratio fractional programs have been discussed. The aforementioned problems are basically single-objective optimizations, no matter whether the objective is single-ratio or multi-ratio. However, in some occasions, we need to maximize many different EEs simultaneously and in this case, single-objective optimizations are not effective enough. For example, if jointly maximizing both the EE of users (uplink) and the EE of base stations (downlink), neither WSEE nor WMEE could provide a good metric for the optimization.

In light of the above, new optimization theory and approaches involving multi-objective optimization are demanded. Therefore, in this part, we will introduce the basic formulation of multi-objective optimization problems and their solutions.

Different from the single-objective optimization, multi-objective optimization aims to maximize many objectives simultaneously, or equivalently, an objective vector. The multi-objective optimization problem (MOOP) can be mathematically formulated as [34]

$$\max_{x \in \mathcal{S}_0} (f_1(x), f_2(x), \ldots, f_N(x)). \tag{A2.12}$$

In general, the solution to a MOOP is not unique since all objective functions cannot always be maximized simultaneously. The solutions to a MOOP are usually known as the Pareto-optimal solutions, which can be defined as follows.

Definition 5 (Pareto-optimal solution to the MOOP). A solution x^* is Pareto optimal to the MOOP defined in (A2.12) if and only if there does not exist any another x, such that $f_i(x) \geq f_i(x^*), \forall i$ and $f_j(x) > f_j(x^*)$ for any index j.

The Pareto-optimal solution for the problem in (A2.12) means that a single objective function cannot be increased without decreasing any other objective functions.

To directly solve a MOOP is very challenging due to the existence of multiple Pareto-optimal solutions. Instead, a MOOP can be effectively solved by converting it into some single-objective optimization problems, e.g., scalarizing the multiple objectives into a single one. There are many scalarization methods, including the weighted-summation of the objective functions, minimizing the maximal of the objective functions, etc. In the following discussion, we will introduce another scalarization method, namely the weighted Tchebycheff method, to solve MOOPs. In fact, the weighted Tchebycheff method can provide the complete Pareto-optimal solutions no matter whether the objective functions are concave. Also, in many occasions, the weighted Tchebycheff method also has a merit of low computational complexity.

To solve the MOOP in (A2.12) using the weighted Tchebycheff method, it is essential to introduce the concept of the utopia point, f_i^u, for each objection function f_i. The utopia point is also known as the idea point, which is defined as the maximal attainable point of each objective function, as

$$f_i^u = \max_{x \in \mathcal{S}_0} f_i(x).$$

By solving the above single-objective optimization problem, the utopia point is often attainable. Having the utopia point for each objective function, a scalarized single-objective problem can be formulated as

$$\min_{x \in \mathcal{S}_0} \left\{ \sum_{i=1}^{N} w_i [f_i^u - f_i(x)]^p \right\}^{\frac{1}{p}}, \tag{A2.13}$$

where w_i is the weight for the i-th objective function with $\sum_{i=1}^{N} w_i = 1$ and $w_i > 0, \forall i$, and $p > 0$. The above optimization problem aims to minimize the norm or distance between the solution point and the utopia point, while the weight vector $\mathbf{w}$ stands for the relative importance of each objective function. According to [34], the single-objective problem in (A2.13) provides sufficient and necessary conditions for the Pareto optimality of the original problem in case that its solution is unique.

The weighted Tchebycheff method is expressed as

$$\min_{x \in \mathcal{S}_0} \max_i \{f_i^u - f_i(x)\}, \tag{A2.14}$$

which is a special case of the problem in (A2.13) if $p \to \infty$.

References

[1] D. Feng, C. Jiang, G. Lim et al. "A survey of energy-efficient wireless communications," *IEEE Commun. Surveys Tuts.*, vol. 15, no. 1, pp. 167–178, Jan. 2013.

[2] L. Suarez, L. Nuaymi, and J. M. Bonnin, "An overview and classification of research approaches in green wireless networks," *EURASIP J. Wireless Commun. Netw.*, vol. 2012, no. 1, pp. 1–18, Jan. 2012.

[3] T. M. Cover and J. A. Thomas, *Elements of Information Theory*, John Wiley & Sons, 1991.

[4] G. Yu and Y. Jiang, "Energy-efficiency region for multiple access channels," *Electron. Lett.*, vol. 50, no. 13, pp. 959–961, Jun. 2014.

[5] C. Xiong, G. Y. Li, S. Zhang, Y. Chen, and S. Xu, "Energy- and spectral-efficiency tradeoff in downlink OFDMA networks," *IEEE Trans. Wireless Commun.*, vol. 10, no. 11, pp. 3874–3886, Nov. 2011.

[6] R. K. Sundaram, *A First Course in Optimization*, Cambridge University Press, 1996.

[7] W. Yu and R. Lui, "Dual methods for nonconvex spectrum optimization of multicarrier systems," *IEEE Trans. Commun.*, vol. 54, no. 7, pp. 1310–1322, Jul. 2006.

[8] S. Boyd and L. Vandenberghe, *Convex Optimization*, Cambridge University Press, 2004.

[9] Z. Song, Q. Ni, K. Navaie et. al., "Energy- and spectral-efficiency tradeoff with α-fairness in downlink OFDMA systems," *IEEE Commun. Lett.*, vol. 19, no. 7, pp. 1265–1268, Jul. 2015.

[10] S. Huang, H. Chen, J. Cai, and F. Zhao, "Energy efficiency and spectral-efficiency tradeoff in amplify-and-forward relay networks," *IEEE Trans. Veh. Technol.*, vol. 62, no. 9, pp. 4366–4378, Nov. 2013.

[11] J. Wu, G. Wang, and Y. R. Zheng, "Energy efficiency and spectral efficiency tradeoff in type-I ARQ systems," *IEEE J. Sel. Areas Commun.*, vol. 32, no. 2, pp. 356–366, Feb. 2014.

[12] W. Zhang, C. X. Wang, D. Chen, and H. Xiong, "Energy–spectral efficiency tradeoff in cognitive radio networks," *IEEE Trans. Veh. Technol.*, vol. 65, no. 4, pp. 2208–2218, Apr. 2016.

[13] J. B. Rao and A. O. Fapojuwo, "On the tradeoff between spectral efficiency and energy efficiency of homogeneous cellular networks with outage constraint," *IEEE Trans. Veh. Technol.*, vol. 62, no. 4, pp. 1801–1814, May 2013.

[14] O. Amin, E. Bedeer, M. H. Ahmed, and O. A. Dobre, "Energy efficiency–spectral efficiency tradeoff: A multiobjective optimization approach," *IEEE Trans. Veh. Technol.*, vol. 65, no. 4, pp. 1975–1981, Apr. 2016.

[15] L. Deng, Y. Rui, P. Cheng et al., "A unified energy efficiency and spectral efficiency tradeoff metric in wireless networks," *IEEE Commun. Lett.*, vol. 17, no. 1, pp. 55–58, Jan. 2013.

[16] D. Tsilimantos, J. M. Gorce, K. Jaffres-Runser, and H. V. Poor, "Spectral and energy efficiency trade-offs in cellular networks," *IEEE Trans. Wireless Commun.*, vol. 15, no. 1, pp. 54–66, Jan. 2016.

[17] L. Zhou, R. Hu, Y. Qian, and H. H. Chen, "Energy-spectrum efficiency tradeoff for video streaming over mobile ad hoc networks," *IEEE J. Sel. Areas Commun.*, vol. 31, no. 5, pp. 981–991, May 2013.

[18] Y. Chen, S. Zhang, S. Xu, and G. Li, "Fundamental trade-offs on green wireless networks," *IEEE Commun. Mag.*, vol. 49, no. 6, pp. 30–37, Jun. 2011.

[19] W. Dinkelbach, "On nonlinear fractional programming," *Manage. Sci.*, vol. 13, pp. 492–498, Mar. 1967.

[20] D. W. K. Ng, E. S. Lo, and R. Schober, "Energy-efficient resource allocation in multi-cell OFDMA systems with limited backhaul capacity," *IEEE Trans. Wireless Commun.*, vol. 11, no. 10, pp. 3618–3631, Oct. 2012.

[21] D. W. K. Ng, E. S. Lo, and R. Schober, " Energy-efficient resource allocation in OFDMA systems with large numbers of base station antennas," *IEEE Trans. Wireless Commun.*, vol. 11, no. 9, pp. 3292–3304, Sep. 2012.

[22] K. T. S. Cheung, S. Yang, and L. Hanzo, "Achieving maximum energy-efficiency in multi-relay OFDMA cellular networks: A fractional programming approach," *IEEE Trans. Commun.*, vol. 61, no. 7, pp. 2746–2757, Jul. 2013.

[23] A. Zappone and E. Jorswieck, "Energy efficiency in wireless networks via fractional programming theory," *Found. Trends Commun. Inf.*, vol. 11, no. 3–4, pp. 185–396, Jun. 2015.

[24] Y. Jong, "An efficient global optimization algorithm for nonlinear sum-of-ratios problem," May 2012. [Online]. Available: www.optimization-online.org/DB_FILE/2012/08/3586.pdf.

[25] J. P. Crouzeix and J. A. Ferland, "Algorithms for generalized fractional programming," *Math. Program.*, vol. 52, no. 2, pp. 191–207, Oct. 1991.

[26] I. Barros, J. B. G. Frenk, S. Schaible, and S. Zhang, "A new algorithm for generalized fractional programs," *Math. Program.*, vol. 72, no. 2, pp. 147–175, Feb. 1996.

[27] G. Yu, Q. Chen, R. Yin, H. Zhang, and G. Y. Li, "Joint downlink and uplink resource allocation for energy-efficient carrier aggregation," *IEEE Trans. Wireless Commun.*, vol. 14, no. 6, pp. 3207–3218, Jun. 2015.

[28] S. He, Y. Huang, L. Yang, and B. Ottersten, "Coordinated multicell multiuser precoding for maximizing weighted sum energy efficiency," *IEEE Trans. Sig. Proc.*, vol. 62, no. 3, pp. 741–751, Feb. 2014.

[29] E. Boshkovska, D. W. K. Ng, N. Zlatanov, and R. Schober, "Practical Non-Linear Energy Harvesting Model and Resource Allocation for SWIPT Systems," *IEEE Commun. Lett.*, vol. 19, no. 12, pp. 2082–2085, Dec. 2015.

[30] L. Xu, G. Yu, and Y. Jiang, "Energy-efficient resource allocation in single-cell OFDMA systems: Multi-objective approach," *IEEE Trans. Wireless Commun.*, vol. 14, no. 10, pp. 5848–5858, Oct. 2015.

[31] J. P. Crouzeix, J. A. Ferland, and S. Scbaible, "A note on an algorithm for generalized tractional programs," *J. Optimiz. Theory App.*, vol. 50, pp. 183–187, 1986.

[32] M. Avriel, W. E. Diewert, S. Schaible, and I. Zang, "Generalized concavity," in *Mathematical Concepts and Methods in Science and Engineering*, vol. 36, Plenum Press, 1988.

[33] M. Sion, "On general minimax theorems," *Pacific J. Math.*, vol. 8, no. 1, pp. 171–176, Mar. 1958.

[34] R. T. Marler and J. S. Arora, "Survey of multi-objective optimization methods for engineering," *Struct. Multidiscip. O.*, vol. 26, no. 6, pp. 369–395, Apr. 2004.

[35] H. W. Kuhn, "The Hungarian method for the assignment problem," *Nav. Res. Logist. Q.*, vol. 2, no. 1–2, pp. 83–97, Mar. 1955.

[36] T. S. Han and K. Kobayashi, "A new achievable rate region for the interference channel," *IEEE Trans. Inf. Theory*, vol. 27, no. 1, pp. 49–60, Jan. 1981.

[37] L. T. H. An and P. D. Tao, "The DC (difference of convex functions) programming and DCA revisited with DC models of real world nonconvex optimization problems," *Ann. Oper. Res.*, vol. 133, no. 1–4, pp. 23–46, 2005.

[38] G. R. Lanckriet and B. K. Sriperumbudur, "On the convergence of the concave-convex procedure," *Neur. Inf. Process. Syst.*, pp. 1759–1767, 2009.

[39] M. Hast, K. J. Astrom, B. Bernhardsson, and S. Boyd, "PID design by convex-concave optimization," in *Proc. European Control Conference,* Jul. 2013.

3 Green and Soft Network Design

3.1 Green and Soft Wireless Communication Network Design

The 5G network is anticipated to be reconfigurable with a software-defined network. A soft network is envisioned to bring agility into the implementation of each network element from core network to access network, as well as the building blocks of the air interface. The network function and resource virtualization should be the core of a soft network. It decouples software and hardware, control and data, uplink and downlink to facilitate a converged network, as well as information technology and communication technology convergence, multiple radio access technology (RAT) convergence, radio access network (RAN) and core network (CN) convergence, content convergence, and spectrum convergence. This enables a super-flat architecture that achieves cost-efficient network deployments, operation, and management.

In a soft network, the computing, storage, and radio resources are virtualized and centralized to achieve dynamic and user-centric resource management, matching service features. Soft networks are expected to build on a telecom-level cloud platform to enable network-as-a-service with the features of open network capability and network sharing. This makes it possible to achieve network flexibility and scalability and provides users with a massive variety of services and consistent quality of experience. Meanwhile, this will lead to much greener network designs and operations from the bottom up.

In this chapter, the design principle of the end-to-end (E2E) network architecture for 5G is first presented, including the considerations on the core network, transport network, and RAN. Then the cloud radio access network (C-RAN) architecture is elaborated as an enabling technology for a green and soft 5G network. With the development and the increasing convergence of big data (BD) analytics, communication technologies, and information technologies, BD-enabled wireless communication networks are motivated, which are anticipated to provide satisfactory services in diversified scenarios with globally optimized resource allocation and maximal extent of software configurations of the E2E network architecture. BD-enabled wireless network design is further investigated, including the potential impact on the network architecture, protocol stack, signaling, and physical (PHY) layer operation. Finally, the benefits of applying BD analytics to the wireless communication network are examined, for the case of mobility management and cross-layer transmission control protocol (TCP) optimization.

3.1.1   E2E Network Architecture for 5G

Besides the performance improvement of mobile network and support of highly diversified applications [1–3, 22, 23], the most important requirements of E2E 5G network architecture design are green and soft. Toward green network functions, interfaces, and protocols can be further simplified and converged, e.g., LTE and 5G network can converge at the RAN side to avoid excessive handover overhead; low-cost deployment, especially under ultra-dense network, efficient flow forwarding, and optimized flow routing, can be implemented to reduce E2E latency and avoid congestion. Besides, part of baseband processing can be centralized based on different fronthaul conditions to improve collaboration capability and pooling gain. In the meantime, the latency can be significantly reduced especially for URLLC services [4–8] by introducing mobile edge computing. Toward soft networks, the transition is inevitable from fixed network entities and deployment, as well as static connection in 2G/3G/4G to NFV (network function virtualization) by introducing functions virtualization and flexible E2E function orchestration to implement dynamic configuration, flexible connections, etc. Besides, operator's revenue can be improved via providing customized services per slice granularity.

5G E2E network architecture has the following characteristics, compared with 4G:

- E2E network slicing: to satisfy diverse requirements from vertical industries, E2E network slices are required to provide guaranteed quality and customized services from UE, RAN, CN, transport network, etc.
- Service-based and componentized core network: service-based function design can facilitate flexible customization and aggregation of network functions. In addition, user plane can be simplified to achieve efficient forwarding by splitting control and data forwarding.
- Flexible and smart RAN: centralized management and collaboration gain from RAN-side can be achieved by separating the central unit and distributed unit. With non-stand-alone deployment, 5G new radio and LTE evolution can be converged at RAN-side [10] to maximize legacy network investment. In addition, by introducing smart service processing on the RAN side to enable mobile edge computing, E2E latency can be greatly reduced, and more local optimization is possible.

In summary, 5G overall architecture design principles of RAN, CN, and transport network perspectives are as follows: for RAN, a two-level architecture with CU/DU split to support flexible deployment is preferred; for CN, a service-based network architecture is motivated; for transport, the fronthaul needs to support CU/DU split, low latency, and high throughput. In addition, the following key issues may be related to E2E architecture and will be further described in the later sections: E2E network slicing, flow-based QoS, non-stand-alone deployment, and edge computing.

3.1.2   Next-Generation Core Network

The traditional core network [14] is generally based on the tightly coupled design of network entities functions, interfaces, and signaling flows between entities. However, this concept does not adapt well with rapid network upgrading, fast time-to-market feature introduction, and system performance scalability. The telecom industry learns from the internet industry the service-oriented architecture (SOA) [31], a similar concept, which is relatively loosely coupled and popular in internet architecture design and can be introduced in the 5G core network design.

A 5G core network is service-based to achieve a flexible combination of network functions, rapid time-to-market, agile usage, and independent scalability. The service-based architecture has the following two major characteristics:

- Micro-service-based network function design: 4G network designs are based on dedicated hardware and network functions are tightly coupled; however, the 5G network is redefined to decouple the network functions of the control plane, which allows for flexible combinations and independent evolution.
- Service-based interface model: Lightweight communication protocol is enabled between network functions to allow easy function invocation; network functions may be invoked on-demand via standard lightweight protocols, which tend to improve efficiency and reduce development complexity.

An example of 5G system reference architecture excerpted from 3GPP TS23.501 [13] is illustrated in Fig. 3.1. A number of network functions, such as NSSF (network slice selection function) and NEF (network exposure function), are defined as types of services, which can be easily tailored for diversified service requirements (e.g., invoking certain functions or expanding capacity on-demand).

These descriptions are all mainly focused on the RAN. The concept of soft design should also be reflected in the evolution of core networks (CNs).

In current LTE networks, there are still many challenges and obstacles for operators to maintain and/or upgrade their services. For example, the EPC entities like MME, SGW, PGW, and PCRF are typically based on customized hardware. This is always cost-inefficient and inflexible for network management. Recently, software-defined networks (SDNs) [27–30] and NFV are being generally identified as the most promising technologies for next-generation CN architecture. SDN decouples the control plane and

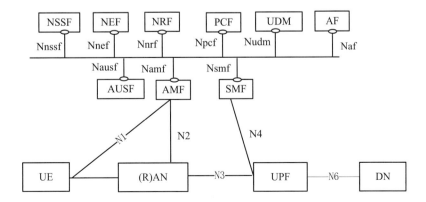

Figure 3.1 5G system reference architecture.

the data plane to reduce the complexity of distributed networking control protocols by using a centralized controller. NFV decouples the hardware and the software, where proprietary hardware network elements are replaced with virtual applications running on low-cost, general-purpose hardware platforms. The integration of SDN and NFV facilitates the deployment of cloud computing with network virtualization, allowing for ubiquitous, convenient, on-demand network access to a shared pool of configurable computing resources. This brings flexibility, reconfigurability, service elasticity, and vendor independence, and also shortens the time-to-market of new applications and services.

Meanwhile, the emergence of SDN and NFV is a great driving force of the development of network slicing technique, which has been envisioned as a key element to meet the diverse demands of 5G use cases and underlying cost requirements. In network slicing, CN is considered the most critical part of providing network services for tenants and their end users. To meet the diversified demands of vertical industries, service customization and on-demand deployment are the key concepts that need to be reflected in CN design. The softness and flexibility inherited in SDN and NFV will enable operators to set up services quickly, and move them around as virtual machines in response to network demands.

3.1.3   Next-Generation RAN

Under ultradense networks, collaboration and centralized mobility control need to be reinforced, since interference and handover signaling overhead may be rather severe. Multi-connectivity seems to be an important way to realize high throughput and ultra reliability. Besides, a number of factors, such as multi-RAT convergence, centralized SON, and NFV/SDN may motivate a new design of RAN architecture, i.e., CU (central unit) and DU (distributed unit) split with an open interface to support interoperability. More specifically, as shown in Fig. 3.2, the functionalities of CU and DU are as follows:

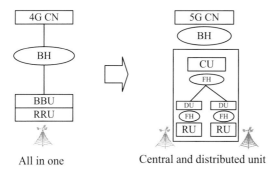

Figure 3.2 From all in one to central and distributed.

- CU mainly deals with non-real-time processing of RAN protocol stack functionalities, which is composed of L3 (e.g., RRC) and partial L2 [16]. CU is required to be centralized and enhanced to support multi-leg management, such as multi-leg reordering, flow control, retransmission, etc.
- DU consists of L1 and partial L2, which is latency-sensitive and mainly performs upon a TTI level, such as ARQ in RLC, and segmentation upon MAC request.

A number of split options have been discussed in 3GPP RAN3 [12]. Option 2 (PDCP and RLC) has been selected for standardization of high layer split [15] and low layer split is still under discussion.

CU is the RAN controller and is in charge of data distribution, which is able to implement multi-connectivity, seamless mobility management, and efficient spectrum usage, which is not a counterpart like RNC in 3G network.

Regarding supporting dual/multi-connectivity, CU is able to serve as an anchor to support data distribution between multiple DUs, which is beneficial to reduce backhaul pressure for data distribution between base stations without anchor points. Additionally, it is also easier to plan the bandwidth of transport network. Looking forward to 5G, CU can be the common anchor for 4G and 5G, which can accelerate data distribution between multi-RATs. In addition, if the transport network between CU and DU is reliable and latency is under millisecond level, multicell radio resource management (RRM) can be introduced in CU to implement relatively rapid radio resource scheduling and coordination of multicells and multi-RATs.

On one hand, regional rapid RRM in CU may play an important role in future networks since it is expected that ultradense networks would be deployed with condensed cell radius and highly overlapped coverage from multiple cells. Under such cases, RRM by the granularity of cells and TTI may not be optimal anymore.

On the other hand, regarding the support of E2E network slicing, it is crucial to implement air interface slicing to satisfy SLA (service level agreement) requirements of diversified services since air interface may be the bottleneck of E2E QoS guarantee. CU and DU split can by nature support E2E slicing. CU can be enhanced to support slice isolation. DU is able to support differentiated configurations, flexible air interface scheduling with customized slicing strategies.

The deployment of CU and DU is required to consider a number of factors, such as transport network conditions, RAN equipment complexity, pooling gain, collaboration gain, and maintenance costs. If the transport network between CU and DU is ideal with high throughput and low latency, such as dark fiber, real-time processing of RAN protocol stack functionalities can be centralized to achieve maximum collaboration gain. However, if the transport network is restricted with limited bandwidth and a long latency, only non-real-time processing is centralized. In addition, the deployment can be adjusted based on service requirements. For example, CU functionalities can be further split into CU-C (central unit – control) and CU-U (central unit – user plane). CU-C can be placed to support large-scale radio resource management and control, and CU-U can be deployed close to UE to support low latency requirement.

3.1.4   Next-Generation Transport Network

5G transport network design is aimed at providing large bandwidth and low latency [12]. Firstly, the fronthaul transport network needs to satisfy the transport requirements from RAN CU and DU split architecture. If high layer split is adopted, delay-tolerant and relatively low-throughput fronthaul transport network is needed. If low layer split is adopted, delay-sensitive and high-throughput fronthaul transport network is needed. In addition, synchronization with high precision is also required to support CU and DU separation. Secondly, the advantage of SDN can be taken to realize flexible configuration and even slicing of network resources.

Meanwhile, due to potentially explosive traffic demands, locally processed data in particular may increase exponentially with the introduction of mobile edge computing, layer 3 IP switch capability would be pushed down to transport aggregation ring to enable local IP flows processing.

3.1.5   Key Issues of E2E Network Architecture

Non-Stand-Alone (NSA) and Stand-Alone (SA) Deployment

In order to take advantage of good coverage capability from well-established LTE low-frequency networks, and in the meanwhile 5G new radio capabilities, non-stand-alone (NSA) [17] deployment is introduced in the first phase. Under this mode, UE relies on LTE BS to provide control plane signaling between UE and CN [25]. Currently, there are two options, as shown in Fig. 3.3:

For option 1, LTE provides signaling connection to 5G NR base station via 4G CN (EPC, evolved packet core). 5G NR (new radio) base station is expected to only provide user plane data transmission to boost capacity. The capability to support dual connectivity of user planes [26] via LTE and 5G NR is needed for terminals.

Option 2 is different from option 1 in that LTE is expected to evolve to support NG Core (next-generation core), which supports new functionalities, such as service-based architecture, network slicing, flow based-QoS, etc. The signaling to support NR data transmission is still anchored to LTE, and the user plane of NR can be either connected to LTE or NG core for boosting capacity.

Regarding the upgrade cost, for option 1, upgrade of 4G RAN and EPC is required to support dual connectivity [21]; for option 2, upgrade of 4G RAN, to support dual connectivity, and connection to NG core is required. EPC upgrade is not required. For both options, if the data plane is split at RAN, upgrade of 4G BBU hardware is required to support 5G NR high throughput. The cost may be relatively high, but higher performance of data aggregation with RAN control may be achieved. In contrast, if data split is expected at the core, 4G BBU hardware upgrade is not required.

NSA mode is expected to be an interim phase and eventually would be replaced by SA deployment. For SA mode, the NR base station works independently and UE connects with the NG Core (e.g., registration and authentication) via NR base station. There are still two options for SA mode: options 3 and 4 in Fig. 3.4. For option 3, a 5G NR system

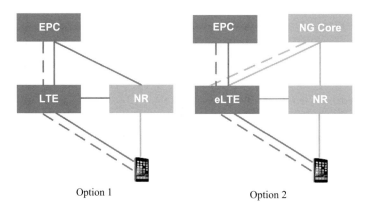

Figure 3.3 NSA options.

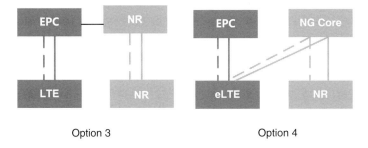

Figure 3.4 SA options.

is required to interwork with LTE from irrespective core networks, i.e., NG core and EPC. For option 4, a 5G NR system is expected to interwork with LTE via NG Core.

For option 3, 4G RAN upgrade is not required, and only EPC is required to be upgraded to support interworking with a 5G NR system. The workload for a network upgrade is relatively low; however, interworking performance may not be as good as expected.

For option 4, both hardware and software upgrades for 4G RAN are needed, but EPC upgrade is not required to support the connectivity to NG core.

E2E Network Slicing

It is foreseeable in 5G that E2E network slicing is indispensable in satisfying emerging diversified services. A network slice refers to a group of logical network functions to satisfy specific quality requirements from vertical industries, which require physical and virtualized resources as needed, including RAN, CN, and even the support of the transport network, or IP network.

From the NGMN paper "Description of network slicing concept" [11], the network slicing concept consists of three layers: service instance layer, network slice instance layer, and resource layer. The service instance layer represents the services (end-user services or business services) that are to be supported. Each service is represented by a service instance. Typically services can be provided by the network operator or by third parties. In line with this, a service instance can either represent an operator service or a third-party provided service.

A network operator uses a network slice blueprint to create a network slice instance. A network slice instance provides the network characteristics that are required by a service instance. A network slice instance may also be shared across multiple service instances provided by the network operator.

The network slice instance may be composed of none, one, or more subnetwork instances, which may be shared by another network slice instance. Similarly, the subnetwork blueprint is used to create a subnetwork instance to form a set of network functions, which run on physical/logical resources.

To implement "network slicing as a service," it is important to design optimum network slicing solutions for each specific use case from vertical industries, including network topology, functions, and protocols. Logically independent network slices may coexist in a shared physical infrastructure. Some functions may be shared by multiple slices, while others need to be customized via parameter reconfiguration or redesign. As shown in Fig. 3.5, an example of RAN slicing is described. Some common network functions (e.g., RRC) are implemented, and slice-specific network functions are available as well, for example, RRC, PDCP, RLC specially designed for URLLC and mMTC. For the low layer, the radio resource is shared by all types of slices to achieve maximum spectrum efficiency.

More specifically, some key issues should be addressed to implement an efficient RAN slice [9].

Efficient radio resource management between slices is crucial for improving spectrum efficiency. Since network physical resources (e.g., computation and storage) could be virtualized and customized for dedicated services, radio resources would similarly be virtualized as time/frequency/space granularity from the following perspectives.

- Spectrum planning: relatively static spectrum allocation (e.g., as a granularity of carrier) for different slices could be implemented. In such a way, sliced QoS

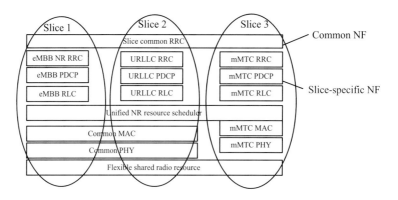

Figure 3.5 An example of RAN slicing.

can be easily guaranteed as expected since radio resource blocks are reserved beforehand, however, the degradation of spectrum efficiency is unavoidable. In addition, the spectrum adjustment is allowed and could take hours or days.

- RT (real-time) scheduling: To maximize spectrum efficiency, radio resource management could perform as a granularity of TTI level, just like common MAC scheduling for different slices. As a result, the radio resource collision may probably happen in the case of bursty traffic in some slices to degrade SLA of other coexisting slices.

- Non-RT scheduling: Similarly, the radio resource management could perform on a larger time scale, for example, a few hundreds of TTI level. The collision probably could be reduced accordingly.

- Access control: Radio resource allocation could also be done via access control per UE level. Some UEs in one slice may be barred from accessing the network since more radio resources are required for another slice.

Different ways of radio resource management may be adopted for diverse scenarios by taking into account network deployment and service requirements. In addition, different levels of adjustment shall be allowed for these ways. BD analytics could be used to make an optimized estimation of radio resource fluctuation so that the corresponding adjustments could be performed in a timely way.

Since the KPIs of eMBB, URLLC, and eMTC are highly diversified and most likely may not be supported over all cells of the RAN. Thus it is important to consider slice availability all cross different cells especially for initial access, handover, and secondary node selection in dual/multi-connectivity.

Flow-Based QoS

4G QoS granularity is based on the E2E bearer, which is too rough to satisfy 5G flow-based service requirements. Moreover, the overhead of E2E bearer establishment or modification is too high to meet rapidly varying QoS requirements of 5G. Flow-based QoS is introduced in 5G to implement finer granularity and faster control.

QoS flow is the granularity of QoS differentiated treatment. During session setup, the CN is able to notify RAN and UE of flow-based QoS profiles, so that flow level treatment can be implemented in RAN and UE.

An in-band QoS control mechanism is also introduced to allow real-time QoS adjustment [19]. In-band QoS marking in a user plane can be modified without signaling involvement, so that flow-based QoS treatment can be reinforced at any time to facilitate real-time processing, which avoids excessive signaling interaction.

Mobile Edge Computing

With the rapid growth of data traffic and diversified services booming, the application of mobile edge computing (MEC) [26] is becoming more and more popular [20]. For example, in the short term, high data traffic demand in enterprises and campuses is expected. In the midterm, low latency and high throughput services, such as AR (augmented reality) may explode in the mobile network. In the long term, auto-drive may become reality with the support MEC.

In general, MEC is adopted not only for performance improvement, such as latency reduction, but also edge intelligence enhancement to boost new business models.

For enterprises, MEC can be used to replace WLAN networks to provide reliable, high-speed, secure, low-cost data access for mobile office automation, industrial control, and IoT (internet of things). For campuses, in order to avoid excessive occupation of a backhaul transport network, MEC can allow users to directly access local networks without traversing the operator's core network.

In the meanwhile, the platform deployed close to UE can process local data and provide user profiles to facilitate more business opportunities. Video surveillance is another good example of MEC's application. Usually most video content is useless and viewed as a waste of backhaul traffic. If a MEC platform can process the local video content with video and picture recognition capabilities, not only can the backhaul pressure be relieved, but efficiency can be improved as well. AR and auto-drive service can also be viewed as typical scenarios of MEC deployment. Normally, AR and auto-drive are required to implement real-time interaction. MEC is crucial to realize low latency and intelligent local data processing.

Small scale of data centers can be deployed at RAN side to offload specific service flows. In the meantime, some computation-intensive processing can be implemented at the edge in a distributed way to reduce backhaul traffic and relieve the processing burden of centralized data centers. For example, image and video reorganization is being processed at the edge, and only recognized information is required to be backhauled to the central server [26].

Meanwhile, RAN network capabilities can be visible to the third-party applications hosted at the edge to improve its performance or incubate new business models. For instance, the information on UE available bandwidth can be gathered and supplied to the application server to enable adaptive transmission, e.g., TCP congestion window adjustment, or for the video server to adjust coding rate under varying wireless conditions.

3.1.6 Summary

To meet the challenging requirements of 5G services, 5G network architecture is anticipated to have the following distinctive characteristics: service-based core network, flexible RAN deployment with CU/DU split, network slicing with flexible orchestration of redefined network functions, flow-based QoS control, edge computing, etc. In 3GPP standardization, RAN CU/DU high-layer split has already been adopted for specification, and low-layer split options and further split between CU control plane and CU user plane are being studied. Regarding slicing, E2E networks (including CN, RAN, and UE) is required to be aware of E2E slicing for slice SLA guarantee. Service-based CN is agreed as the only option for CN evolution. Edge computing is also being standardized in SA2, and context awareness in RAN to accelerate content delivery has also been studied in RAN3. The concept of SDN/NFV has been well reflected in 5G network architecture design, which allows 5G network to progress towards "Green" and "Soft."

3.2   C-RAN: Revolutionary Evolution of RAN

3.2.1   Introduction

The telecom industry has been witnessing a traffic explosion in recent years. It is estimated that internet traffic is expected to increase over 1,000 times by the year 2020, with over 50% of the traffic volume in file sharing. Unfortunately, operators will not see a proportionate increase in revenue. Instead, mobile operators have to invest in more infrastructure just to keep up with such data explosion, significantly increasing the total cost in addition to increasing the total cost of ownership (TCO), while complicating network maintenance, as there are legacy 2G, 3G, and 4G components all coexisting with each other. In addition, system upgrades will become even more challenging when 5G is introduced [32].

The traditional RAN architecture is facing various challenges in the 4G era and beyond. First, traditional network deployment usually requires a separate room per site with supporting facilities such as air conditioning to accommodate the base station (BS) or baseband unit (BBU). This form of deployment is becoming increasingly difficult since the available real estate is becoming scarcer and rental costs are increasing. Furthermore, it could be foreseen that this issue would become more severe when heterogeneous networks with a high density of small cells begin to prosper.

Second, interference problems in the current LTE networks are much more severe than in 2G and 3G networks due to a larger number of small cells in order to facilitate higher data capacity. In order to mitigate this interference, collaborative radio techniques, such as coordinated multipoint (CoMP) [33], have been proposed. However, efficient CoMP algorithms, such as joint transmission (JT), cannot achieve maximum performance gain with the traditional X2 interface with LTE architecture due to high latency and low bandwidth [34, 35]. Furthermore, in 5G the network would be much denser than 4G and therefore the interference issue is more critical.

Last but not least, power consumption is a great concern for operators as both the carbon footprint and energy costs of the network increase. From [36, 37], a large percentage of power consumption in mobile networks comes from RAN. As a result, saving energy in the network's RAN directly lowers the operation expense (OPEX) of the network.

Centralized, collaborative, cloud, and clean RAN [36, 38] (C-RAN), proposed by China Mobile Communications Corporation (CMCC), is a new type of RAN architecture to help operators address the aforementioned challenges. A C-RAN system centralizes different baseband processing resources to form a single resource pool such that the resource can be managed and dynamically allocated on demand. C-RAN has several advantages over traditional BS architecture, such as increased resource utilization, lower energy consumption, and decreased interference due to better support for CoMP implementation.

There have been many studies on C-RAN in the literature [39, 40]. In this chapter, we will briefly recall the original definition of C-RAN as well as its features and advantages. Then we will focus on the evolution of C-RAN, i.e., how C-RAN is evolving to become the essential element of 5G to meet its diverse requirements.

3.2.2   C-RAN Basics

C-RAN was proposed as a key RAN architecture by CMCC in 2009. The basic idea of C-RAN is to centralize the processing resources (e.g., baseband processing resources) in the pool and, further, virtualize them to realize on-demand allocation. Its original definition includes the following four key features.

- Centralization: Instead of requiring one equipment room for each base station, as in the traditional deployment, a certain number of BSs would be centralized in a bigger room in C-RAN. With centralization, first, site selection becomes much easier, since there is no need to find many sites. This is particularly important given that LTE systems require many more sites than 2G/3G due to higher frequency, and the expense on real estate is getting higher and higher in current society. Furthermore, network construction time could be reduced and network deployment could be sped up. In addition, using a big central room to accommodate several BSs means that the BSs could share the same facilities, such as power, air conditioning, etc., which then contributes to the saving on OPEX. In fact, the benefits from centralization have been continuously verified in CMCC's commercial networks. For example, it was observed that power consumption could be reduced by 41% in one trial due to shared air conditioning.
- Cooperation: The idea behind it is to use high-speed and low-latency switches to connect the BSs in the same central room so that the BSs could cooperate with each other. In this way, it is expected the system performance could be improved with the support of cooperative technologies, such as CoMP. C-RAN's capability to facilitate CoMP technologies has been verified in commercial networks, where improved system performance has been observed.
- Cloudification: Instead of traditional equipment that is developed based on specialized hardware such as a digital signal processor (DSP), field programmable gate array (FPGA), etc., the ultimate C-RAN system is targeting to adopt general-purpose hardware, e.g., standard IT servers to realize the mobile communication functionalities. In fact, the idea of cloudification is in line with the philosophy of Network Functions NFV.
- Clean system: C-RAN systems are expected to enjoy much more energy saving than the traditional architecture for several key reasons, namely, facilities centralization, which helps save energy with such tactics as shared air conditioning, and the adoption of NFV [41].

C-RAN realization is a stage-by-stage process with different features realized at different stages. Centralization is the first step, while cloudification is the ultimate goal of C-RAN systems.

3.2.3   Evolution of C-RAN towards 5G

The concept of C-RAN is evolving in the past few years as research and technology progresses. In particular, the concept of centralized unit/distributed unit (CU/DU) function re-split and next-generation fronthaul interface (NGFI) are introduced in C-RAN. Combined with such new concepts and new features, evolving C-RAN has become the essential element of 5G.

5G BBU will be further divided into a CU and a DU. The principle of CU/DU functional re-split lies on real-time processing requirements from different functions. A typical CU mainly includes non-real-time RAN high-layer processing, functions migrating from CN and MEC services. Accordingly, a DU is mainly responsible for physical layer processing and real-time processing of layer 2. In order to lessen transport requirement between the RRU and the DU, partial physical layer processing can be moved from the DU to the RRU. From the equipment point of view, CU equipment can be developed based on general purpose platform, which supports RAN functions, functions migrating from the CN and MEC services. DU equipment can be developed based on a customized or hybrid platform, which supports intensive computing. With NFV infrastructure, system resources, including the CU and the DU, can be orchestrated flexibly via management and orchestration (MANO), a SDN controller, and traditional operating and maintenance center (OMC), which could support operators' requirements on fast service rollout.

In order to solve the transport challenges among the CU, the DU, and the RRU, NGFI [42, 43] is proposed. As shown in Fig. 3.6, the NGFI switch network provides the connection between the CU and the DU, as well as between the DU and the RRU. With the help of the NGFI, the CU and the DU can be flexibly deployed according to multiple

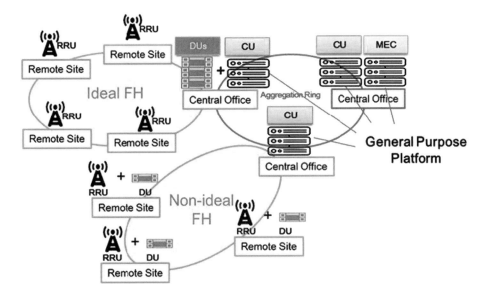

Figure 3.6 CU/DU and NGFI-based C-RAN architecture.

scenarios. In case of ideal fronthaul, the deployment of the DU can also be centralized, which could support physical layer collaboration. In case of nonideal fronthaul, the DU could be deployed in a distributed way. Therefore, the C-RAN architecture based on the NGFI supports not only DU centralized deployment but also DU distributed deployment.

The concepts of the CU-DU structure and the NGFI will be detailed in the following sections of this chapter.

5G C-RAN is based on the CU/DU architecture, the NGFI, and NFV infrastructure, which has been viewed as a promising 5G RAN architecture. Compared with 4G C-RAN, the main features of 5G C-RAN are still centralization, collaboration, cloudification, and clean (green), with each evolving in the context of 5G.

- Centralization: For 4G C-RAN, all the baseband functions are centralized in the central office. With the introduction of CU/DU and NGFI concept, centralization in a 5G C-RAN has two-fold meanings. First, BBU equipment could be physically centralized, which has been demonstrated to have such advantages as fast network deployment, facility sharing, etc. Second, from the function perspectives, with CU/DU re-split, partial high layer functions could be realized on the same central platform.

- Collaboration: There are two kinds of collaboration in C-RAN. First, the central BBU site serves as a wireless service anchor of the control plane and the user plane, which could support multicell high-layer collaboration, multi-connection, seamless mobility management and cooperative spectrum sensing. Second, when both the CU and the DU are centralized in the center BBU site, the SE (spectral efficiency) of cell edge and average throughput can be improved further by introducing physical layer collaboration, such as CoMP, D-MIMO, etc.

- Cloudification: There are two key aspects of the concept of cloudification. One is function abstraction. The other is decoupling of processing resources and applications. Traditionally, processing resources are allocated to a single BS. For C-RAN cloud, processing resources would be allocated in a resource pool, which is good for processing resource multiplexing, reducing system cost, and flexible network function deployment. For example, MEC, which is viewed as one of cloud RAN's flexible deployment use cases, will be well supported in C-RAN. Moreover, the radio resource can also be abstracted as a resource in the resource pool. Decoupling of the radio resource and radio access technology (RAT), on-demand network capacity adjustment, and service customization can be supported by C-RAN. For example, a specific radio resource can be configured for a designated area in order to meet the requirement of group customers. In general, processing resources and radio resources can be dynamically adjusted according to traffic load, user profile, and service requirements in C-RAN. Therefore, operators' requirement for fast service deployment will be better supported via C-RAN on-demand network capacity.

- Clean (Green): Based on centralization, collaboration, and cloudification, the number of BBU sites, air conditioning, and other facilities can be reduced, and the TCO can be saved by centralization. Moreover, on-demand network capacity and processing resources adjustment are supported. Therefore, overall network efficiency is improved, and the clean (green) target is reached.

3.2.4   NGFI: Next-Generation Fronthaul Interface

Traditional 2G/3G/4G BSs consists of BBUs and RRUs with fibers to connect each other. The link between a BBU and a RRU is called fronthaul (FH). With centralization, different BBUs are clustered in the same room. Thus, from the central room, a lot of fibers would go out to connect the RRUs in the remote sites. The more BBUs are centralized, i.e., the larger the centralization scale is, the more fibers are required. It is unfortunate that fibers have become scarce. The FH issue has become a key obstacle to C-RAN centralization for operators who are lacking fiber resources. There has been some study on the FH solution in literature, for example [36], with several proposed schemes including various compression techniques, wavelength division multiplexing (WDM), optical transport networks (OTN), microwave transmission, and so on. In the later part of this book, we will present field trial results to verify the feasibility of passive WDM FH solutions. In general, it is widely agreed that WDM-based FH solutions are mature enough to effectively save fiber consumption in support of C-RAN large-scale deployment for 2G, 3G, and 4G systems.

When it comes to 5G, the FH issue becomes much more severe and more challenging, given some key 5G features and technologies. One typical example is the massive MIMO technology. In 5G massive MIMO with at least 64 antennas and 100 MHz bandwidth is expected. In this case, the required FH data rate between the BBU and the RRU would be around 100 Gbps, which is unaffordable for even dark-fiber FH connections due to the high cost of 100Gbps optics modules. If C-RAN is further taken into account, the problem would be more critical. There must be a revolutionary design on the FH solution to replace traditional CPRI interface. In fact, this is a consensus in the industry and the answer lies on a new concept, called next generation fronthaul interface.

The basic design principle for NGFI is to repartition the baseband functions within the BBU so that partial baseband functions would be moved from the BBU to the RRU side. As the result, the BBU has been divided into two logical parts and the NGFI connects them.

It is clear the NGFI is depending on used function split schemes, yet from a design perspective, the ideal NGFI is expected to have the following features [43].

- Its data rate should be traffic-dependent and therefore support statistical multiplexing.
- The mapping between the BBU and the RRH should be one-to-many and flexible.
- It should be independent of the number of antennas.
- It should be packet-based, i.e., the fronthaul (FH) data could be packetized and transported via packet-switched networks.

The key way towards NGFI is the function split between the BBU and the RRU. There has been an extensive study on the comparison for various split options. Please refer to Fig. 3.8 in the next subection for an overview. It is not until recently that 3GPP has finally decided to adopt split option 2 as the split scheme. Option 2 is the split between the packet data convergence protocol (PDCP) and the radio link control (RLC) layer. In other words, with option 2, the PDCP functions would remain inside the BBU, while the RLC layers, together with those layers below the RLC, such as MAC and PHY, would be moved to the remote side and become a new entity. There are more details regarding the split in the next part (Section 3.2.5) of the CU-DU structure.

Function splitting is just the first step for the NGFI. When it comes to the FH networks in the context of C-RAN, there is a radical change compared with the original WDM or other existing FH solutions. Thanks to the packet-based features, it is expected to use packet switching networks to transport the NGFI packets when the ethernet can come into play. Thanks to its ubiquity, low cost, high flexibility, and scalability, the ethernet should be adopted as the NGFI FH solution. There are several benefits. First, an ethernet interface is the most common interface on standard IT servers, and the use of ethernet makes C-RAN virtualization easier and cheaper. Second, the ethernet can fully make use of the dynamic nature of the NGFI to realize statistical multiplexing. Third, flexible routing capabilities could also be used to realize multiple paths between the BBU pools and the RRH.

The main challenges for the ethernet as an FH solution lie on the latency. Traditional ethernet would have the latency capacity of several dozen to several hundred milliseconds, which is unacceptable for 5G scenarios. Low latency is one of the key features of 5G. Not to mention the ultrareliable low-latency communication (uRLLC) application, which requires 1 ms E2E latency. Even for eMBB, the user-plane latency requirement is around 10 ms, which, using traditional ethernet, is hard to meet.

Other challenges include high timing and synchronization requirements imposed by the NGFI interface. Although the exact NGFI has so far not been specified, it is possible that the NGFI may keep some requirements of the CPRI, such as synchronization requirements. The allowable radio frequency error for a CPRI link is 2 parts per billion (ppb). Synchronization is another concern. In order to meet the timing requirements, both the BBUs and the RRUs should be perfectly synchronized, which therefore requires a very accurate clock distribution mechanism. Potential solutions may include any combination of the global positioning system (GPS), IEEE 1588, and synchronous ethernet (Sync-E).

3.2.5   CU-DU Architecture for 5G

CU/DU Definition

The development and standardization of 5G have brought new challenges and opportunities to RAN. Since 5G envisions supporting much wider service types than 4G, the current RAN architecture makes it hard to satisfy all requirements adopted by 5G. As a result, an innovation of RAN architecture with high capability, flexibility, and scalability is expected to free this gap, which leads to the adoption of CU/DU architecture.

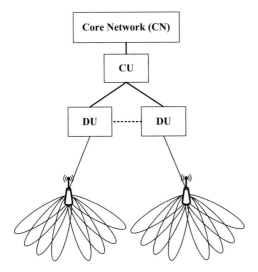

Figure 3.7 CU/DU architecture.

Figure 3.7 shows a conceptual CU/DU architecture with one CU and several DUs. In a CU/DU architecture, the CU can support multiple DUs and each DU belongs to only one CU. In general, CU is responsible for data processing as well as controlling the context of users, the profiles of services, and the connections between the DUs. The DU is responsible for scheduling radio resources and physical layer numerologies.

More specifically, CU controlling could be classified into the intercell level and user level. For intercell-level controlling, the CU keeps track of the capacity per cell and informs the DU to perform interference coordination, load balancing and power control among cells that ensure the wireless coverage area (note that each DU is in charge of at least one cell, and each cell only belongs to one DU). For the user level controlling, the CU selects a suitable DU for UE based on its channel quality and DU capacity. By CU scheduling, no data forwarding is needed within a single CU, which is advantageous in Ultradense Network (UDN) scenarios.

DU scheduling is based on signal quality, user mobility, radio link robustness, and air interface numerologies. Under the CU controlling, the DU should anchor to a CU and be mastered by the same. In addition, the DUs might be logically interconnected with each other. The functions fulfilled on the DU can be split into two parts, which are implemented by two functional modules. The functional module for radio resources scheduling performs allocations of control channels, resource blocks (RBs), and power, achieving the signaling procedures and timing relationship over the air interface. The functional module for physical layer numerologies fulfills the fundamental processing operations, such as modulation, multiplexing, channel coding, and channel measurement. Besides, the DU should report status information to the CU and cooperate with the CU to fulfill the maintenance of the CU/DU architecture, which is easy to be extended according to performance provisions.

A 5G base station, called a gNodeB (gNB), is comprised of the CU, the DU, and the RRU. Compared to LTE eNodeB (eNB), which only contains the BBU and the RRU, new RAN architecture further divides the BBU into a CU and one or multiple DU(s). Most of the controlling functionalities are centralized on the CU while the fast scheduling on the air interface is realized on the DU. In fact, different functions for the CU and the DU reflect the more precise division of responsibilities than the original BBU, in order to achieve high capability, flexibility, and scalability. The CU is able to pile up the resources that enable the manipulation of a number of DUs, and the DU is able to handle multiple numerologies that indicate different processing requirements; thus, the high capacity can be achieved. Furthermore, the CU and the DU can be deployed independently according to the requirements of services and the capacities of the hardware; thus, high flexibility can be achieved. Thirdly, the CU is suggested to use general hardware, which is easy to be extended, and the number of the DUs connected to one specific CU depends on which is on demand and dependent on the capacities of the CU and the DU. Thus, high scalability can be achieved.

Feasible Functional Split Options

The functional split of the protocol stack is the key to CU/DU architecture design. In the study item for a new radio (NR), 3GPP RAN3 has investigated several possible functional splits based on E-UTRA protocol stack. As shown in Fig. 3.8 [12], a total of eight split options have been identified.

Roughly speaking, the more parts of protocol stack reside in the CU, the higher pooling gain the system can achieve. However, this situation may introduce the transmission latency of the interface between the CU and the DU. As a result, the judgment of the feasible functional split should concentrate on data processing capability, implementation complexity, and transmission latency.

3GPP RAN3 has divided the options into two categories, i.e., higher layer split and lower-layer split. More specifically, options from 1 to 3 are classified as higher-layer split while options from 6 to 8 are lower-layer split. Option 4 is not taken into consideration since its benefit is unforseen. In addition, the classification of Option 5 is controversial and will be investigated independently.

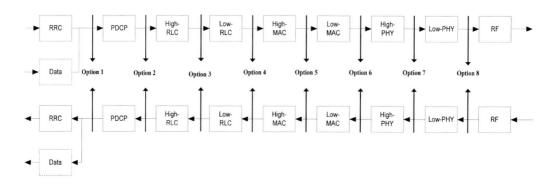

Figure 3.8 Functional split between CU/DU [12].

Based on the current process of 3GPP RAN3, Option 2 (PDCP/RLC split) is suggested as the normative higher-layer split after a down selection between option 2 and 3–1 (intra-RLC split, as specified by 3GPP). The reasons to standardize option 2 are listed as follows:

1. Option 2 adopts the standardized LTE dual connectivity (DC) split as the baseline, which requires relatively small normative work compared to option 3–1.
2. The ARQ at the RLC layer of option 2 is located on the DU, which achieves lower retransmission latency compared to option 3–1, for which the ARQ resides in the CU.
3. With certain enhancement, option 2 can achieve comparatively fast centralized retransmission of lost RLC PDUs compared to option 3–1.

However, for lower-layer split, further study is needed to justify the benefits of each option. Although the determination of lower-layer split is still an open issue, several common views can be summarized as follows:

•   Generally speaking, apart from option 8, which has been used in LTE and will not be detailed in new radio (NR), all the other lower layer split options suffer from the transmission latency between the CU and the DU. Therefore, in our opinion, an ideal FH condition should be premised. Otherwise, the lower layer split brings no benefit.
•   One of the most important advantages for the lower-layer split is the realization of centralized scheduling, which is a key function to achieve the intercell level controlling as specified earlier.

In summary, the determination of the functional split should be based on the following two aspects. On the one hand, most of the controlling and centralized scheduling functions are suggested to be located on the CU while the fast scheduling functions are suggested to be located on the DU. On the other hand, the difficulty of implementation and standardization should be kept within a scope while ensuring the capability for future optimization.

Interfaces

After determining functional split, the interfaces should be defined. The interfaces related to the CU/DU architecture can be classified into two categories. The RAN-core network (CN) interfaces, and the interfaces between the CU and the DU. For RAN-CN interfaces, the CU, DU, and RRU can be treated as one logical node, namely gNB, which is regarded as the termination point of the NG interfaces. Based on the current agreements of 3GPP RAN3, the standardization of the NG interfaces sets LTE RAN-CN interfaces, i.e., S1 interfaces, as a baseline. Therefore, the NG interfaces won't be detailed in this section. On the other hand, the interfaces between the CU and DU, or F1 interfaces, may adopt the same protocol stack as in LTE for both the user plane and the control plane, since the legacy protocol stack has proved to be sufficiently reliable. However, because of the introduction of functional split and the characteristics of the CU and the DU, more F1 application protocol (F1AP) should be added in order to ensure the normal operations and possible optimizations. Although the detailed descriptions of F1AP are still under discussion, in our opinion, there are two new aspects that need to be considered:

- Since radio resource control (RRC) is operated on the CU, the F1 interface should support the transfer of the RRC messages between the CU and the DU.
- Since the CU is responsible for managing multiple DUs, the F1 interface should consider including management functions to support operations, such as flow control and load balancing among DUs.

Note that the normative work of CU/DU interfaces is expected to be done in 3GPP's specification TS 38.401. We suggest that the reader directly refer to TS 38.401 Release 15 for more details.

Mobility

The implementation of the CU/DU architecture enables RAN to optimize functions of the protocol stack and procedures of operations. One of the most vital advantages of the CU/DU architecture over 4G is mobility. For example, seamless switch, which is one of the objectives that 5G RAN must meet to satisfy requirements of services with low latency, can be realized by means of CU/DU implementation. By adopting the CU/DU architecture, the radio link of UE can be split into the CU-level part and DU-level part.

The CU-level radio link is mainly in charge of the data processing of UE, which means two types of basic information are needed: the context of the radio link and the data transmitted on the radio link. The context identifies a radio link including identifiers of RBs and logical channels, formats of PDUs and SDUs, and the mapping between RBs, logical channels, and transport channels. With the aid of the context of the radio link, the data can be accurately transmitted and received.

The DU-level radio link is mainly in charge of the transmission and reception of transport blocks (TBs) over the air interface, which means two types of basic functions are needed: the function of selecting an available physical channel and the function of assembling suitable TBs. The scheduler performs the selection and assembling, and matches TBs and physical channels dynamically.

Given the difference between CU-level and DU-level radio link, the inter-CU and intra-CU mobility management of UEs should be treated separately.

When the inter-CU switch is triggered, both CU-level and DU-level radio links should be switched. For the CU-level radio link, data forwarding is required, and the context of UE should be forwarded to the target CU in order to achieve lossless switch. For the DU-level radio link, the seamless switch cannot be achieved because both the CU-level and the DU-level radio links need to be reestablished. However, by switching the CU-level and the DU-level radio links separately, the interruption time over air interface could be minimized. First, a cloned CU-level radio link can be established on the target CU while maintaining the whole radio link on the source node. Second, the DU-level radio link is quickly switched from the source DU to the target DU by CU scheduling. Note that it takes a longer time for the CU-level switch because of CU-level data forwarding.

When the intra-CU switch is triggered, no data forwarding is required and no switching latency is caused. Therefore, the seamless switch of CU-level radio link can be achieved. For the DU-level radio link, the DU can fulfill seamless switch if the DU schedules new users and resource blocks, which are TTI-based. More specifically, if the switch of a partial radio link can be scheduled within one TTI, the seamless switch is achieved within a DU; while the seamlessness is nearly achieved among DUs if the switch of a radio link can be scheduled within several TTIs. The value of TTI can be dynamically configured by the scheduler according to each radio link switch.

In summary, the partition of the CU-level and the DU-level radio links is expected to simplify the switch procedure and reduce the switch latency, especially for the intra-CU switch scenario. Intra-CU switch achieves seamlessness by integrating radio link switch into scheduling operation, which is a significant advantage of the CU/DU architecture.

3.2.6   Rethink Protocol Stack for 5G: MCD

Motivation

ITU-R has defined three key scenarios for 5G including eMBB (10 Gbits/s), URLLC (99.999%, 1 ms) and mMTC (1 million/km^2), each of which provisions stringent requirements in different aspects. From the aspect of RAN, the classical 4G RAN is not capable of satisfying all of these demands simultaneously. For massive data scenarios and the deployment of more dense nodes in 5G, multi-RAT for physical layer (PHY) and BD computing capability based on cloud platforms are proposed. As a generalization, five innovative R&D themes have been proposed in [32] for 5G RAN. As indicated in [32], with the introduction of UDN in 5G, the importance of flexible air interface is highlighted, which is especially reflected by rethinking of Ring and Young, which represents an effort to review traditional cell-centric network design and put forward user-centric design by adopting the concept of "no more cells" (NMC). With NMC, the available radio resources from multiple access points could be jointly scheduled dynamically for each user and the selection of control/user plane and uplink/downlink (UL/DL) channels respectively could be done separately. With the development of the 5G RAN architecture, the emergence and the adoption of C-RAN by the industry has facilitated the realization of the concept of NMC.

From the perspective of the protocol stack, the signaling interaction in protocol stack architecture is complex, although the framework of the traditional LTE protocol stack architecture is clear. More specifically, in traditional LTE/LTE-A, the basic element of communication network is "cell" that manages the radio resources and the users connected with it. As shown in Fig. 3.9 in the traditional LTE protocol, the UE context can only be established based on a specific cell. Even in carrier aggregation (CA) scenarios, the UE context is established based on the primary cell (PCell) rather than secondary cell (SCell). The SCell only provides channels for data transmission/receiving. In the procedure of cell handover, signaling interaction between cells is slow, and the duration is on the order of seconds or even minutes, which is in the same case as the signaling interaction in some technologies, such as intercell interference coordination (ICIC).

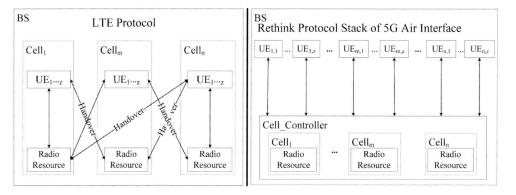

Figure 3.9 Rethink protocol stack.

In 5G, a user-centric network (UCN) is introduced [44] to solve the problem of explosive growth of data traffic and increasing density of deployed BSs. The signaling procedure in the UCN reflects the concept of CP/UP decoupling. According to the quality of the air interface, the network should provide corresponding radio services in order to maintain the connection of CP/UP and the transmission of signaling and data [45]. To improve the quality of the air interface, many new air interface technologies have been introduced to 5G networks, e.g., full duplex, hybrid beamforming PHY, and so on. However, those new technologies bring new challenges to the 5G air interface, requiring the network to provide corresponding services to UE to meet the demand for data rate and channel quality in every TTI. In order to support the requirement, the coordination among cells should keep real-time in each TTI, which greatly increases the difficulty of processing on the network side.

As a result, it is necessary to rethink the protocol stack of air interface for the requirements of 5G and the status of the traditional networks. The traditional network, which is characterized as cell-centric, has been proved to be a simple and practical method of radio resource management [46]. The protocol stack of the 5G air interface should inherit the advantages of the traditional networks, which will be reconsidered to meet the requirement of 5G network and coexist with the traditional network protocol stack.

In summary, the traditional LTE protocol stack architecture is not suitable to support the UDN scenario because of the signaling overhead caused by frequent handover, which motivates us to rethink the protocol stack for 5G air interface. Following the concept of NMC, the protocol stack architecture should be "user-centric" to provide flexible air interface and reduce frequent radio resource control (RRC) signaling transmission. Meanwhile, the innovative protocol stack should take full advantage of "cloud" with strong computing capability. Considering the high density of users and cells, the protocol stack architecture should implement the optimized configuration of air interface resources, including frequency domain resources, time domain resources, and air space resources. As an outcome of rethinking the protocol stack, MCD design logic has been proposed, aiming at achieving the goal of NMC for 5G NR.

The MCD protocol stack is a unified and seamless mobility-enabling framework, which is able to fulfill the same functions as other multicell frameworks adopted by 3GPP, including CA [45] and dual connectivity (DC) [46], with the help of the innovative modification on MAC scheduling and RRC signaling. First, the MCD protocol stack achieves the seamless handover by means of MAC scheduling, which is different from the CA case. Furthermore, the MCD protocol stack achieves the seamless handover without the establishment of an additional RLC entity, which is different from the DC case. Consequently, the MCD protocol stack reflects the concepts of green and soft. On one hand, the data forwarding procedure during the handover within the same BS, is not required, which saves the energy of the BS and the idea of green is achieved. On the other hand, the functions of each layer can be dynamically adjusted by different service requirements, which embodies the idea of soft.

More Details in Differences

According to the analysis just discussed, we have proposed the concept of MCD protocol stack of 5G air interface, in which the UE makes decision independently. In our proposal, both "UE" and "cell" are the basic elements of communication networks, and cells become the dedicated radio resource management elements.

The difference between the protocol stack "rethink" and the traditional LTE protocol stack is shown in Fig. 3.9. In the traditional LTE protocol stack, all the signaling and context of UE treat the connected cell as the only key label. For example, the scope of cell radio network temporary identifier (C-RNTI) for each cell is 0-65535 [47]. The data radio bearer (DRB), signaling radio bearer (SRB) (and the mapping process of SRB/DRB to E-UTRAN Radio Access Bearer (E-RAB) of UE) are both allocated and managed in the scope of one cell. With CA, UE can use the resource of more than one cell. However, as a supplement technology of the LTE protocol stack, CA cannot assist the LTE network to solve the problem in 5G since CA is still cell-centric. Generally speaking, the adoption of the cell-centric scheme is a double-edged sword. On one hand, with "cell" as the key label, the LTE protocol stack simplifies the process of radio resource management for the network [48]. On the other hand, such protocol stack increases the complexity of management and leads to high delay in UE mobility, which is hard to fit into the needs of 5G.

In the MCD protocol stack, "UE" is also a basic element as well as "cell." On one hand, as the element of the protocol stack, UE is responsible for the management of all its information, including its context, the mapping process from DRB/SRB to E-RAB, its channel quality, and the dedicated radio resources allocated, etc. On the other hand, as the only element of the original protocol stack, the cell manages all the radio resources that are not allocated to any users, including ICIC radio resources [49]. As shown in Fig. 3.9, the Cell_Controller allocates the radio resources from different cells to fulfill the requirement of a specific UE according its specific requirements. The radio resources allocated to the UE become its specific attribute, and the UE returns radio resources to cells when the transmission process ends [50]. Consequently, the resource allocation and release are just related to changes of the UE attribute, which works in the same way as the UE context setup/modification. Such an operation avoids the changes of DRBs and logical channels when radio resources change. As a result, the procedure of handover is replaced by the modification of UE attributes for radio resources, which is expected to be carried out in a much faster way than in the traditional structure. In addition, the modifications of semi-static links only occur when the inter-gNB handover is required, while for the intra-gNB inter-cell switch, no modification is needed, which dramatically reduces the frequency of necessary handover, especially for the scenarios with high-density cells.

The MCD Protocol Stack

The functional blocks of a cell and UE in the MCD protocol stack for 5G air interface is shown in Fig. 3.10. The functions of a cell include the following two parts. The Cell_Signaling Controller module is in charge of the air interface signaling while the Cell_UU (UE to UTRAN) Controller module is responsible for the radio resource allocation. The functions of UE include the following two parts. The UE_Packet Processing module is in charge of the data packet processing while the UE_Channel Measurement module is responsible for the collection and analysis of the measurement results [51, 52].

The Cell_Signaling Controller module corresponds to the traditional RRC protocol in the LTE protocol stack [53], which manages the radio resource at a time interval much

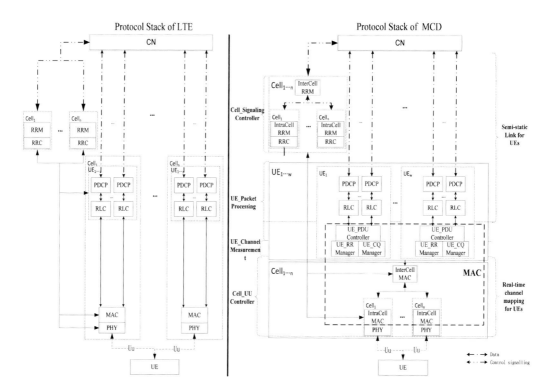

Figure 3.10 "Independent UE" protocol stack.

larger than TTI. More specifically, the Cell_Signaling Controller module is comprised of the RRC module and the radio resource management RRM) module. In the MCD protocol stack, the RRM module is further divided into InterCell RRM and IntraCell RRM.

The Cell_UU Controller module corresponds to the PHY protocol [54] and most part of the MAC protocol in the traditional protocol stack. It fulfils the radio resource allocation between InterCell MAC and IntraCell MAC and the mapping process from UE data to the physical channel. It should be noted that the division of the original RRM and MAC modules into intercell and intracell parts is a reflection of the sinking of functions related to the cell.

The UE_Packet Processing module corresponds to the original PDCP/RLC protocol in LTE, and the UE_PDU Controller. The PDCP/RLC module achieves the data packet processing and provides the service access point (SAP) of the UE context while the UE_PDU Controller module controls the collection and dissemination of PDUs for concurrent PDU streams [55, 56].

The UE_Channel Measurement module corresponds to the channel control part of the MAC protocol, which includes the UE_CQ (channel quality) Manager and the UE_RR (radio resource) Manager modules. The UE_CQ Manager module is in charge of monitoring, modification, and computation of the channel quality, which provides the information to support the MAC scheduling. The UE_RR Manager module is responsible for the management of UE specific radio resources.

To meet the stringent requirements of three typical scenarios adopted by 5G, the MCD protocol stack provides an innovative pattern for the link control, which is the collaborative management of semi-static links and real-time channel mappings.

The semi-static links for UEs is comprised of logical channels, DRB/SRB, and E-RAB links, all of which indicate a specific type of service (ToS) of the UE and work in a semi-static way. To achieve more flexibility, it is necessary to unbundle the ToS with a specific PHY mode, which decouples the UE from the cell. As a result, the semi-static links only need to be modified when inter-BS handover occurs.

The real-time channel mapping for UEs is responsible for the mapping of logical channels, transport channels, and physical channels. The UE provides parameters to MAC including the quality of channels, the buffer state, the request to PHY, and the characteristics of allocated radio resources. According to radio resources of all the available cells and the parameters received, MAC configures cells together with their radio resources as attributes of the UE. By this means, the logical channel matches the appropriate cells at first, and then performs channel mappings within each cell. With real-time channel mappings, UE can receive data from different cells.

With the combination of semi-static links and real-time channel mappings for the UE, the MCD protocol stack optimizes the handover procedure compared to the traditional LTE protocol stack. As shown in Fig. 3.11, in traditional LTE protocol stack architecture, handover is required as soon as the UE moves across cells. In comparison, in the MCD protocol stack architecture, the real-time mapping replaces the handover within the same BS, which enhances the flexibility of the radio resource management

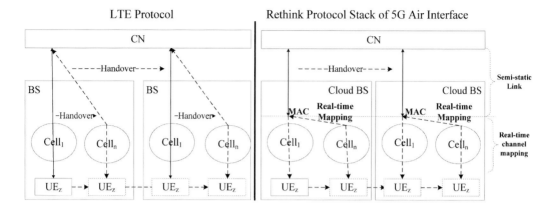

Figure 3.11 The mobility of UE.

while ensuring the stability of the system by redefining the relationship between the cell and the UE.

The Application to C-RAN

As mentioned earlier, the concept of C-RAN has been widely accepted by the industry for 5G. One of the most important characteristics of C-RAN is the adoption of the cloud platform, which in theory provides sufficient hardware processing capabilities for the management of the air interface. However, in reality, the hardware resource of the cloud platform is relatively limited, and so is the network scale of 5G wireless network; thus, the mapping between the hardware resource and the network scale needs more research.

The cloud platform is able to achieve the strong computing capability and dynamically adaptive hardware management capability. The MCD protocol stack can make the most of the cloud platform in terms of intercell management for the performance optimization.

As indicated previously in the section, the MCD protocol stack redesigns the architecture of the protocol stack into the centralized and distributed levels, which achieves the centralized link control and the real-time collaboration of the air interface across cells. On one hand, the centralized functions can make full use of the strong computing capability of the cloud platform. On the other hand, the distributed functions make dynamical adjustments according to the load of the network, which makes full use of the adaptation ability of the hardware resource management of the cloud platform.

A possible design for the MCD protocol stack with the introduction of the cloud platform is shown in Fig. 3.12. From Fig. 3.12, the MCD protocol stack isolates the PHY and IntraCell MAC from InterCell MAC, and only the InterCell MAC is implemented on the cloud platform. As a comparison, PHY and IntraCell MAC are still implemented on the dedicated hardware devices. As a result, the original functions of MAC can be split into two parts: intercell control for UEs (all the other blocks in the MAC module except IntraCell MAC in Fig. 3.12) and IntraCell MAC. On one hand, InterCell control for UEs realizes seamless handover and performs operations including flow control

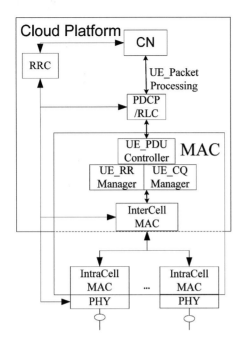

Figure 3.12 The cloud platform for the MCD protocol stack.

and load balancing across cells. On the other hand, IntraCell MAC schedules UEs within a cell and performs operations, including transport/physical channel selection, service mapping, and priority control, which are under strict TTI constraint over the air interface.

The Progress of 5G NR Protocol Stack in 3GPP

The progress of 5G NR protocol stack is mainly obtained in [57]. According to discussions and agreements in 3GPP, NR protocol stack has been enhanced in the following aspects:

- New AS layer

  In NR, a new AS layer is added on top of the PDCP layer that is in charge of the mapping from QoS flow to DRB. The new AS layer is called the service data adaptation protocol (SDAP) layer. Originally in LTE, there is only one-to-one mapping between E-RAB and DRB, and each E-RAB only corresponds to one set of QoS parameters; while in NR, E-RAB is replaced by QoS flows, and multiple QoS flows can map to the same DRB. As a result, 5G NR can acquire the characteristics of service streams more precisely compared to LTE. In addition, the SDAP entity is configured per PDU session, which reflects the idea of user-centric design logic.

- PDCP

  As mentioned in the previous section, according to the current progress of RAN3, the functional split option 2 has been adopted as the normative split for the CU/DU architecture, which splits between PDCP and RLC. As a result, the functions that are suitable to be operated on the cloud platform are centralized to the PDCP layer. Therefore, in 5G NR, the reordering function has been moved from RLC to PDCP. And the duplication function, which is newly introduced to improve the reliability of services in NR, is added at the PDCP layer. Moreover, the dual connectivity adopted by NR has set PDCP as the anchor of the data split bearers, which results in the introduction of other functions, including data dispense and flow control at the PDCP layer.

- RLC

  As indicated by NR, one of the main objectives of the 5G protocol stack is to minimize the processing delay at the RLC layer. Consequently, the reordering function is moved from RLC to PDCP, and the concatenation function at RLC is merged to the multiplexing function at MAC. In addition, in order to further optimize the processing delay, the concept of preprocessing is introduced to both RLC and MAC layers.

- MAC

  Since the concept of the numerology is introduced at PHY, 5G MAC is enhanced accordingly compared to LTE. Besides the TTI length, the mapping relationship between logical channel group (LCG) and the numerology should be reconsidered, and Buffer status report (BSR), scheduling request (SR), logical channel prioritization (LCP), and discontinuous reception (DRX) in NR will also be impacted. In addition, from the perspective of random access (RA), multiple beams will support different set of RA parameters, such as backoff and power ramping. Moreover, in order to achieve the fast activation of the duplication function at the PDCP layer, a new MAC control element (CE) is introduced at the MAC layer.

  The detailed functions for L2 layers in NR are shown in Fig. 3.13. Note that the main enhancement for the protocol stack is marked with circles. In summary, all these modifications and enhancements reflect the thought of decoupling UE from cell. Combined with the adopted CU/DU architecture by 3GPP, CU fulfills the more UE-specific centralized control, while DU achieves fast scheduling functions that are more cell-related. Therefore, the current progress of 3GPP on the protocol stack matches the MCD design logic proposed in this book.

Besides, in order to further decouple UEs from cells, 3GPP has introduced the concept of beam at the air interface. Beams are mainly operated at the MAC and the PHY layers. A specific UE can be served by a dedicated beam during a time period based on the beam tracking, and the fast beam switch can be achieved at TTI level. In addition, the division of traditional RRM is under discussion in 3GPP RAN3. As indicated by the latest progress of RAN3, 5G NR architecture is striving to sink part of RRM functions, which are more cell-related, down to DU. For example, 3GPP is discussing to achieve C-RNTI

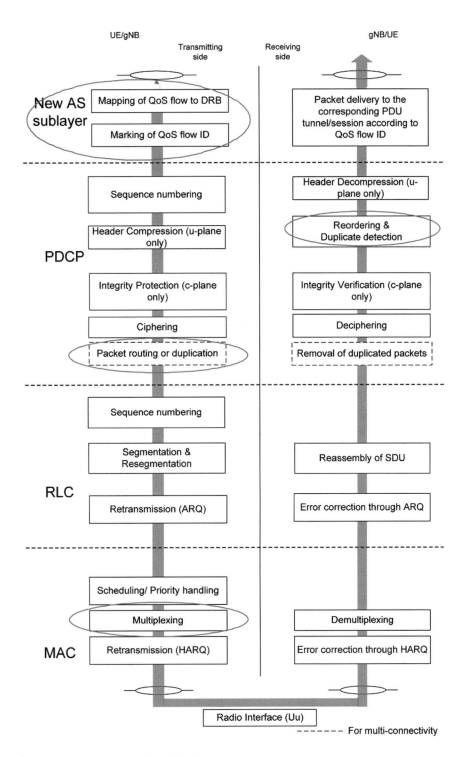

Figure 3.13 The 5G NR protocol stack [57].

allocation in DU since C-RNTI is a cell-specific parameter. As a generalization, the beam management and the sinking of RRM functions again reflect the idea of the MCD design logic.

3.3   Big-Data-Enabled Mobile Network Design

There is a broad agreement about the remarkable potential of big data to lead innovation, boost commerce, and drive progress, among leaders in the industry, academia, and government. The term "dig data" can be literally understood as the surge of data in today's networked, digitized, and information-driven world. With vast data resources, people can tackle questions previously out of reach. Additionally, an agreement is also reached on the inevitable fact that BD will overwhelm traditional analytical approaches, architecture, information management, and data transport. In this chapter, we begin with the consensus on some important basic issues with BD, such as definition, characteristics, history, and mainstream technology. Then, it is followed by a special focus discussion on wireless BD and its applications into the mobile network, seen from a network operator's perspective. And last but not the least, we present the BD-driven mobile network design, as well as some initial investigative results.

3.3.1   Background of Big Data

The rate of growth of data generated and stored has been increasing exponentially. The 1965-born Moore's Law has been applied to almost every aspect of the computer industry, from integrated circuits to memory, whereas the growth rates of data volumes are estimated to be faster than Moore's Law, i.e., more than doubling every eighteen months.

This data explosion sparks new ways of gathering and utilizing data to extract value, meanwhile providing significant challenges due to the size of the data being manipulated and studied. Another significant shift stems from the increasing amount of unstructured data. Historically, enterprise data analytics has been focusing on structured data and using relational data models to capture data characteristics, and discarding the fragmented and unstructured "noise." With the soaring of the quantity of unstructured data, such as web pages, texts, images, and videos, there now is a strong desire to obtain additional value from this heterogeneity nature of today's data. The ability to process not only a large amount of data but also various types of data leads to the current revolution to parallel scalability in the architecture and subsequent data handling methods.

To understand this BD revolution, the following four aspects should be considered: the characteristics of the datasets, the historical milestone events, the relation between internet technology and data technology, and the state-of-the-art scientific and technological tools. The following sections will briefly discuss these aspects.

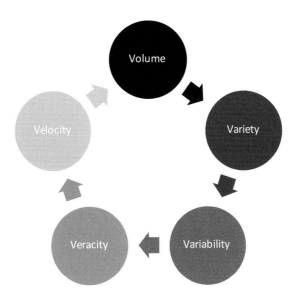

Figure 3.14 5V characteristics of big data.

Understanding Big Data

BD is a broad concept where different industries have different understandings and thus different definitions [58]. Nevertheless, BD can be distinguished by the following five characteristics as in Fig. 3.14 :

- **Volume**, the most commonly recognizable feature, i.e., the presence of the extensive data sets available for extraction of value. The underlying assumption here is that the larger the volume the greater the value.
- **Velocity**, a contextual reference to the speed at which the data is generated and processed, often in real-time, or near real-time. It is opposite to the data at rest.
- **Variety**, the need to process data from multiple repositories, types and domains and the need for analytics across a range of logical models, timescales, and semantics.
- **Variability**, the changes in a dataset's characteristics; whether it is infrastructure, interface, flow rate, or volume, all have impacts on data processing (whereas volatility refers to the changing values of the actual data elements in a data set).
- **Veracity**, the varying quality and creditability of the data, which comes from multi-sourced and multidimensional inputs and affects the accuracy of the analysis.

To reference the National Institute of Standards and Technology (NIST) Big Data Working Group (NBD-WG) in the United States, it describes BD as the datasets that are so large or complex that even some current data processing application software and data analytics cannot adequately deal with. Scientific and technological advances have been outrun by the rapid growth of data. Therefore, innovative technical approaches in information processing are needed to reap the fruits of BD, such as enhanced insight, decision-making, and process automation.

Historical Journey

The BD industry has gone through roughly three main development stages:

1. 2003–2006: BD development was still in its infancy. The explosion of unstructured datasets such as text and video that do not fit into predefined relational data models, and dataset research gave rise to scientific and technological advances. The milestone event was that Google published three influential papers that introduced its Google File System (GFS), MapReduce, and BigTable to the world. All three core technologies are distributed in nature, and able to process huge amounts of unstructured data using cost-effective hardware and software implementations. The most notable application was search engines.
2. 2006–2009: BD saw some major breakthroughs. Parallel computing and distributed processing became the pillars to BD's fast-paced advances. With reference to Google's technical architecture, Hadoop, an open-source software framework used for distributed storage and processing, emerged and was soon used by IT companies to build data processing systems. By leveraging users' web browsing behavioral data, IT companies were generating user profiles that in turn guided targeted advertising and promotions.
3. 2010–now: With the widespread adoption of mobile phones, mobile data usage skyrocketed. BD development has entered a new era, when the data is more fragmented, distributed, and optimzed for streaming media than ever before. There are many novel BD analytics tools that have originated from the open-source community, for instance, Storm, S4, and Spark. Innovative applications also flourished, ranging from real-time marketing and internet credit investigation to vertical application solutions for industries such as tourism, real estate, transportation, and business.

IT vs. DT

Information technology is all about data processing and manipulation, and it provides the software tools for people to use predefined procedures to conduct business. The analogy is that IT has extended the reach of human beings' arms and legs. Data technology focuses on data analytics and provides the means to extract information and produce new knowledge. The analogy is that DT is expanding human beings' brains. Therefore, the difference here is not only in the methodology, but it is rather a paradigm shift.

It can be predicted that there is a revolution in the near future where data defines and determines the technological progression, resource distribution, fund flow, and talent flow in society. It will fundamentally shake traditional social structure and cooperation, leading to more effective and efficient modes of production and economic operation. The IT industry's motto is to own, to manipulate, to transport, and to control; DT industry's slogan is rather to open, to interwork, to experience, and to share. DT is taking over IT as the fundamental driver of social progress.

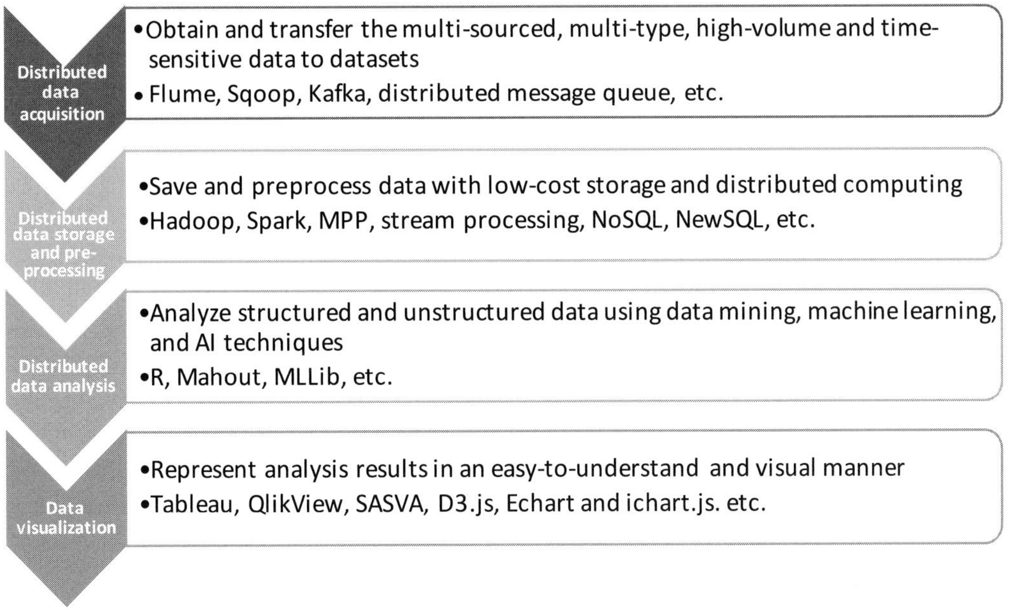

Distributed data acquisition
- Obtain and transfer the multi-sourced, multi-type, high-volume and time-sensitive data to datasets
- Flume, Sqoop, Kafka, distributed message queue, etc.

Distributed data storage and pre-processing
- Save and preprocess data with low-cost storage and distributed computing
- Hadoop, Spark, MPP, stream processing, NoSQL, NewSQL, etc.

Distributed data analysis
- Analyze structured and unstructured data using data mining, machine learning, and AI techniques
- R, Mahout, MLLib, etc.

Data visualization
- Represent analysis results in an easy-to-understand and visual manner
- Tableau, QlikView, SASVA, D3.js, Echart and ichart.js. etc.

Figure 3.15 Big data life cycle and main analytical tools.

Mainstream Technology

In order to meet the challenging requirements of BD applications and massive datasets, the technical community has come out with numerous data analytics and solutions. From the perspective of the life cycle of data in the system, there are four processes: data acquisition, data storage and computing, data analysis, and data visualization, as shown in Fig. 3.15.

1. **Distributed data acquisition**

   This process includes obtaining the source data and transferring them to datasets. Due to the multi-sourced, multi-type, high-volume, and time-sensitive features of the captured data elements, it is critical to have cloud computing and a distributed system architecture. The mainstream techniques are Flume, Sqoop, Kafka, and distributed message queue.

2. **Distributed data storage and computing**

   The amount of data is growing at such a rapid rate that it demands extremely low-cost storage capability to handle the diverse forms of data, as well as the massive computing capability to preprocess the raw data. Traditional CPU and parallel computing can no long meet the requirements on speed, scalability, and cost. Distributed computing becomes the dominant method, including Hadoop, Spark, MPP, stream processing, NoSQL, NewSQL, etc.

3. **Distributed data mining**

   Data-mining techniques are used to discover new knowledge and new relations from the chaos of data, which is the primary focus of any BD system. Traditional data mining was performed on structured and small datasets, where predefined experience-based relational models were established and then used in data analysis. However, little prior knowledge is known about unstructured and multi-type data, let alone used for building an explicit mathematical model. More intelligent and distributed non-relational data platforms should be investigated. Some of the analytic tools are R, Mahout, MLLib, etc.

4. **Data visualization**

Representing the data analysis results to the user in a straightforward and meaningful manner is seen as a crucial segment in decision-making scenarios. The biggest challenge is how to make the complex outcome easy to understand, and there are a few current techniques, including Tableau, QlikView, SASVA, D3.js, Echart, ichart.js. etc.

3.3.2 Wireless Big Data

The global success of the wireless industry is undeniable, namely cellular networks, internet of things (IoT), Wi-Fi, satellite, and sensor networks, as well as many private and dedicated communication networks, such as smart grid and intelligent transportation networks, that are traditionally wired communication networks but are migrating to the wireless domain.

The explosion of wireless data and the increasing number of types of data has demonstrated that the wireless communication industry has entered the age of BD. Wireless big data (WBD) applications are borne out by the realization that there is an enormous amount of system data generated each day (estimated more than 14 Petabytes/day from a dozens of CMCC data centers alone) and should be properly analyzed and used for improving the ecosystem comprised of users and the network.

- **Definition, Sources, and Classification**

There is no universally agreed definition of WBD as of today, yet it is generally described as the massive data consumed by wireless service users, and the data generated by the wireless system when providing such services. This may relate to wireless spectrum, transmission, access, and network, etc.

With respect to the sources of the WBD, we initially proposed three categories: raw WBD, derived WBD, and trial WBD. Raw WBD is the data, old and new, that has passed through wireless communication systems, originating from content providers, end users, and the network itself. Derived WBD is the statistical analysis that is beneficial to service delivery, such as spectrum utilization distribution, small-cell deployment sites statistics, and radio resource allocation statistics. Trail WBD is the acquired performance data for any new trial and testing effort, either for new transmission techniques or innovative functional and architectural proof points. Alternatively, WBD can be divided into four categories according to [59], as shown in Table 3.1, i.e., application data, user data, network data and link data. The rationale is explained in details for each category with the means of acquisition. The application data usually entails the statistics of content popu-

Table 3.1 Four categories of wireless big data

Data Category	Contents	Big data vs. Traditional data
Application data	Content popularity, service type, etc.	Big data analytics helps to obtain these data, which can span across multiple layers of the network and be time-sensitive.
User data	User preference, location, mobility, online behavior, etc.	Traditional network optimization did not use application and user data.
Network data	Cell configuration, signal strength, traffic load, outage rate, inter/intracell interference, signaling, UE capability, etc.	Big data analytics is able to sense the wireless environment and the network status, thus providing globally optimized solutions.
Link data	Physical channel information such as path loss, shadowing, channel statistics, etc.	Traditional network optimization is generally confined to per radio link/user/cell optimization, or simple intercell coordination.

larity. The user data implies user behavioral profiles, such as location, mobility, and preference. Both categories can be obtained using BD analytics, for instance, through deep inspection of the user plane packets. User's location and mobility can either be obtained via GPS or network measurements. Note that BD analytics and traditional data processing may work together to obtain certain information, which may involve multiple layers in the network and have stringent time constraints. The network data includes the network configurations, such as the coordinates of the physical BS, its antenna configuration, and KPIs, such as traffic load and outage, UE capability, and various signaling interactions between UE and network. The link data concerns with the wireless links between the users and the BSs, generally obtainable via downlink and uplink measurement and reporting.

- **Potential value**

  Many mobile operators, such as CMCC, own vast ranges of WBD, including verified user information, user application usage statistics, and network data, as well as private service data. WBD is present in business, operation, and management (B-/O-/M-) domains, and on value-adding service platforms. B-domain WBD contains user's basic/personal information, user device specs, contract plans and more, which are structured data and often superior in data quality.

O-domain WBD consists of control signaling data, network performance logs, application usage breakdown, etc. It encompasses user's online behavior, and is thus much larger in volume than B-domain data, and more difficult to process, but with greater value. M-domain WBD comprises the operator's ERP, financial reports, human resources, and alike; luckily these are normally structured data, and can provide the most valuable insights into the company's strategic financial decisions.

Operator's internal WBD has many advantages against the data from over the top (OTT) companies: 1) the data is more trustworthy since it is mandatory that mobile numbers are verified through national ID associations; 2) user profiles are multifaceted, i.e., a user's internet browsing history can be analyzed to label the individual with certain behavioral traits while a user's voice call records can reflect his/her social ties; 3) operators possess real-time information about users, such as location data; 4) data volume is unprecedented, to be specific, DPI that has captured data's daily increase is in terabytes scale.

Operators can also leverage two additional forms of data, i.e., publicly open data on the internet through web crawler systems or API on websites, and vertical industry's internal data through bilateral cooperation and data sharing.

There is also great value for terminal users through sharing information and helping with the formation of BD in order to allow the operators and service providers to provide better services and experience. Naturally, WBD is of great value to third party applications, who can leverage both internal and external data, and promote cross-association and data-mining analysis. Such cooperation can help to break free of the data island and further develop new and innovative applications and business models for the next generation. WBD analytics also provides important support for public safety. When monitoring the flow of people in real time, especially large-scale spatial location movement, it helps to find out the potential safety risk. The BS signaling data can also record the moving trajectories of users. By using user mobile sequence detection technology and key location mining algorithms along with the matching algorithm between mobile device spatial-temporal trajectories and real-time road networks, we can compute and show real-time urban traffic and provide valuable suggestions for urban road construction. Finally, wireless communication research could also benefit from WBD, in terms of physical layer transmission scheme design, channel modeling and emulation, network management and orchestration, radio resource management, cell deployment and optimization, and more.

In the past 50 years, we have seen the success of the business model of "IT+CT," which has created applications such as targeted advertisements and internet credit systems. The next 50 years will be the era of "IT+CT+DT," where BD and artificial intelligence (AI) bring new momentum to cellular industry development in terms of network optimization, capacity improvement, customized services, and better user experience, giving rise to more innovative and disruptive technologies.

3.3.3   Artificial Intelligence in Wireless Networks

To support a range of diversified scenarios and service requirements, the 5G network is becoming softer and more agile. Unfortunately, the complexity of the network optimization problem becomes increasingly challenging if traditional methods are used. The 5G network needs to embrace new and cutting-edge technologies, such as AI, to efficiently boost both SE and EE. In one respect, ever-increasingly complicated configuration issues call for smart algorithms to replace manual changes according to prior experiences. In another respect, the network needs to intelligently adapt to the environment (e.g., traffic load, service characteristics, user behavior, etc.) to fulfill the evolving service requirements. In the following discussion, the AI concept and AI applications in wireless networking are briefly discussed.

AI is the science and engineering behind machines that behave like humans, and has long been applied to optimize communication networks [60–74]. Generally, AI encompasses multi-disciplinary techniques, such as machine learning, optimization theory, game theory, control theory, and meta-heuristics [60]. Machine learning belongs to one of the most important subfields in AI and is typically classified into three categories [75–78]: supervised learning, unsupervised learning, and reinforcement learning (RL), as shown in Fig. 3.16. Supervised learning's goal is to obtain an optimal model/function though analyzing a set of training examples, each of which consists of an input object and a desired output value. Such inferred model/function is then used to map new inputs in a seemingly intelligent way. Unsupervised learning is mainly used when the expected output is not known and the system has to learn by itself. RL works similarly to the unsupervised scenario, where a system must learn the expected output on its own, but with the help of a reward mechanism. If the decision made by the system was good, a reward is given, otherwise the system receives a penalty [79]. This reward mechanism enables the RL system to continuously update itself to maximize some notion of cumulative reward. The emerging hot topic, deep learning methods, can also fit perfectly into one of the three types [80, 83]. Figure 3.16 shows a general view of the different learning schemes [81].

Recently, ML and AI applications in cellular network domains have been heatedly discussed. In [66], a survey of ML techniques for self-organizing cellular networks was provided. In [65], typical AI algorithms were investigated to enhance cellular network performance. With the exploitation of WBD and the rapid development of learning algorithms, AI will play an unprecedented role in future networks. It helps to sense, mine, predict, and reason in the context of wireless systems, enabling fast and optimal adaptation and configuration of various network parameters/processes. AI is envisioned to be the next disruptive element for 5G and beyond to improve system performance in terms of both network capabilities and user experience. To employ AI technology for wireless communication systems, it is necessary to pay great attention to three main problems: the construction of wireless data sets, real-time requirement of network functions, and the universal applicability of algorithms. These issues need to be always kept in mind when studying any specific use cases.

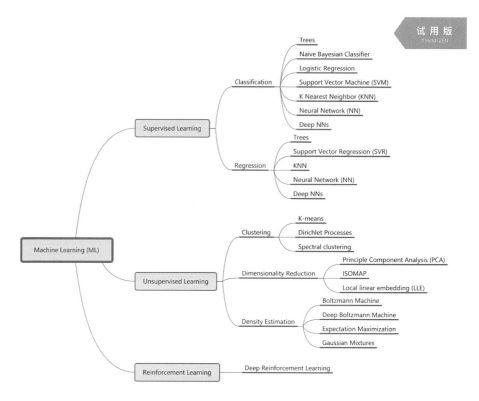

Figure 3.16 An overview of machine learning algorithms.

3.3.4   Application of WBD and AI into Mobile Network

The fundamental drive of operators' engagement in WBD research is to ensure coverage/connectivity, improve management efficiency, alleviate operational cost, maintain competitiveness, and offer superior services to customers. Traditional solutions for achieving those goals are very limited if not nonexistent, in terms of both methodology, algorithm, and implementation. With the advent of WBD and AI concepts and methods, we suddenly have a new weapon to tackle these tricky issues with promisingly high efficiency and great results. In addition, there are also AI-enabled applications where collaborative effort with third party companies can lead to new revenue for operators. A mobile operator is inherently the owner of the data, and a data service enabler, at the same time, should also strive to be a key technology holder and a data solutions provider. This ambition needs continuous research and trial and error to fulfil.

Traditional research interest in WBD has been focused internally as vital for network operators, and it falls within these four categories:

- Customer service data: Voice calls, web forums, etc. are being collected and solidified into valuable insights about the current quality of and possible improvements to our service chain.
- Sales support: By investigating customers' online behavior and forming user profiles/preferences, the success rate of targeted advertisements can be significantly improved.

- Network optimization: Operators are using WBD analytics to better understand the relations between the network operational status and user's QoE.
- Enhanced management capability: Scientific data-mining techniques are being applied to operation and management data to improve the efficiency and effectiveness of cooperating actions. Specifically, there have been multiple success stories about using internal WBD resources on areas like wireless caching [82], network resource allocation [84], network planning, and management optimization [85], etc.

Note that the goal here has been to improve the network's energy efficiency and spectrum efficiency in the long run. For instance, the authors in [86] made an attempt on implementing a signaling-based network optimization scheme on 4G LTE networks, where such intelligent operation only happens at the mobile network management plane and is a long-term action.

A more recent development is the new approved study item in 3GPP on network data analytics (NWDA) [87], which is a network entity that is designed to provide network analytics feedback (currently only slice-specific network status) to the policy control function (PCF) at the CN. NWDA is envisioned to identify service characteristics to form a BD model, which is then used to classify incoming traffic to allow customized and improved service delivery.

At the RAN side, Fig. 3.17 illustrates some use cases ranging from PHY-layer optimization, as well as protocol and signaling simplification, to network resource orchestration, where sensing, prediction, and intelligent decision-making form the BD analytical process.

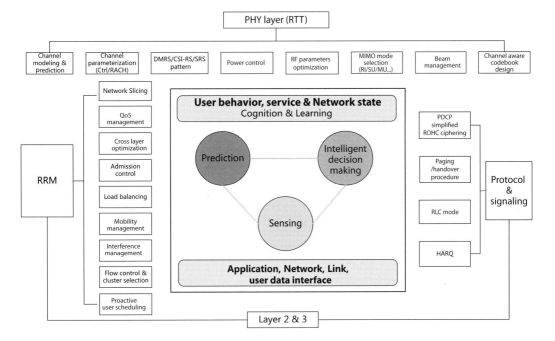

Figure 3.17 The collection of RAN optimization uses cases.

Furthermore, various forward-thinking characterizations and use cases of WBD have been identified in [88, 89], and provided a more integrated picture of the currently fragmented research in the field, albeit mainly focused on the core network again. In [90], a BD-aware wireless network has been proposed and some state-of-the-art signal processing techniques adaptable for managing WBD traffic have been identified. Authors in [91] have further analyzed problems, such as resource management, cache server deployment, QoE modeling, and monitoring in heterogeneous networks, and proposed solutions in a wireless BD-driven framework. However, the authors have not considered how to integrate the framework into the real network. In the next section, a WBD-enabled wireless communication network is proposed, with WBD implemented in both the CN side and RAN side.

In the upcoming sections we will try to minimize the research and application gap between the CN and the RAN by introducing BD and AI concepts in many use cases and scenarios on the RAN side, along with the architecture that supports such innovation. In terms of use cases, they can be roughly categorized into four types (illustrated in Fig. 3.18).

- Intelligent NMS/MANO: Automatic network management and control, such as coverage enhancement (interference management and coverage hole discovery); QoE monitoring and optimization through CP/UP signaling and active measurements; implementation of energy-saving schemes that rely on sensing the network traffic and user profiles; cooperation and control in heterogeneous networks and among multiple RATs; cross-layer service optimization, etc.
- Intelligent MEC: MEC enables a range of new applications that require massive connectivity, huge data volume, ultra-low latency and high reliability by ways of moving the computing and radio resources closer to the users; the success of MEC applications, such as proactive caching, depends on intelligent management that is based on the understanding of the various service requirements, network status, user profiles, etc.
- Intelligent RRM: Traditional RRM includes power control and allocation, channel allocation, handover, access control, traffic load balancing, end-to-end QoS control. All can benefit greatly with intelligent setting of the control parameters and operational models based on WBD.
- Intelligent RTT: This category lies at the most fundamental level of the wireless communication system, for instance, channel modeling, spectrum mapping, signal detection, automatic MCS selection, and rank selection; the insertion of intelligence will directly improve the air interface throughput.

3.3.5   Green and Soft Network Architecture with WBD

Overall Vision of the Architecture

To enhance the system performance of wireless communication networks, we have proposed a BD-enabled network architecture in [59]. As shown in Fig. 3.19, this architecture takes different network layers into consideration, i.e., the access network, the

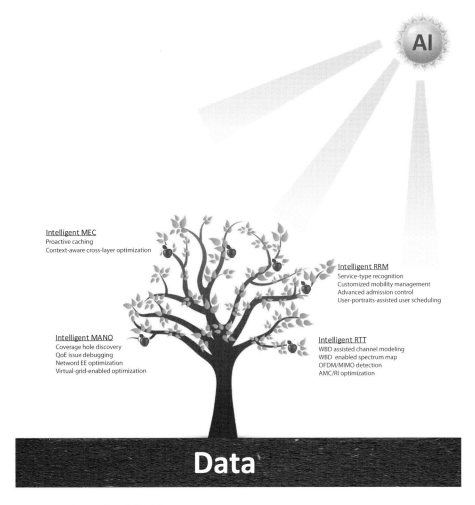

Intelligent MEC
Proactive caching
Context-aware cross-layer optimization

Intelligent RRM
Service-type recognition
Customized mobility management
Advanced admission control
User-portraits-assisted user scheduling

Intelligent MANO
Coverage hole discovery
QoE issue debugging
Netword EE optimization
Virtual-grid-enabled optimization

Intelligent RTT
WBD assisted channel modeling
WBD enabled spectrum map
OFDM/MIMO detection
AMC/RI optimization

Data

Figure 3.18 Categories of AI-driven uses cases

core network, and the IP backbone network. The distributed RAN with integrated BSs coexists in this architecture with Cloud RAN, which has a flexible functional partition between the CU and DU. Each network node, for instance, core network gateway (CN-GW), CU, DU, integrated BSs, and data-only BSs, which are equipped with low data-rate fronthaul and can only provide local data services, possesses proper storage and processing capabilities.

The BD platform is responsible for processing large volumes of data and providing useful information for resource optimization and RAN optimization. It can be deployed at either the CN or RAN side. For RAN optimization, the BD platform is recommended to be deployed at the RAN side and possibly in the CU. For CN optimization, the platform needs to be deployed at the CN. There is a possibility to define an interface between CN BD and RAN BD for information exchange. For RAN BD, the processing duty may be split between the CU and the DU. In phase 1 of 5G RAN standardization,

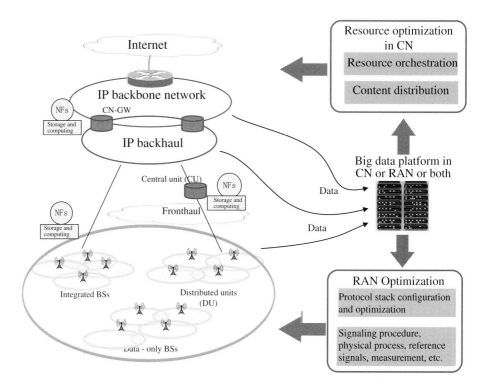

Figure 3.19 BD-enabled mobile network architecture.

the CU-DU architecture to support centralized deployment has been introduced to improve collaboration and pooling gains. Under such a centralized deployment, it is expected to deploy gNB-CU with more powerful processing and storage capabilities. Therefore, it is natural to host computation and storage intensive functionalities inside the CU, such as data collections, data storage, and data processing, including BD model training and distribution of well-trained models, etc. In the meantime, some relatively less demanding functionalities, e.g., data reporting, training-model-based decision-making, and execution, can be placed at DUs to support the real-time control. More details on the BD-empowered RAN architecture are investigated in [92, 93].

By leveraging data mining and data analytics, BD technology can help predict user mobility, traffic behavior, network load fluctuation, channel state variations, link level, and system-level interferences. This facilitates efficient resource assignment, flexible network capabilities distribution, flexible protocol stack configuration and optimization at each network node, signaling procedure, and physical layer optimization. BD changes the network design from the reactive BS-centric paradigm to the proactive user-centric paradigm.

According to the service requirements and network conditions analyzed by the BD platform, network functions (NF) implementation can be either centralized or distributed at the edge. To be more specific, the new features in BD-enabled networks are listed as follows.

- **On-Demand Resource Orchestration**

  In conventional networks, a lot of resources may be wasted in light-traffic scenarios. By utilizing WBD, it is possible to predict users' future service requests, location, mobility, and the network conditions. With such useful information, a proper amount of resources can be provisioned, guaranteeing high resource utilization efficiency and reducing network cost by avoiding over-provisioning. As an example, the processing resources at the CU can be intelligently configured according to the predicted traffic fluctuation to promote more pooling gain. In addition, some access resources can be turned off if no traffic is predicted in the corresponding coverage area.

  Network slicing is a key new feature of 5G networks, defined to suit highly diversified 5G services. In order to guarantee QoS of each specific network slice, some relatively static isolation of radio resources may be adopted, which may lead to insufficient resource utilization. Different slices may have diverse characteristics, e.g., distinct peak or idle traffic patterns. There exists a possibility that the usage of radio resources for different slices may compensate for each other. For example, a slice is at its busy hour, and the other slice is at its idle hour. Such traffic patterns of different slices can be identified via BD technologies, leading to a better utilization of radio resources among slices.

- **Flexible Content Distribution**

  With the assistance of the BD technologies, the network is able to predict users' traffic patterns with more accuracy. With adequate storage and computing capabilities, the network edge nodes could pre-fetch the predicted popular contents beforehand during idle hours, instead of fetching contents upon users' request via potentially overloaded backhaul links.

  Obviously, the closer the content is to the users, the less response latency they see. However, storage and computing at higher network levels mean more pooling gain and less maintenance cost. Different traffic patterns, such as content popularity distributions, may require content deployment in different levels of network nodes. For example, when the traffic is low, or the traffic trend is of low predictability, it is more attractive to place content storage and computing at a higher level in a cost-efficient way. However, when the traffic trend can be accurately predicted, it is beneficial to improve user's experience by moving the corresponding content storage and computing functions to network nodes that are closer to users.

- **Protocol Stack Configuration and Optimization**

  With flexible function split between CU and DU, the corresponding protocol stack configuration is required. Also, there may be the scenario where DC or CA is implemented in the network. The protocol stacks at the CU, DU, integrated BS, and data-only BS need to be flexibly configured. For example, with ideal fronthaul, MAC functions can be configured at the CU, optimally allocating resource between different DUs. BD analytics helps to optimize the protocol stack configuration and processing in various scenarios.

- **Signaling Procedure Optimization**
  Signaling procedure, as specified in various standards, stipulates the signaling flows between the UE, eNB, MME, SGW, PGW, PCRF, and home subscriber server (HSS), to use the terminologies of LTE as an example. With BD analytics, many signaling procedures can be simplified, which brings much-reduced operation complexity and latency and is urgently motivated in ultra-low latency applications.

- **Physical Layer Procedure Optimization**
  When the protocol stack is configured and the processing at each layer is optimized based on the service type and the scenario, the application data will be passed down, layer by layer, across the stacks. Traditional physical layer processing, such as synchronization, modulation and coding, multiple access, multiple antenna precoding, duplex mode selection, numerology configuration, reference signals, measurements and feedback, and power control, etc., can also be significantly enhanced via BD technology.

Data Driven Intelligent RAN Architecture

To leverage the WBD and AI capabilities for smart 5G, the data analytics functionalities need to be introduced into the network architecture [92]. The data analytics functionalities are mainly responsible for WBD collection, analysis, feature extraction, and model training, and provide intelligent network guidance for management and control decisions. It is quite different from the current network architecture, which mainly focuses on the communication part.

In order to meet both long-term network optimization and approximately real-time predication requirements for RRM/RTT algorithms, the architecture of the intelligence engine should be hierarchical and distributed. The overall reference architecture is illustrated in Fig. 3.20.

Multilevel BD processing and intelligent learning functions are supported. BD analytics functionalities are envisioned to be located at the operating support system (OSS)/MANO, the CN and RAN side. BD analytics will be naturally supported by the OSS/MANO for long-term network planning and network management, e.g., coverage hole discovery. The core network introduces NWDA to network data analytics (NWDA) to the 5G standard, which is used to collect massive data about each function of the core network, including user mobility data, service flow data, billing information, and network entity status data. NWDA can analyze the collected data, and improve QoS through policy control function (PCF), to enhance the user's service experience. Correspondingly, RAN domain should also define data analytics elements to optimize radio networks and coordinate with NWDA. We term this RAN data analytics element as RDA.

RDA is primarily used to support data-driven intelligent radio resource management and PHY-layer or higher-layer optimization in the RAN. The RDA should interact

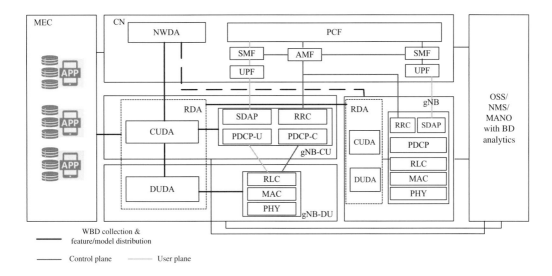

Figure 3.20 Data-driven intelligent RAN architecture

with both the control plane and user plane of RAN to realize the data collection and policy configuration. It also needs to provide data subscription services for NWDA and business and operation support system (BOSS)/OSS/MANO, and upload preprocessed subscription data to NWDA and BOSS/OSS for further BD operations and services. RDA can also subscribe to the NWDA results for the RAN-side service optimization. RDA may include central unit data analytics (CUDA) and distributed data analytics (DUDA).

CUDA is used in quasi-real-time optimization for RRC, SDAP [18], PDCP, and other protocol layers (such as multi-connection, interference management, mobility management, etc.). More specifically, it includes the data analysis, quasi-real-time predicting, decision-making model training, online model prediction, and strategy generation and configuration based on the predict results, in order to provide DUDA data features and model subscription distribution. CUDA can support both master and slave modes. The slave CUDA can request the master CUDA to perform some computational tasks, such as model training; while the master CUDA can conduct some computationally intensive model training for the slave CUDA, and offer some network-level collaborative optimization recommendations.

DUDA is adopted for real-time RAN data collection and preprocessing, prediction, parameter optimization and training tasks with low computational complexity in DU (e.g., PHY/MAC/RLC). DUDA needs to offer data features needed for training prediction/decision models after preprocessing to the CUDA, while CUDA can assist DUDA to conduct some computationally intensive model training tasks. Assisted by CUDA, the trained model can be sent to the DUDA for installation, perform real-time prediction/decision-making based on the real-time collected data, and generate corresponding strategy based on the prediction results, in order to perform real-time closed-loop control for the DU's process (such as scheduling, link adaptation, etc.). Note that the architecture should also expose north interfaces to the third parties and vertical industries, and training data should be comprised of E2E network data and application data in order to provide differentiated services.

Hierarchical and distributed architecture can significantly reduce data transmission costs, since the vast amount of collected data can be analyzed and trained for local needs without being uploaded to the centralized BD analytics. Additionally, online training and prediction for real-time applications (e.g., 1ms $\sim$ 10ms) can be distributed at the DU instead of the centralized nodes with large delay backhauling, which helps to guarantee the real-time needs for the RRM/RTT use cases.

3.3.6   Big-Data-Enabled Automatic Network Management and Operation

Coverage Hole Discovery

Network KPIs (key performance indicators) can be clarified into accessibility, retainability, availablity, mobility, traffic, and radio quality. Operators use KPIs to monitor how well a network is performing. Bad radio coverage can impact network KPIs significantly. If a coverage hole could be efficiently recognized and optimized, it will improve the KPIs. It's time and man power costly to find coverage hole with drive testing. WBD platforms make coverage hole recognition easy and efficient. It's done via statistics (clustering) based on information collected in WBD platforms, such as call trace, UE measurement reporting, base station measurement results, location, and other input.

From a network point of view, it's not easy to locate UE position precisely without a positioning feature enabled. It's a fact that most networks don't enable a positioning feature. Generally, triangulation is the way used to estimate UE location. It uses base station position (preconfigured by the network operator) and direction together with UE-reported RSRP/RSRQ to do the calculation. It will work but the precision couldn't be guaranteed. Using novel machine learning algorithms, which combine elements from supervised learning-based RF fingerprinting and particle-filter-based hidden Markov model learning used for robot path-tracking [96], operators can achieve median accuracy of 20–30 m. It shows significant improvement compared to no machine learning, which is about 100 m. With this more accurate positioning info, together with call drop events, handover failure events, and RRC reestablishment events, it can quickly identify the area of weak coverage, or coverage hole. Exhibiting the result in the map, it will guide maintenance teams to perform optimization.

As shown in Fig. 3.21, red-colored dots exhibit coverage holes calculated based on UE location, UE reported RSRP, and call drop/handover failure events.

QoE Issue Debugging

Good KPIs could indicate that the network element is working at its best, but they cannot ensure meeting the end user expectations. Modern operators shift to QoE monitoring and

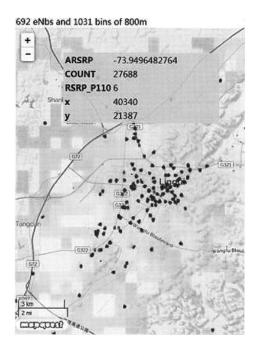

Figure 3.21 Coverage hole exhibition based on location.

optimization, which aims to improve end user satisfaction. QoE is more of a subjective experience. Factors such as personal mood could impact user perception. From the network's point of view, in order shift to QoE optimization, some KQIs (key quality indicators) are defined. KQIs could be MOS for voice, web page download time, round trip time, delay, jitter, packet loss rate, etc.

In WBD platforms, there are collections of traces from RANs and CNs, including control plane signaling, UP datas, and a variety of real-time measurement data. For a given user, it's possible to correlate information from different network elements. For example, using the unique identity of the interface in the network element, known as the time stamp, it can form an overview of a user. With all such history data and using ML algorithms, the performance problem of a given user could be automatically identified by the system, e.g., using decision trees, and the results are clustered and classified by BD analytics.

When a user is surfing the internet, if a requested web page is returned within 1s, the user feels good. If the same page were to return in more than 5s, the experience would be awful. A WBD platform can recognize all users experiencing web page download rates below a given threshold, e.g., 64 kbps for DL or 32 kbps for UL. All such users can be categorized to identify network coverage problems, device problems, terminal problems, or application server problems through decision trees, as shown in Fig. 3.22. All possible badly behaved user records are summarized and clarified; maintenance and development engineers can take this input for further checking and optimization.

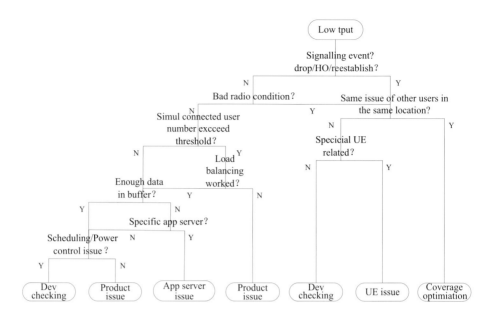

Figure 3.22 Decision tree used for low user throughput classification.

Network Energy Efficiency Optimization
Dynamic Enable/Disable Secondary Carrier

KPI data of network data volume and radio resource utilization with time of day statistics reveal that there are daytime busy hours and nighttime idle hours. It's consistent with the human's work-and-rest rhythm. There is also a significant tidal effect between office buildings and residential areas. In order to improve network coverage and user experience, operators deploy multiple carrier frequencies for hot spot. LTE advance features such as carrier aggregation can improve throughput significantly.

Power saving is one efficient means to reduce the OPEX for operators. On WBD platforms, various cycles can be achieved by the analysis of historical resource utilization data, traffic volumes. As there is also detailed application traffic information, more application behavior can be classified with periodicity and trends. With the location information, it can identify different types of coverage area for clustering. Based on the current collection of network resource usage and periodicity of change, it can provide guidance of switching ON/OFF the secondary carrier frequency. Compared to the base station method, which is based on real-time resource utilization, the number of online users to manage the extra carrier switch ON/OFF, BD platforms can more accurately predict the demand for resources and reduce the unnecessary ping-pong effect.

Figure 3.23 shows network traffic volume usage of a day for hour of a day. The time period between 23:00 and 6:00 are not shown. The secondary carrier could be dynamically switched ON/OFF based on the real network resource usage.

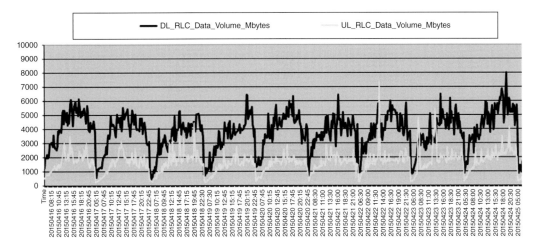

Figure 3.23 Hourly network traffic volume.

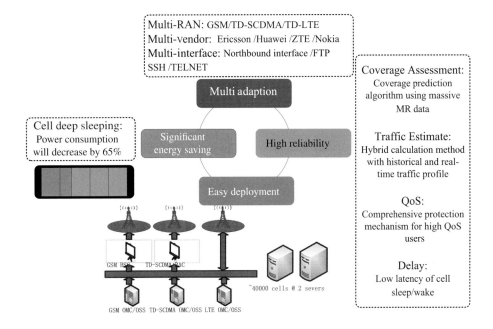

Figure 3.24 Illustrations of MCES.

Multi-RAT Cooperation Energy-Saving

To cope with the severe energy challenges caused by incessant mobile network expansion, the multi-RAT cooperation energy-saving system (MCES) was developed to effectively improve the energy efficiency of mobile networks. As shown in Fig. 3.24,

MCES interacts with the RAN in a real-time manner and can support multiple vendors' 2G/3G/4G RAN equipment. MCES identifies and coordinates cells with overlapping coverage to achieve network energy savings. Through BD analytics of massive MR data and traffic profiles, MCES can find the energy-saving cells and their compensating cell, and predict their traffic load trends. When the traffic load meets the criteria, for example, falling below some pre-defined threshold for some time, MCES will migrate its traffic to the compensating cell and place the energy-saving cell into a sleep state. With a real-time monitoring function, MCES can also turn on the sleeping cell before high-volume traffic arrives abruptly.

Up to now, MCES has been deployed for over 70,000 cells across ten provinces in China. The average annual electricity saving is 400,000 kwh in a 10,000-cell area.

3.3.7   Big-Data-Empowered MEC

Proactive Caching

Proactive caching predicts the contents that users may request and cache them at nodes of the wireless edge, e.g., base station (BS) and user equipment, before users send requests. The major approaches of proactive caching are caching at BSs, cache-enabled D2D communications, and pushing. When users send requests, the requested contents can be found from the cache of nearby nodes (e.g., macro BSs, pico BSs, or even user equipment) immediately and then were sent to the users, which can improve user experience [97], reduce E2E delay, and boost overall network performance efficiently [98–101].

To implement proactive caching, we need to resort to BD analytics to predict the contents to be proactively cached. In traditional wired networks, contents are stored at the content server, usually in a reactive manner. Since the content server has large storage size and covers a large number of users in a large time–space range, it is possible to predict the content popularity by BD analytics. However, in wireless networks, due to the limited size of storage and coverage area of BSs, designing caching policies based on the content popularity cannot improve caching gain efficiently. This is because the content popularity as a demand of multiple users cannot reflect the demand of each individual user, which results in lower cache-hit ratio and degrades the gain of improving network performance and user satisfaction. Therefore, to improve the gain of proactive caching, we need to predict the probability of each user requesting each file (i.e., user preference) and active level of each user. How to quickly learn user preference and active level for a large number of users with low complexity in practical environment with dynamically generated contents is an urgent task to be solved. Due to the limited number of user requests obtained at the BSS, the predicted file request probability is not accurate. Designing cache policies under these imperfect factors is also an inevitable problem.

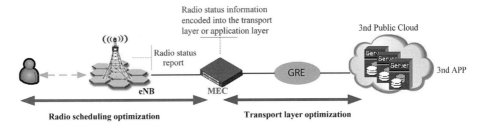

Figure 3.25 Illustration of cross-layer optimization.

RAN-Assisted Application Optimization

While the radio condition of the mobile network can fluctuate on the order of milliseconds and may occasionally result in packet losses even without network congestion, the application is adjusted on a larger timescale on the order of seconds and tends to attribute packet loss to network overloading. As a result, there is a misalignment between the RAN and the application, which may lead to noneffective usage of available radio resources and a degraded user experience. Therefore, it is highly desirable to introduce cross-layer optimization between the RAN and the application. The RAN may inform the application of the real-time radio air-interface channel status, based on which the application adjusts its transmission data rate. An illustration of the cross layer optimization is shown in Fig. 3.25.

Take TCP optimization for example, with some useful RAN information, e.g., buffer size, load of the base station (BS), the link throughput, and service type, packet error rate (PER), the TCP congestion window can be optimized and predicted to better match the radio channel variations. However, it is extremely hard to find a mathematical cross-layer model to determine the optimal TCP window with so many affecting factors. In this case, BD-assisted ML-based optimization offers an effective solution. With well constructed training data, a supervised learning-based model can be trained for the TCP window prediction. The BD-assisted learning approach allows for a good match between the TCP window and the wireless channel condition. This will significantly improve system throughput and buffer utilization.

The RAN may also identify the traffic features or traffic priorities within the application sessions via the big data analytics. Accordingly, the RAN can dynamically reconfigure its network protocols and parameters and performs prioritized scheduling strategies to guarantee traffic transmission of higher priority.

Intelligent Service Provision

The interworking between the 5G and 4G networks will be ever more tightly, especially with the introduction of NSA mode, making the network service flow more complex. How to control the flow of service and QoS, make full use of network resources, and protect the user's service experience will be some of the key factors affecting the network performance. Intelligence control of various services can be based on real-time data collection (including equipment and cell size of the resource occupancy rate, user distribution, user priority, traffic distribution, and other information) together with data-mining techniques to achieve 4G and 5G network load balancing, and QoS based on user granularity.

For the NSA scenario, the intelligent control center learns the traffic distribution characteristics via data-mining algorithms, automatically allocates the traffic load on the 4G or 5G base station according to the QoS requirements of different types of services, and promotes higher throughput and resource utilization and ensures the users' experience. In addition, the intelligent control center determines through user behavioral analysis and prediction, different QoS guarantees for each user, and reserves enough transmission resources for high-priority users in advance, which encourages user loyalty. At the same time, proactive edge caching can be achieved through intelligent service control. By using MEC and coupling it with the wireless network status, the platform can learn the service profile and the best practices so as to better allocate and schedule resources according to the service priority, timing, and resource availability and further enhance the content hit rate and resource utilization efficiency.

3.3.8 Big-Data-Assisted Protocol Stack and Signaling Procedure Optimization

Protocol Stack Configuration

The traditional protocol stack for the normal integrated BS is shown in Fig. 3.26a, where the data processing is through PDCP, RLC, MAC, and physical processing, with the management of RRC. While in DC, as shown in Figs. 3.26b and 3.26c, the PDCP function resides at the master eNB (MeNB). The secondary eNB (SeNB) is only responsible for RLC, MAC, and PHY. For CA operations, joint MAC scheduling is feasible, leaving the SeNB responsible only for PHY processing, as shown in Figs. 3.26d and 3.26e. For both DC and CA, there is only one RRC at the MeNB. As shown in Fig. 3.26f, when the data-only BS is deployed with low data rate fronthual to provide local data services, TCP and IP processing can be conveniently omitted. The PDCP processing can be further simplified, e.g., IP header compression is no longer needed. With BD analytics, BSs may receive recommendations on the operation mode, e.g., CoMP, DC, or CA. The protocol stack is also configured accordingly. For example, with the information of user's geographical distribution and/or the intercell interferences, CoMP schemes may be adopted in a certain area for better performance.

In the CU/DU architecture, protocol stack partition between CU and DU can also be optimally configured via BD analytics based on service type, fronthaul capability, frequency band, user mobility, QoE, etc. Some exemplary cases are envisioned and explained as follows:

- For low latency services, fewer functions will be allocated to CU, e.g., full eNB protocol stack at DU.
- For high-frequency bands, such as mmWave, more processing is required at DU to alleviate the burden of fronthaul from extremely high data rate. One stack

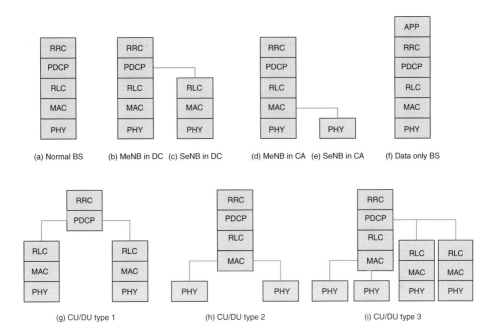

Figure 3.26 Protocol stack configuration.

configuration for cases 1 and 2 is shown in Fig. 3.26g, where RRC and PDCP reside at CU, while RLC, MAC, and PHY are handled at DUs.

- For low-frequency bands where severe intercell interference exists, more processing will be motivated at CU for efficient interference mitigation.

- With ideal fronthaul, MAC function can be configured at CU, optimally allocating resource between different DUs. While with nonideal fronthaul, MAC function at the central unit is not necessitated since cross-cell fast scheduling and resource allocation may not be well supported due to much longer fronthaul latency. In this case, only RRC and PDCP can be allocated at the CU. One stack configuration for cases 3 and 4 is shown in Fig. 3.26h, where RRC, PDCP, RLC, and MAC reside at the CU while PHY is handled at DUs. A hybrid CU/DU protocol stack is shown in Fig. 3.26i, where the CU is controlling DUs with different protocol stack configurations.

- For high-mobility users, integrated BS with a full protocol stack implementation is not preferred. Cloud RAN architecture with RRC at the CU is more suitable.

Protocol Stack Processing Optimization

- Reconfigurable Compression and Encryption in PDCP
  After the protocol stack is configured, the data processing can also be further optimized based on BD analytics. Taking a look at the robust header compression

(ROHC) mechanism, it is utilized to handle user plane data flow that has large packet headers; however, the incurred delay takes up 20.01% of the L2 total delay. BD analytics can be used to identify data packets or data flows with the same service types. The identified latency-insensitive IP packets can then be aggregated into a large data packet that shares one IP header, resulting in a much reduced ROHC's processing delay.

Traditional UP and CP packets all need to be encrypted when passing through the PDCP layer. The delay caused by ciphering process accounts for 59.16% of the L2 total time delay. If the service type of the packets can be analyzed and identified by BD processing, differentiated ciphering processing can be selected accordingly. If some services are not private, they do not need ciphering, thus reducing the processing delay and complexity. For example, e-commerce transactions and news browsing definitely have distinct privacy requirements. Therefore, ciphering over the air should be adaptable for different service categories to avoid unnecessary overhead. Besides, powerful BD analytics is capable of identifying potential security attacks, monitoring and eliminating potential dangers, and may effectively reduce the necessity of data ciphering.

- Optimized Transmission Mode in RLC
  Traditional RLC modes are configured by CN according to service types. Audio and video flow services basically use unacknowledged mode (UM), but acknowledged mode (AM) can be used if delay requirements are not very strict, under which reliability will be greatly improved. The UM mode transmission is more suitable for small-packet traffic as well, which usually has a small number of segments and generally does not need ARQ. BD analytics is capable of identifying traffic delay sensitivity and accurate identification of small packet traffic, bringing much-reduced delay and processing complexity.
- MAC Hybrid ARQ
  Through statistical analysis of the channel and traffic characteristics, maximum retransmission numbers can be dynamically configured by hybrid ARQ. This will bring overhead reduction and better resource utilization.

Signaling Procedure Optimization

The implementation of BD platform in wireless communication networks will naturally lead to much changes of the signaling procedure.

Taking handover as an example, we will investigate how BD helps to simplify the system operation and brings performance improvement. The basic procedure of the X2 handover [94] is illustrated in Fig. 3.27 (not including the gray color procedures with arrows and the BD center), which is very complicated. The handover is generally triggered by the eNB, based on the measurement report feedback from UEs if certain criteria of the measured channel conditions in the adjacent cells are met. The measurement process is complicated and the signaling overhead is large.

Based on BD analytics, one possible handover approach is proposed as follows (shown also in Fig. 3.27 with bold gray arrow lines). Note that the BD center in the CN is depicted in this figure, which coordinates adjacent eNBs for possibly more efficient

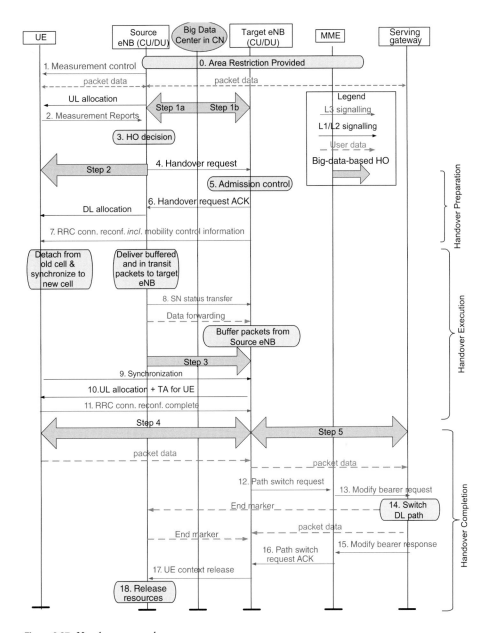

Figure 3.27 Handover procedure.

handover. For other handover cases, e.g., the CU controlled intra-DU handover, the procedure can be modified accordingly.

- Step 1a: The BD center in the CN determines when each UE should handover to which target eNB based on an accurate prediction of UE's location and movement. It sends a handover command to the source eNB, with information of the target eNB.

- Step 1b: Meanwhile, the BD center sends a handover command to the target eNB, possibly with the sequence number (SN) status. Note that without BD capabilities, the handover needs multiple signaling exchanges between the source and target eNB, e.g., handover request, ACK, SN status transfer.
- Steps 2, 3, and 4 (steps 7 to 11 in the traditional handover): These can be the same as the traditional handover.
- Step 5: Path switch (steps 12 to 16 in the traditional handover) and context release can be made before steps 2, 3, and 4 finish, especially when the BD center is sure of the success of the handover.

Compared with the traditional handover procedure, BD-based handover has the following advantages: 1) Signalling overhead is reduced, e.g., handover request, acknowledgment (ACK), admission control, and possibly SN status transfer can be omitted; 2) UE's measurement and feedback efforts can be reduced; 3) Handover interruption time can be reduced due to reduced signaling procedure and to possible concurrent UE access to the target eNB and path switch process; and 4) Ping-pong handover can be effectively avoided.

This methodology can be extended to many other signaling procedures. For example, in the "attach" procedure, if it is not initial access, and the time interval between the current attach action and the previous one is small (e.g., several minutes), the identity, authentication, and security procedures can be significantly simplified or even omitted.

3.3.9   Big-Data- and AI-Enabled Radio Resource Management

Network Slicing Optimization

Next-generation (5G) wireless networks shall support various services with different characteristics. In addition to what has been supported in LTE networks (e.g., mobile broadband, VoLTE), it would support ultra-reliable and low-latency communications (URLLC) and massive machine-type communications (mMTC) services. Different services have different requirements over wireless networks, e.g., real-time video service and voice service is more sensitive to delay but has relatively loose requirements over packet loss rate; intelligent meter, like IoT devices, has strict requirement over reliability but loose requirement over bandwidth and delay. In order to meet the diverse requirements of various services, 5G introduces network slicing technique into the same physical network instruments. Network slicing is a logical concept, achieved by assigning services with the same network requirements to a slice. Different network slices are logically independent of each other and physically share the same network resources.

Per experience of traditional network planning, engineers should build traffic models and then configure resource reservation of each slice. The difficulty here is that there is no such data available, and traffic model cannot reflect real resource requirement. On WBD platforms, there are collections of user traffic data (e.g., via DPI) and location info. With statistical analysis of packets, it can cluster services based on application behavior and its resource requirements, devices distribution, and periodicity. It can predict slice resource requirements by using ML and feedback to the network element for auto tuning.

Customized Mobility Management

The 5G registration area will be similar to the 4G tracking area. The registration area list is maintained in both UE and core access and mobility management function (AMF), and 5G RAN still needs to broadcast registration area code to UE. There may exist three mobility categories for UE, namely, no mobility/restricted mobility/unrestricted mobility.

Without big data, it shall be very difficult to track and categorize UEs' mobility patterns. It is beneficial to leverage BD analytics to mine the collected network information so as to precisely predict a particular UE's mobility pattern. As shown in Figure 3.28, the associated UE could track, for instance, gNB lists or cell lists per time-of-day and then feedback the earlier data analytics to the network. This allows AMF to page the UE via the reported data and therefore bring down the paging load in gNB and save corresponding processing resource in gNB.

The handover parameters have a close relationship with different network propagation environments (such as building occlusion), interference, load, and service types. If the handover parameters are set unreasonably, premature handover, late handover, or a ping-pong handover may occur, which deteriorates the handover performance and load balancing performance, and finally results in poor user experience. BD analytics can be used to collect and learn handover data in different network environments, such as different cells, slices, user types, and service types, and then to predict cell interference, as well as load and user service performances, and adaptively set handover parameter configurations.

In general, traditional methods always make handover decisions based on instantaneous measurements. BD analytics algorithms could fully exploit the historical

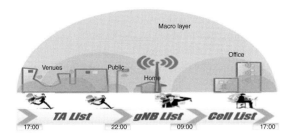

Figure 3.28 Customized mobility management.

information and predictive information during handover procedures, which would help make intelligent decisions.

Predictive Resource Allocation

Traditional wireless systems are designed for E2E communications, which assume that the required contents, the request time, the user location, and the user trajectory are random. In these scenarios, the static resource allocation scheme is considered in wireless radio access. In other word, there is a fixed matching between the association resources (e.g., radio spectrum and BS) and RAT. As a result, the resource allocation scheme cannot be adjusted adaptively according to the dynamic tendency of spectrum and traffic load. In the network that is designed based on such a principle, the unbalanced space–time distribution of traffic causes the BSs in hot spots to overload, but there are a lot of idle resources in other BSs or in idle time.

Predictive resource allocation makes the plan for radio resources allocation in advance based on the prediction of the network environment, channel quality, and traffic demand. For example, for real-time traffic (e.g., phone calls and video conference), it is possible to ensure QoE by forecasting the traffic demands and reserving resources. For non-real-time traffic (e.g., video on demand [VoD] and file download traffic), it is possible to borrow future residual resources to improve overall network resource utilization with the individual QoE guarantee, according to user-required content, QoS requirements, and the prediction of user trajectory, channel statistics, and interference statistics. For predictive resource allocation, the related context information includes application-level information (e.g., QoE of VoD, video conference, voice, and other traffic), network-level information (e.g., congestion status, spectrum, and interference environment), user-level information (e.g., mobility trajectory and the average channel gains in corresponding locations), and device-level information (available storage capacity and remaining battery capacity of mobile terminals). So far, the existing research on predictive resource allocation considers user-level or network-level context information to allocate resources in advance, which can improve network throughput [102], reduce transmission costs, enhance user experience [102, 103], and lower network congestion [104–106].

Effective predictive resource allocation requires the prediction of user behavior, network traffic, and spectrum situation. The predicted information relies on a mass of comprehensive data from low-level layer to high-level layer, which raises a higher demand to the network function than before. On the existing network configuration, it is necessary to deploy the radio spectrum data collection module on the edge network, introduce the network traffic acquisition and recognition module at the convergence layer (e.g., the gateway), and add the matching module between the traffic information and the traffic data in the CN. Moreover, since the highly dynamic characteristic of the wireless environment causes stringent requirements on data collection, feedback, and processing cycles, it is necessary to introduce high-rate, high-reliability, and low-latency control information exchanging channels in existing network systems. In addition, the massive data generated by the dense data collection module causes serious challenges on the control network, hence it is also necessary to solve the problem of all kinds of data collectors deployment.

Intelligent Network Access Control

The existing strategies of wireless device access to different-mode networks and different access points (AP) rely on a single instantaneous indicator, such as the received signal strength and the theoretical rate. The limitations of traditional access strategies in the future network have become increasingly obvious. For example, in a mmWave communication network, due to the small coverage of the network, the traditional access strategies lead to frequent switching among different APs (i.e., ping-pong effect), where redundant switching severely degrades the user's throughput. In addition, for IoT applications, diverse service-quality requirements and special device characteristics make it difficult for massive devices to access the networks quickly and easily [107]. In order to realize the intelligent access of wireless devices, instantaneous information will no longer be the only basis for decision-making; historical information and forecasted information will also be taken into account. In particular, the historical data to be utilized include network information such as received signal values, device status information, device geospatial information, throughput, and latency.

WBD contains information closely related to network behavior and the user behavior, which brings great opportunities for intelligent wireless access. With the help of BD analytics, we can extract the space–time variability and relevance of wireless service/user/network behaviors, and provide the basis for decision-making to realize intelligent and efficient wireless access [108]. There are still many problems to solve for BD-enabled wireless access, including the assessment of the service quality of different networks and the establishment of a unified assessment system. Also, the process for handling the huge feedback generated by massive devices to assist decision-making is not clear. It is also important to transfer the extracted knowledge from offline analysis of massive historical data to specific real environment for real-time online decision-making.

In order to realize WBD-enabled intelligent wireless network, the network structure needs to integrate the following modules. Firstly, an offline analysis module is required in the cloud to achieve joint mining and complex analysis of BD. High-level semantic knowledge is extracted through the complex mining of multi-modal data and will be used for the whole network. Secondly, a semi-real-time analysis module is required in the local BSs (group) to localize the off-line extracted knowledge. The module will combine the high-level semantic knowledge provided by the cloud and the local information to derive knowledge suitable for local learning. Then, an online real-time analysis module is required in edge devices to achieve real-time matching of highly dynamic networks. The module will combine the localized semantic information and real-time measured data to make quick decisions. Finally, to realize the sharing and regeneration of knowledge in the whole network, a separate knowledge flow network is required to add to the existing data transmission network, for continuously improving the intelligence of the whole network.

Coverage and Capacity Optimization

Coverage and capacity optimization (CCO) is one of the typical operational tasks of the RAN. CCO aims to provide the required capacity in the targeted coverage areas, to minimize interference and maintain an acceptable QoS in an autonomous way. To achieve these targets, antenna power and configuration (pilot power, antenna down-tilt, antenna azimuth, or massive MIMO pattern in 5G) play a critical role, as they affect the direction of the antenna radiation pattern. Fixed RF parameters could not bring the best network performance for the ever-changing radio network environment. CCO can be used to improve the received signal strength in its own cell as well as to reduce the interference to neighboring cells by selecting appropriate RF parameters.

The network is very complicated, and it is not possible to find definite function to map between the RF parameters and the target coverage and capacity performance. The main reason is that the set of configurable RF parameters is multidimensional, and each RF parameter has a wide range of values, leading to very large number of possible options.

Reinforcement learning with the neural network is used to adjust RF parameters for CCO. The algorithm can leverage ML to analyze and learn what the proper action is for each current network state. A neural network is used to build the mapping between network states, RF parameters, and the network performance. Compared to the traditional Q-table method, it has better generalization ability and can respond to changes in the wireless network in a good way. The algorithm also punishes KPI violations. This policy guarantees the network KPI remaining stable during the process of CCO.

RF optimization based on CCO can be used in many scenarios, such as carrier aggregation (CA), energy saving, and SFN. A CA scenario is present in Fig. 3.29 by way of example. Nowadays, more and more terminals begin to support multiband CA characteristics, which makes CA terminals fully use multiband network resources, improving terminal throughput and other performance. Because the different frequency bands are not related to coordination or antenna pattern, the CA terminal of the CA wireless environment is different. For example, some terminals are in the poor CA area (low-quality CA area) in multiple frequency bands and RSRP, although giving multiple frequency bands resources, the overall throughput of current CA users is not very large.

CCO-based RF optimization is a scenario-oriented CA feature-enabling algorithm, which improves the capacity upper bound of CA characteristics in the multiband region, and forms a good complement to both the RRM and CA characteristics.

The future of CCO-based intelligent RF will involve scenario-oriented RRM characteristics to constantly improve their theoretical upper bound.

Virtual-Grid-Enabled Network Performance Optimization

Compared to traditional geographic grid, Virtual Grid does not need to divide the grid according to actual locations. Instead, the system measurements (e.g., RSRP) of multiple cells are used to define the grid. Historical KPI statistics were stored in Virtual Grid, and used for grid-level radio performance optimization. The concept of virtual grid is illustrated in Fig. 3.30, where the grid is defined by the RSRP measurements from three adjacent cells.

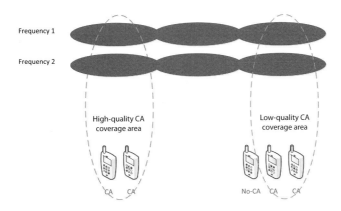

Figure 3.29 AI RF enabling CA roof.

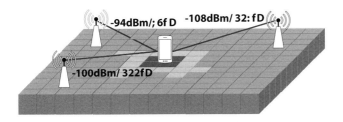

Figure 3.30 Virtual raster example.

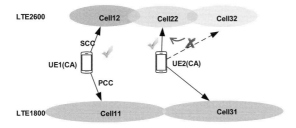

Figure 3.31 SCC selections in CA.

In the inter-frequency, inter-systems and multifrequency network scenarios, changing the granularity of the current cell-level algorithm to the grid level can improve the performance of many features such as CA, MLB, inter-frequency and inter-systems handover CSFB, and SRVCC. For example, in a CA scenario, it usually configures SCC in a blind way to reduce the GAP measure of inter-frequency, however, the carrier or PCI selected may not be the best, and sometimes may even fail to configure. As shown in Fig. 3.31, UE1 and UE2 will select Cell12 as SCC when using blind configure, but UE2 may fail. UE1 and UE2 will select Cell12 and Cell22 as SCC respectively when using the Virtual Grid method.

Figure 3.32 Wireless network user portraits.

For BD-driven methods, the system measurements of multiple cells act as wireless fingerprints, which are used to associate the statistics and measurements stored in the grid. After that, the model can be built and then be used to predict desired statistics and measurements given the inter-frequency measurements. By doing so, suitable policy and action can be selected to improve the performance of many radio features.

User-Portraits-Based User Scheduling

User portraits refer to the classifiaction of users by given features based on real data. Through user classification by wireless network features and data, the wireless network can be targeted on-demand to provide the appropriate services to improve the user experience and network utilization.

There are many dimensions that describe user features in the wireless network. For example, as illustrated in Fig. 3.32, RSRP, RSRQ, and CQI could describe the characteristics of the wireless environment; terminal type, chip type, and transmission capacity could describe the characteristics of the user type; service type and buffer length could describe features of user demands; some features like location, movement speed, and trajectory could be used for further analysis.

For different wireless applications, we can choose different multidimensional feature combinations to classify users. For example, you can sort the user transmission priorities by the service type and buffer length; assign the appropriate cell access and presence policy to the user through the service type and the movement speed; user location, moving speed, motion trajectory, or business type to find appropriate user pairing.

3.3.10   Big-Data-Assisted High-Efficiency Physical Layer Operation

Channel State Information (CSI) Acquisition and Feedback

The availability of CSI at BS and UE is crucial for wireless communication systems. To this end, reference signals are widely adopted in various wireless standards for estimation of fast-changing wireless channels. For time-varying channels, the wide-sense stationary uncorrelated scatter (WSSUS) model is widely used in theoretical analysis. Typical channel responses in a multiple path environment in time domain (i.e., $h(t, \tau)$, where, t and τ denote time and delay spread, respectively) and in frequency domain (i.e., $h(t, f)$, where f denotes frequency) are all very dynamic, especially in high mobility scenarios.

The density of reference signals in frequency division duplex (FDD) systems should be high enough to capture the channel's characteristics in both time and frequency domains. Generally, this density is selected to accommodate high mobility and large delay spread, especially for the cell's common reference signals. In cases where most users within the cell are moving slowly, a high density of the reference signals leads to unnecessary overhead, which is over-provisioning. The situation is more problematic with a large number of antennas at the BS, since the overhead of reference signals scales with the number of antenna ports.

One efficient way to minimize the overhead of the reference signals is to resort to WBD analytics. With the prediction of users' mobility and wireless channels' statistics, the BS is able to schedule common reference signals accordingly. If there are no high-mobility users, very sparse reference signals may suffice. If high-mobility and low-mobility users coexist, the BS can schedule dense reference signals in some frequency bands and sparse reference signals in other bands, thus effectively minimizing the overall overhead.

In time division duplex (TDD) systems, the CSI can be obtained by exploiting the channel reciprocity using uplink pilots. However, due to limited training sequences, pilot contamination may dramatically degrade downlink performance, since the uplink channels are contaminated by intercell interference from the adjacent cells. In order to reduce pilot contamination, with the help of BD, joint user and pilot scheduling schemes can coordinate the non-orthogonal pilot sequence among users of weak mutual interference, and arrange orthogonal pilot sequences to users with strong mutual interference.

Transform-Domain Signal Processing for Semi-Static Scheduling

Dynamic channel variations in the time and frequency domain necessitate dense reference signals, fast feedback, adaptive modulation and coding, and fast scheduling, thus posing tough challenges for wireless communication system design. Fortunately, via Fourier transformation with respect to variable t, the dynamic $h(t, \tau)$ can be transformed to a stationary $h(v, \tau)$ in delay and Doppler domains, where v is Doppler frequency. In contrast to the time and frequency domain channel responses, the delay and Doppler domain channels are more stable, depending merely on the multipath channel structure (angular distribution and the power delay profile) and mobility. Therefore, it is almost static if the channel structure and mobility does not change.

The direct impact of the transform-domain signal processing is the alleviated difficulty in tracking the time-varying fading. This is particularly useful in high-speed train communications. The significantly increased coherence time of the effective channel in transform domains brings abundant opportunities for the simplification of wireless systems in both the standardization and the implementation. For example, reference signals can be designed with very low overhead and the channel feedback need not to be fast. Channel coding schemes can also be simplified. The well-studied AWGN codes may perform sufficiently well over the effective channel, thus alleviating the burden of traditional adaptive modulation and coding. Moreover, it enables FDD massive MIMO in moving applications due to easy CSI estimation. Most importantly, the slow variation of the effective channels in transform domains can significantly facilitate BD analytics since analyzing the statistical channels may be enough for satisfactory PHY and MAC operations.

Flexible Frame Structure Configuration

Since there may be many use cases emerging in 5G and beyond, it is very important for operators to deploy one network to support all deployment scenarios and use cases. Toward this end, it is critical to adopt one unified and flexible air interface framework to meet diverse requirements of the key 5G scenarios, e.g., eMBB, URLLC, and mMTC. The unified framework of SDAI [95] may include flexible frame structure, waveform, duplex mode, multiple access, MIMO, coding and modulation, and corresponding layer 2 and layer 3 signaling. For efficient operation of SDAI, the key is a flexible frame structure, e.g., the numerology can be dynamically configured; the time resource within each subframe can be flexibly partitioned between downlink and uplink; the duration of the transmit time interval is adaptive; and the period of uplink feedback of ACK/NACK is configurable. The practical implementation of flexible frame structure at each BS is very difficult. For example, flexible downlink and uplink transmission may bring severe intercell and intracell interference. However, mitigating interference, especially the cross-link interference, can be very challenging. Another example is that the frequent uplink feedback for the latency-sensitive service may cause severe interference to the downlink of the latency-nonsensitive service. Thanks to BD technologies, a lot of useful information can be utilized to optimize the frame structure, e.g., UE's service types, QoE, location, traffic volume, mobility, channel information, and inter-user interference.

Power Control

In mobile communication systems, the power control mechanism is mainly used in uplink to adjust UEs' transmission power for the purposes of compensating time-varying path loss and reducing mutual interference. The power control procedure of LTE uplink data transmission generally consists of two parts, basic open-loop operating point and dynamic power offset.

The basic open-loop operating point is determined by UE bandwidth M, preconfigured cell nominal power P_{o_PUSCH}, estimated downlink path loss PL and path loss factor a. The BD technique could potentially aid the value setting of certain parameters. For instance, the path loss factor is chosen to balance the edge UEs' performance.

In the downlink, eNB marginalizes the functionality of power control, and only adopts power allocation between downlink traffic data and control signaling. By utilizing network load data, we can optimally allocate the power ratio to maximize the network energy efficiency under the constraint of various QoS requirement of UEs. Furthermore, benefiting from the accurate estimation of channel state information and characteristic of deployment scenarios, a more flexible downlink power control scheme may be also considered in future mobile network design, e.g., interference map-based power control.

Machine-Learning-Based MIMO Transmission

In those large-scale MIMO-equipped future wireless communication scenarios, the computational complexity in determining optimal transmission using matrix calculations becomes unfeasible. In particular, when combined with multi-carrier technologies such as OFDM, the optimal transmission mode cannot be obtained in closed form. Applying ML on massive MIMO link adaptation can deeply exploit the inherent connection among channel characteristics, transmission mode, and error performance. Based on data-driven ML methods, the mapping relation can be effectively learned by observing the channel data and the corresponding error performance so as to select the optimal transmission mode for each channel realization. It is of vital significance to apply the combination of deep learning and classification algorithms in ML on feature extraction of the channel state information so that it can improve the adaptive learning ability and universality of the MIMO link, further reducing the computational complexity.

3.3.11 Big Data Platform Capabilities/Environment

Platform Requirements and Definitions

To support fast-evolving, data-driven wireless services, the following five capabilities are required for the WBD platforms: 1) BD clusters management capability; 2) BD analytics capability; 3) wireless networks, wireless transmissions, and wireless applications support capability; 4) self-optimization, iterative update, and continuous integration capability; and 5) the AI/ML framework support capability.

The key features of WBD platforms are summarized as follows:

1. **Simple and unified management interface:** The management interface is expected to effectively support the WBD platform construction, operation, and maintenance services. It needs to support forming a distributed WBD platform via the management interface. In addition, different BD software and an integrated BD capacity platform can be built according to the different hardware resources of components. For the user interface, powerful engineering management capabilities and unified operating interface are desired.

2. **Efficient off-line data analysis:** To efficiently deal with the massive amount of structured, semi-structured, and unstructured data, the WBD platform needs to be able to do unified processing for different types of different systems. In addition, specific efforts are also needed for fast and stable processing of WBD.

3. **Real-time online data analysis:** For some wireless applications, the platform needs to provide real-time analysis capabilities to make timely decisions. The timescale for the real-time decision can be by the hour, minute, second, or even millisecond.

4. **Support off-line algorithm:** Data analysis usually provides an elementary statistical analysis. For a deeper analysis, a ML algorithm may be needed to build a dedicated decision model. The WBD platform needs to provide the off-line algorithm for the original off-line data analysis, e.g., model training. The off-line

algorithm mainly refers to non-incremental algorithms. Taking the wireless network optimization use cases for example, instead of building complex mathematical models through manual experience, the WBD platform helps to build a data-driven model, which is expected to obtain a more accurate and practical mapping, and automatically provide suggestions.

5. **Support online algorithm:** Online data analysis provides real-time feedback for fast and timely response and decision-making. The online algorithm mainly focuses on the incremental algorithm or the reinforcement algorithm that can make the previously constructed model evolve in real time. It helps to reduce the human intervention and avoid waste.

Framework of Wireless Big Data Platforms

A WBD platform can be built based on the cloud or physical servers. The framework of the WBD platform is shown in Fig. 3.33. The platform is divided into three layers, namely the data acquisition and preprocessing layer, the BD computing layer, and the application/algorithm layer. It uses the corresponding interface between the different components for isolation.

For the data acquisition and preprocessing layer, we mainly consider the collection of the commercial network data, laboratory simulation data, and third-party data. Data transmission includes both off-line transmission and online transmission according to the application requirements. As the data form is very diverse, a unified representation

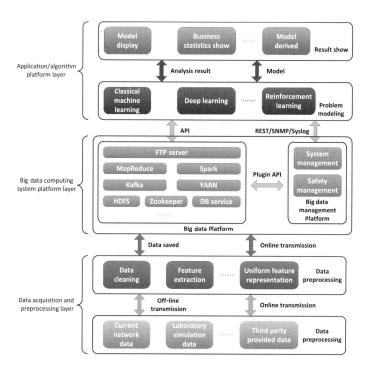

Figure 3.33 Framework of a WBD platform.

is desired. Moreover, we need to preprocess the data by cleaning and feature extraction before using it for specific applications. With regard to specific use cases, the data can be stored in a BD database or directly input to the algorithms.

The BD computing system platform layer is composed of the BD analytics part and management part. The management part is mainly responsible for BD platform operation and maintenance, platform capacity expansion, etc. As for the BD analytic part, we need to take account of the wireless services' requirements on the platform, and the capability of the different physical servers. For example, a machine that computes superior performance can deploy Spark services to support online and off-line computing. While machines with sufficient storage space and better performance exist, Spark and HDFS services can be deployed to support off-line processing of massive amounts of data. As communication between servers often limits the real-time capabilities of cluster computing in BD clusters, it is recommended to use fiber, optical switching, and other more rapid communication media to speed up the communication between the nodes.

For the application layer/algorithm platform, the following three aspects should be considered. The first is the project management capabilities. It is required to be friendly to users with different backgrounds. The second is the algorithm. Many outstanding ML libraries can be integrated. The ML algorithm can be divided into classical ML algorithms, incremental learning algorithms, deep learning, and reinforcement learning. For classical ML algorithms, the needs of stand-alone and distributed clusters must be taken into account, blending in such tools as scikit-learn, H2O.ai, Spark MLlib, and so on. Incremental learning can refer to the StreamDM system developed by Huawei Noah Labs. For deep learning algorithms, the platform can be integrated with Tensorflow, Caffe, Theano, and other deep learning systems. For reinforcement learning algorithms, Caffe2, DeeR, Tensorflow, and other existing system frameworks can be considered. The third is the efficient communication to support real-time online analysis and algorithm construction for wireless service.

Data Representation in Wireless Big Data Platforms

The data processing performance of the platform heavily relies on the data representation. It not only affects the storage of data, but also affects the efficiency of data processing. Here, we provide some guidance for the data representation of different data structures.

In general, data can be divided into structured data, semi-structured data, and unstructured data. A two-dimensional table is usually used to express the structured data. Call history record, words, KPI data from the commercial network, or laboratory simulation can be regarded as structured data. It is not convenient to present unstructured data as a two-dimensional logic table. It generally includes all formats of office documents, text, pictures, images, audio/video, and so on. Compared with structured data, unstructured data usually has unknown data semantic information and requires a large amount of space for single-record storage. Therefore, it is difficult to store and retrieve the unstructured data. Semi-structured data exists between fully structured and completely unstructured data, such as HTML documents. It is generally self-describing; the structure and content are mixed together.

For unstructured data, it is recommended to use a table containing the metadata of three fields to represent: number, description (varchar (1024)), content index (blob). Metadata uses structured data representation to manage the source data. Through the content description, we can know the meaning of the data. Through the content index, we can obtain the corresponding raw data. For the storage and representation of semi-structured data, two widely used methods are suggested:

1. Decompose the semi-structured data into structured data by extracting the the required fields. If the data does not contain certain attributes, the vacancy or default value can be used. The rest of the data can be added to the memo field as the note information. This method is convenient for querying. But it lacks flexibility of expansion, which is unable to handle the extra information beyond the preliminary design.
2. Use XML format to organize the unstructured data and save to the CLOB field. Information of different categories can be stored in different nodes of XML. The advantages of this approach are: it is easy to expand the information by simply changing the corresponding DTD or XSD. Disadvantages are the relatively low query efficiency.

3.3.12   Enhanced System Performance with WBD

In the previous three sections, we have discussed how WBD can impact wireless communication network design, from different perspectives, such as network architecture, protocol stack, signaling procedure, and physical layer operations. A BD-enabled network architecture has been proposed, along with design considerations on protocol stack configuration and simplified signaling procedures such as handover, simplified stack processing in PDCP, RLC, and MAC layer, low overhead reference signals, and flexible frame structure. The potential impact of transform domain signal processing on system design is also discussed, which facilitates the application of WBD. WBD, which is generated in mobile networks and is seemingly a burden to the network, nevertheless can be eventually transformed to a blessing, enabling much simplified network operations and standardization. In this section, the enhanced system performance with WBD is investigated with some preliminary simulation results. We will present three typical use cases in the wireless domain [92], i.e., mobility management, TCP window adjustment, and beam sweeping procedure, where BD and AI algorithms are used to optimize simulated system performance in a heuristic manner.

Big-Data-Enabled Efficient Mobility Management

With BD, user behavior such as trajectory pattern can be accurately predicted, which helps to improve the efficiency of the mobility management in cellular networks. In LTE, mobility management includes the paging procedure in the idle state and the handover procedure in the connected state. In the following section, the efficient mobility management scheme based on the BD analytics is discussed for handover procedures.

When UE is at the connected state, mobility management is realized by the handover procedure. Handover optimization has two sometimes conflicting objectives: minimizing the unnecessary handovers and minimizing the likelihood of dropped calls. The state-of-the-art optimization method is mainly based on two tunable parameters: time to trigger (TTT) and handover hysterisis value (Hys) [109], which is also known as the mobility robustness optimization studied in the self-organized network (SON). The handover parameters optimization is a complicated task since the coverage areas of BSs are usually irregular and the signal quality and noise vary rapidly due to the variation of shadowing and multipath propagation in realistic wireless environments. When TTT and Hys are set too small, unnecessary handover may occur and induce ping-pong effect. When TTT and Hys are set too large, it will increase the probability of radio link failure.

Handover Parameters Optimization
The traditional handover is manually carried out by network operators and based on prior experience. Recently, BD analytics has been introduced to facilitate the automation of the mobility robustness optimization in SON. Via BD analytics on the historical handover KPI, the optimized cell-specific TTT and Hys can be found via either statistical analysis or advanced reinforcement learning algorithms. To further improve handover performance, the parameters can be further enhanced to be a user-location-specific rather than the rough-cell-specific basis. As shown in [110], the number of dropped calls and the number of unnecessary handovers can be significantly reduced via location-specific handover parameter settings. Through an unsupervised neural network, the simulated scenario of the specific indoor environment is efficiently learned, and the handover parameters are finely tuned.

Another promising handover optimization direction is to rely on BD analytics to automatically find service-specific handover parameter configurations. Now 4G RAN is configured with two fixed handover thresholds for VoLTE and data respectively. However, operators have to perform field tests to get the voice handover threshold, which results in high OPEX as well as long time to market. Besides VoLTE, more new services such as V2X will be supported and likely to have their own handover threshold preference. Moreover, service coverage could be very different due to service usage pattern, e.g., temporal factors. During the daytime more traffic (e.g., web browsing) shows up in work places and transportation hubs, and during the night more traffic (e.g., video) appears in residential areas. To perfectly match the traffic and service variations, BD analytics is in need. By leverage BD analytics on the coverage and service performance, RAN can derive appropriate service handover threshold accordingly. Thus, the occurrence of ping-pong handovers and handover failures will drop, leading to better user experience and reduced signaling overhead.

Intelligent Handover Decision Policy
In some specific scenarios, the conflicting demands of minimizing unnecessary handovers and the likelihood of dropped calls cannot be satisfied using the traditional handover method, i.e., adjusting TTTs and Hys of the serving cell and the neighbor cell.

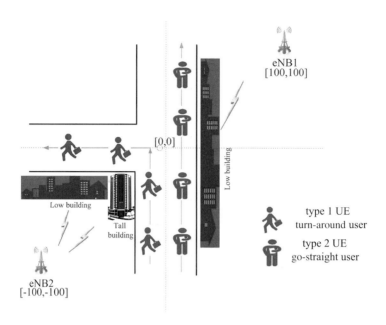

Figure 3.34 Example of the handover scenario.

A handover scenario is illustrated in Fig. 3.34. Two eNBs are respectively deployed at $[-100, -100]$m and $[100, 100]$m at a crossroad. One tall tower stands at the turn corner, and two low buildings are along the street. In this example, we deployed two types of users, type 1 UE (UE1, turn-around) and type UE 2 (UE2, go-straight). At the initial stage, UE1 and UE2 were both connected to eNB2 due to higher reference signal receiving power (RSRP). For UE1, his eNB2 RSPR experienced a decrease first, followed by an increase after making the right turn. To avoid ping-pong handovers, it was better to set a large TTT. For UE2, his eNB2 RSRP decreased along the route, and the best policy was to handover from eNB2 to eNB1. If TTT is set to a large value, the probability of dropped calls will increase for UE2. In this case, the appropriate TTT settings for UE1 and UE2 are in conflict, which the state-of-the-art approach cannot efficiently resolve. For this case, more intelligent handover decision policy is needed.

The data-driven learning-based algorithm emerges as a potentially fantastic solution. Based on the historical RSRP, reference signal receiving quality (RSRQ), the handover records, and the corresponding handover performance, e.g., throughput during the handover, powerful machine learning techniques are able to learn the realistic propagation environment. An intelligent handover decision policy is proposed by solving a classification problem via supervised machine learning. To do so, a classifier can be built based on the user's RSRP/RSRQ sequences along the route. The RSRP sequence serves as the input vector, and the output label is the best handover cell index based on the judgment of the handover performance. As for the simple example here, labels can be simplified as 1 or 0 for handover from eNB2 to eNB1 or not. Support vector machines (SVM) are used as the ML technique to perform the classification. SVM seek for the optimal hyperplane, which optimally separates the data into two classes in the feature space. The proposed data-driven learning-based handover scheme includes the following steps: 1) training dataset generation; 2) build a training model and model testing; 3) model execution in real-time handover procedure.

1. Training dataset generation

   - Input samples:
     The training input samples are also known as features. The RSRP, RSRQ sequence extracted from the measurement report (MR) can be selected as the training input samples to capture the feature of user's trajectory.
   - Labeling:
     The labels are generated based on the evaluation of the handover performance. In general, the labels are set as the cell index that gives the best handover performance. For instance, if the ping-pong handover is detected for certain input vector, the label will be set as the current cell index to indicate that it is better to stay on the current cell rather than perform handover.

2. Build a learning model
   Using the labeled training data, the SVM classifiers can be obtained.

3. Real-time handover decision making:
   With the trained classifiers and the real-time input of the MR data, intelligent handover decisions can be made.

   In Table 3.2, we show the count of service interruption and the count of the ping-pong handover for 100 simulation instances. In this simulation, the ping-pong handover is defined as a handover from cell B to cell A, then handover back to cell B if the time-of-stay-connected in cell A is less than 1s. The service interruption happens when the user's signal-to-interference-and-noise ratio (SINR) of served cell is less than 3dB during a period of 100ms. In that case, the effective transmission SINR will be zero, due to data transmission failure. For each instance, we generate one user, and that user may go straight or turn around with the same probability. The result shows that the proposed handover scheme can efficiently reduce ping-pong handovers and service interruptions. It indicates that data-driven learning-based optimization can be used to improve users' mobility performance. It will be especially useful in more challenging propagation scenarios. Note that the proposed algorithm can also be easily extended to wider areas and more complex deployment scenarios (e.g., heterogeneous cells).

To further enhance the learning-based handover scheme, the probability of the user's direction of movement can be designed as the additional input of the handover classifier trained by the supervised ML algorithm. To support the acquisition of such probability, the CN BD platform may extract the mobility profile of the user by analyzing his MR data over the duration of days/weeks/months and obtain the probability at each cell with respect to time, then push the user mobility profile to the corresponding CU when that user enters the cell hosted by that CU.

Table 3.2 Counts of service interruption and ping-pong handover for 100 instances

TTT setting (ms)		*200*	*400*	*600*
State-of-the-art method	*Count of service interruption*	7	26	51
	Count of ping-pong handover	51	51	0
Learning-based method	*Count of service interruption*	3		
	Count of ping-pong handover	0	0	0

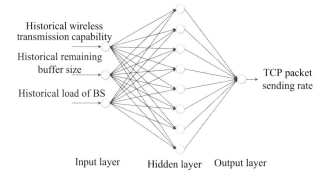

Figure 3.35 The structure of the BP neural network for TCP window optimization.

Big-Data-Assisted TCP Optimization

In current wireless networks, the TCP packet rate cannot be adjusted in a timely and accurate manner due to lack of knowledge on the real-time wireless transmission rate of the L2 layer. The traditional adjustment mechanism of the TCP window is based on trial and error [111, 112], which limits the overall system throughput and wastes wireless resources. With some useful RAN information, e.g., buffer size, load of the BS, the link throughput, service type, and PER, the TCP congestion window can be optimized and predicted to better match the radio channel variations. However, it is extremely hard to find a mathematical cross-layer model to determine the optimal TCP window with so many affecting factors. In this case, BD-assisted ML-based optimization offers an effective solution. With well constructed training data, a supervised learning-based model can be trained for the TCP window prediction.

A back propagation (BP) neural network can be utilized to predict the appropriate TCP window based on the context information of the wireless transmission rate. One example of the neural network is shown in Fig. 3.35. The inputs of the training model include the historic wireless transmission capability, historic remaining space in the buffer, and historic load of BS. The output is the rate at which the TCP should send out data packets. The prediction model was trained with 300,000 training data samples and 20,000 iterations.

The performance comparison of the traditional method and our proposed learning-based method is shown in Fig. 3.36. The x-axis shows the total simulation time of 12s,

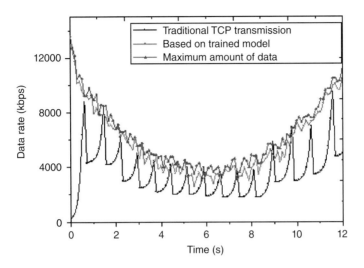

Figure 3.36 Comparison of packet sending rates.

and the y-axis is the data rate in the unit of kbps. The blue curve indicates the maximum amount of data that could be transmitted at each time instance. The black curve is the performance of the traditional TCP transmission scheme. The TCP packet sending rate initially increases to 2^{n+n_0} kbps, where n is the time index. It will drop to half when congestion is detected. Afterwards, it will rise with a rate of 2^{m+m_0}, where m is the time index starting from the last congestion point to the next congestion point. In the simulation, the update interval is 105ms; $n_0 = 8$; $m_0 = 3$. The red curve shows the achievable data rate of the BD-aided learning-based scheme. It can be seen that the BD-aided method is well-matched with the maximum transmission capability, while the traditional method shows great performance loss.

To measure the effectiveness of the BD-aided TCP window optimization, we calculate the similarity between the TCP packet sending rate vector and the maximum wireless transmission rate vector. The similarities for the traditional method and the proposed method are 0.867 and 0.953 respectively. A higher similarity value indicates better wireless resource utilization efficiency; thus the proposed method demonstrates a superior performance.

In summary, the BD-enabled learning-based method facilitates the dynamic optimization of TCP packet sending rate. It allows for a good match between the TCP window and the wireless channel condition. This significantly improves system throughput, buffer utilization, and the potential to reduce the retransmission.

Big-Data-Aided Beam Sweeping Optimization in Initial Access

In 5G NR, hybrid analog and digital beamforming is proposed in MIMO systems, which helps to balance performance and complexity, especially at high frequencies such as mmWave. In the initial access stage under the hybrid framework, beam sweeping

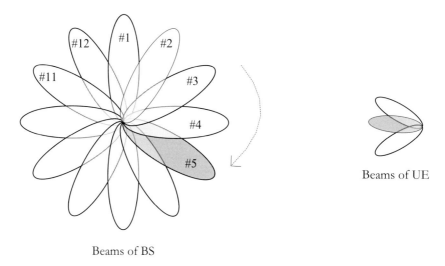

Beams of UE

Beams of BS

Figure 3.37 An example of beam sweeping.

is performed by the BS periodically to allow users to access the network from any direction in the cell. Specifically, the BS periodically transmits synchronization signals in different directions, covering the whole angular space. We assume that if the signal power received by the UE is greater than a threshold, the UE can successfully decode the signal and camp onto the network. As for high frequency communications, narrow beams are usually employed by the BS to compensate the overwhelmingly high path loss, and thus a large number of beam patterns may be needed to cover the whole angular space. If the UE also uses directional analog beams to receive synchronization signals, the exhaustive search for the desired beam pair may be prohibitively time-consuming.

At the conventional initial access stage, since there is no a prior shared information between the BS and the UE, they have to search for the desired beam pair sequentially. An example is shown in Fig. 3.37, wherein the BS transmits the synchronization signals in a beam-sweeping manner from beam #1 to beam #12. Imagine, when the UE roams to the cell, that the BS is transmitting the synchronization signals from beam #5 onwards, and then it has to wait for a long time until the synchronization signals are transmitted from the beam #4 direction again. As a result, the delay for initial access is large. Intuitively, the beam-sweeping process can be optimized if the users' spatial/angular distribution can be known at the BS.

Based on BD analytics, the BS is able to predict the user distribution by exploring the information feedback from UEs, and adjust the beam pattern and/or the beam-sweeping order according to the predicted user distribution. The information for predicting the user distribution includes the initial access delay of old UEs, the beam pair direction- and the received power of the synchronization signals. BD-aided initial access is suitable for the scenario where group characteristics exist among the users. For example, if the BS predicts that there is a high probability that a large number of users are distributed

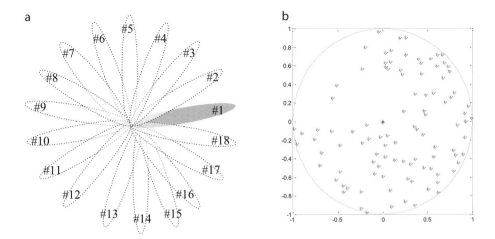

Figure 3.38 (a) Illustration of BS beams; (b) Illustration of user distribution.

in a certain direction, the beam-sweeping order should be adapted to guarantee that the users from these directions can discover the network in a timely manner. Therefore, the average initial access delay can be reduced.

In the following passage, some preliminary evaluation results are illustrated to show the effectiveness of the proposed BD-aided beam sweeping. In the simulation, we assume the BS has 16 beams to cover the whole cell shown in Fig. 3.38a, and the UE has one single beam direction. There are 100 users in the cell, and an example of the user distribution is shown in Fig. 3.38b. The BS has 64 antenna elements, and follows 8×8 uniform panel models with 20dB transmit power. The noise and cell radius is normalized to 1. The channel propagation is model as $h = d^{-3.7}$, where d is the distance between the UE and the BS. The access threshold is set as 0dB.

The following two schemes are evaluated for comparison:

- Conventional sequential beam sweeping: the BS sequentially scans the whole angular space from beam #1 to beam #16.
- BD-aided beam sweeping: The beam sweeping is optimized based on the information of UE access delay and access beam index, where high priority is allocated to the beam direction covering more users and larger access delay in the beam sweeping procedure.

In Fig. 3.39, the comparison of the average beam-sweeping times for the above two schemes are illustrated. It can be seen that the BD-aided scheme has much lower average beam-sweeping times. It indicates that the average access delay is significantly reduced. Besides, the proposed BD-aided scheme is quite robust in regards to the UE distribution. As we can see, with different degrees of user nonuniform distribution, the average beam-sweeping times at the BS are almost the same. The degree of the user nonuniform distribution is defined as the ratio of probability of the user located within the upper semicircle and the lower semicircle.

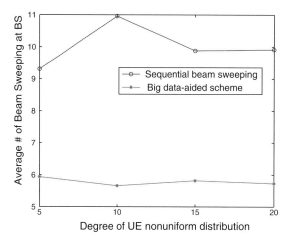

Figure 3.39 Comparison of initial access performance of the conventional scheme and the BD-aided scheme.

Summary

In this section, the benefits of applying WBD analytics to the wireless communication network are investigated. AI algorithms are first surveyed, with the basics briefly illustrated as examples. With BD analytics in the BD-enabled network architecture, e.g., in the CN or RAN or both, some potential performance improvements are expected in mobility management, cross layer TCP optimization, and beam-sweeping enhancement for initial network access. The user behavior, e.g., location and trajectory pattern, helps to improve the efficiency of the paging process in the idle state and the handover process in the connected state. Also, the BD-enabled learning-based method facilitates the dynamic optimization of TCP packet sending rate, achieving a good match between the TCP window and the wireless channel condition, significantly improving system throughput, buffer utilization and potential reduction of the retransmission numbers. Finally, the BD-aided beam-sweeping scheme, which is very important for MIMO operations at high frequency bands, shows much lower average beam-sweeping time. Accordingly, the average access delay is significantly reduced. Meanwhile, the proposed BD-aided scheme is robust against different UE distributions.

References

[1] ITU-R M.2083, "IMT vision-framework and overall objectives of the future development of IMT for 2020 and beyond," Sept. 2015.

[2] NGMN Alliance, "5G white paper," Feb. 2015.

[3] METIS 2020 Project, "The 5G future scenarios identified by METIS: The first step toward a 5G mobile and wireless communications system," Sept. 2013. www.metis2020.com/press-events/press/the-5g-future-scenarios-identified-by-metis/.

[4] M. Chiosi et al., "Network functions virtualisation: Introductory white paper," ETSI, Oct. 2012. www.etsi.org/technologiesclusters/technologies/nfv

[5] N. M. K. Chowdhury and R. Boutaba, "Network virtualization: State of the art and research challenges," *IEEE Comm. Mag.*, vol. 47, no. 7, pp. 20–26, 2009.

[6] B. Han, V. Gopalakrishnan, L. Ji, and S. Lee, "Network function virtualization: Challenges and opportunities for innovations," *IEEE Commun. Mag.*, vol. 53, no. 2, pp. 90–97, Feb. 2015.

[7] P. Veitch, M. J. McGrath, and V. Bayon, "An instrumentation and analytics framework for optimal and robust NFV deployment," *IEEE Commun. Mag.*, vol. 53, no. 2, pp. 126–133, Feb. 2015.

[8] H. Hawilo, A. Shami, M. Mirahmadi, and R. Asal, "NFV: State of the art challenges and implementation in next generation mobile networks (vEPC)," *IEEE Netw.*, vol. 28, no. 6, pp. 18–26, Nov. 2014.

[9] I. Silva, G. Mildh, and A. Kaloxylos, "Impact of network slicing on 5G radio access networks," *European Conference on Networks & Communications*, 2016.

[10] I. Silva, G. Mildh et al., "Tight integration of new 5G air interface and LTE to fulfill 5G requirements", *IEEE 81st Veh. Technol. Conf. (VTC Spring)*, Glasgow, May 2015.

[11] NGMN Alliance, "Description of network slicing concept," Version 1.0, Jan. 2016. www .ngmn.org/fileadmin/user_upload/160113_Network_Slicing_v1_0.pdf.

[12] 3GPP TR 38.801, "Study on new radio access technology: Radio access architecture and interfaces (Release 14)."

[13] 3GPP TS 23.501, "System architecture for the 5G system, (Release 15)."

[14] 3GPP TS 23.401, "General packet radio service (GPRS) enhancements for evolved universal terrestrial radio access network (E-UTRAN) access (Release 8)."

[15] 3GPP TS 38.401, "NG-RAN; Architecture description (Release 15)."

[16] 3GPP TS 38.300, "NR; Overall description; Stage-2 (Release 15)."

[17] 3GPP TS 37.340, "NR; Multi-connectivity; Overall description; Stage 2 (Release 15)."

[18] 3GPP TS 37.324, "Evolved universal terrestrial radio access (E-UTRA) and NR; Service data adaptation protocol (SDAP) specification (Release 15)."

[19] 3GPP TS 38.470, "NG-RAN; F1 general aspects and principles (Release 15)."

[20] 3GPP TR 36.933, "Study on context aware service delivery in RAN for LTE (Release 14)."

[21] 3GPP TR 36.300, "Evolved universal terrestrial radio access (E-UTRA) and evolved universal terrestrial radio access network (E-UTRAN); Overall description; Stage 2 (Release 8)."

[22] 3GPP TR 22.891, "Study on new services and markets technology enablers (Release 14)."

[23] 3GPP TR 38.913, "Study on scenarios and requirements for next generation access technologies (Release 14)."

[24] 3GPP TS 38.410, "NG-RAN; NG general aspects and principles (Release 15)."

[25] 3GPP TS 38.420, "NG-RAN; Xn general aspects and principles (Release 15)."

[26] ETSI, "Multi-access edge computing." www.etsi.org/technologies-clusters/technologies/ multi-access-edge-computing.

[27] Open Networking Foundation, "Software-defined networking: The new norm for networks," Apr. 2012. http://www.opennetworking.org/images/stories/downloads/sdn-resources/white-papers/wp-sdn-newnorm.pdf

[28] J. Wan et al., "Software-defined industrial internet of things in the context of industry 4.0," *IEEE Sensors J.*, vol. 16, no. 20, pp. 7373–7380, 2016.

[29] S. Sezer et al., "Are we ready for SDN? Implementation challenges for software-defined networks," *IEEE Commun. Mag.*, vol. 51, no. 7, pp. 36–43, July 2013.

[30] B. Nunes et al., "A survey of software-defined networking: Past present and future of programmable networks," *IEEE Commun. Surveys & Tutorials*, vol. 16, pp. 1617–1634, 2014.

[31] The Open Group, "Service-oriented architecture standards." www.opengroup.org/standards/soa.

[32] C. L. I, C. Rowell, S. Han et al., "Toward green and soft: A 5G perspective," *IEEE Commun. Mag.*, vol. 52, no. 2, pp. 66–73, Feb. 2014.

[33] 3GPP TR 36.819, "Coordinated multi-point operation for LTE physical layer aspects (Release 11)."

[34] China Mobile Communications Corporation, "Simulation results for CoMP phase I evaluation in homogeneous network," 3GPP R1-111301, Meeting R1-65 contribution, Barcelona, May 2011

[35] Q. Wang, D. Jiang, G. Liu, and Z. Yan. "Coordinated multiple points transmission for LTE-advanced systems," in *Proc. of 5th Int. Conf. on Wireless Commun., Netw. and Mobile Comput., 2009 (WiCom '09)*, Sept. 2009, pp. 1–5.

[36] China Mobile Research Institute, "C-RAN white paper: The road towards green RAN," Jun. 2014.

[37] C. L. I, J. Huang, R. Duan et al., "Recent progress on C-RAN centralization and cloudification," *IEEE Access*, vol. 2, pp. 1030–1039, 2014.

[38] J. Wu, S. Rangan, and H. Zhang, *Green Communication*, CRC Press, 2013.

[39] NGMN, "Further study on critical C-RAN technologies," Apr., 2015.

[40] Next Generation Fronthaul Interface (1914) Working Group, 2015, http://sites.ieee.org/sagroups-1914.

[41] ETSI, "Network functions virtualisation," 2012. http://portal.etsi.org/portal/server.pt/community/NFV/367.

[42] B. A. A. Nunes, M. Mendonca, X.-N. Nguyen, K. Obraczka, and T. Turletti, "A survey of software-defined networking: Past, present, and future of programmable networks," *IEEE Commun. Surveys Tuts.*, vol. 16, no. 3, pp. 1617–1634, Sept. 2014.

[43] C. L. I, Y. Yuan, J. Huang et al., "Rethink fronthaul for soft RAN", *IEEE Commun. Mag.*, vol. 53, no. 9, pp. 82–88, 2015

[44] Y. Liu, and G. Liu, "User-centric wireless network for 5G," in *5G Mobile Communications*, W. Xiang, K. Zheng and X. Shen (eds.), Springer, 2016, pp. 457–474.

[45] 3GPP TS 36.300, "Evolved universal terrestrial radio access (E-UTRA) and evolved universal terrestrial radio access network (E-UTRAN); Overall description; Stage 2 (Release 8)."

[46] 3GPP TS 37.340, "NR; Multi-connectivity; Overall description; Stage-2 (Release 15)."

[47] 3GPP TS 36.410, "Evolved universal terrestrial radio access network (E-UTRAN) S1 general aspects and principles (Release 8)."

[48] 3GPP TS 36.413, "Evolved universal terrestrial radio access network (E-UTRAN) S1 application protocol (Release 11)."

[49] 3GPP TS 36.420, "Evolved universal terrestrial radio access network (E-UTRAN X2) general aspects and principles (Release 8)."

[50] 3GPP TS 36.423, "Evolved universal terrestrial radio access network (E-UTRAN) X2 application protocol (Release 8)."

[51] 3GPP TS 36.321, "Evolved universal terrestrial radio access (E-UTRA); medium access control (MAC) protocol specification (Release 8)."

[52] 3GPP TS 36.211, "Evolved universal terrestrial radio access (E-UTRA); Physical channels and modulation (Release 8)."

[53] 3GPP TS 36.331, "Evolved universal terrestrial radio access (E-UTRA); Radio resource control (RRC); Protocol specification (Release 8)."

[54] 3GPP TS 36.212, "Evolved universal terrestrial radio access (E-UTRA); Multiplexing and channel coding (Release 8)."

[55] 3GPP TS 36.322, "Evolved universal terrestrial radio access (E-UTRA); Radio link control (RLC) protocol specification (Release 8)."

[56] 3GPP TS 36.323, "Evolved universal terrestrial radio access (E-UTRA); Packet Data convergence protocol (PDCP) specification (Release 8)."

[57] 3GPP TR 38.804, "Study on new radio access technology radio interface protocol aspects (Release 14)."

[58] X. Cheng et al. "Mobile big data: The fuel for data-driven wireless," *IEEE Internet of Things J.*, vol. 4, no. 5, pp. 1489–1516, Oct. 2017. DOI 10.1109/JIOT.2017.2714189.

[59] S. Han, C.-L. I, S. Wang et al., "Big data enabled mobile network design for 5G and beyond," accepted by *IEEE Commun. Mag.*, 2017.

[60] J. Qadir et al., "Artificial intelligence enabled networking," *IEEE Access*, vol. 3, pp. 3079–3082, 2015.

[61] X. Wang, X. Li, and V. C. M. Leung, "Artificial intelligence-based techniques for emerging heterogeneous network: State of the arts, opportunities, and challenges,", *IEEE Access*, vol. 3, pp. 1379–1391, 2015.

[62] M. Bkassiny, Y. Li, S. K. Jayaweera et al., "A survey on machine-learning techniques in cognitive radios," *IEEE Commun. Surveys and Tutorials*, vol. 15, no. 3, pp. 1136–1159, 2013.

[63] M. A. Alsheikh, S. Lin , D. Niyato et al., "Machine learning in wireless sensor networks: Algorithms, strategies, and applications," *IEEE Commun. Surveys and Tutorials*, vol. 16, no. 4, pp. 1996–2018, 2014.

[64] C. Jiang, H. Zhang, Y. Ren et al., "Machine learning paradigms for next-generation wireless networks," *IEEE Wireless Commun.*, vol. 24, no. 2, pp. 98–105, Apr. 2017.

[65] R. Li, L. Zhao X. Zhou, et al., "Intelligent 5G: When cellular networks meet artificial intelligence," *IEEE Wireless Commun.*, vol. 24, no. 5, pp. 175–183, Oct. 2017.

[66] P. V. Klaine, A. I. Muhammad, O. Oluwakayode et al., "A survey of machine learning techniques applied to self organizing cellular networks," *IEEE Commun. Surveys & Tutorials*, vol.19, no.7, pp. 2392–2431, fourth quarter, 2017.

[67] S. Hu, Y.-d. Yao, and Z. Yang, "MAC protocol identification using support vector machines for cognitive radio networks," *IEEE Wireless Commun.*, vol. 21, no. 1, pp. 52–60, Feb. 2014.

[68] R. Li, L. Zhao, X. Chen et al., "TACT: A transfer actor-critic learning framework for energy saving in cellular radio access networks," *IEEE Trans. Wireless Commun.*, vol. 13, no. 4, pp. 2000–2011, Apr. 2014.

[69] N. Sinclair, D. Harle, I. A. Glover, J. Irvine, and R. C. Atkinson, "An advanced SOM algorithm applied to handover management within LTE," *IEEE Trans. on Veh. Technol.*, vol. 62, no. 5, pp. 1883–1894, Jun. 2013.

[70] P. Wang, S. C. Lin, and M. Luo, "A framework for QoS-aware traffic classification using semi-supervised machine learning in SDNs," *IEEE Int. Conf. on Services Comput. (SCC)*, San Francisco, pp. 760-765, 2016.

[71] P. Amaral, J. Dinis, P. Pinto et al., "Machine learning in software defined networks: Data collection and traffic classification," *IEEE 24th Int. Conf. on Netw. Protocols (ICNP)*, Singapore, pp. 1–5, 2016.

[72] I. Yahia, J. Bendriss, A. Samba et al., "CogNitive 5G networks: Comprehensive operator use cases with machine learning for management operations," *20th Conf. on Innovations in Clouds, Internet and Netw. (ICIN)*, Paris, pp. 252–259, 2017.

[73] X. Gao, L. Dai, Y. Sun et al., "Machine learning inspired energy-efficient hybrid precoding for mmWave massive MIMO systems," *IEEE Int. Conf. on Commun. (ICC)*, 2017.

[74] J. Joung, "Machine learning-based antenna selection in wireless communications," *IEEE Commun. Lett.*, vol. 20, no. 11, pp. 2241–2244, Nov. 2016.

[75] C. Bishop, *Pattern Recognition and Machine Learning*, Springer, 2007.

[76] K. P. Murphy, *Machine Learning: A Probabilistic Perspective*, The MIT Press, 2012.

[77] E. Alpaydin, *Introduction to Machine Learning*, Second Edition, The MIT Press, 2010.

[78] T. Hastie, R. Tibshirani, J. Friedman, *The Elements of Statistical Learning: Data Mining, Inference, and Prediction*, Second Edition, Springer, 2016.

[79] R. S. Sutton and A. G. Barto, *Reinforcement Learning: An Introduction*, Second Edition, The MIT Press, 2018.

[80] I. Goodfellow, Y. Bengio, and A. Courville, *Deep Learning*, The MIT Press, 2016.

[81] N. Kato et al., "The deep learning vision for heterogeneous network traffic control: Proposal, challenges, and future perspective," *IEEE Wireless Commun.*, vol. 24, no. 3, pp. 146–153, 2017.

[82] D. Liu, B. Chen, C. Yang, and A. Molisch, "Caching at the wireless edge: Design aspects, challenges and future direction," *IEEE Commun. Mag.*, vol. 54, no. 9, pp. 22–28, Sept. 2016.

[83] Wikipedia, "Deep learning." https://en.wikipedia.org/wiki/Deep_learning.

[84] C. Yao, C. Yang, Z. Xiong, "Energy-saving predictive resource planning and allocation," *IEEE Trans. on Commun.*, vol. 64, no. 12, pp. 5078–5095, Dec. 2016.

[85] S. Zhou, D Lee, B Leng, et al., "On the spatial distribution of base stations and its relation to the traffic density in cellular networks," *IEEE Access*, vol. 3, pp. 998–1010, 2015.

[86] C.-L. I, Y. Liu, S. Han, S. Wang, and G. Liu, "On big data analytics for greener and softer RAN," *IEEE Access*, vol. 3, pp. 3068–3075, Mar. 2015.

[87] 3GPP SA2, "Study of enablers for network automation for 5G," Doc number: S2-173827, May. 2017, Hangzhou, China.

[88] X. Zhang et al., "Social computing for mobile big data," *Computer*, vol. 49, no. 9, pp. 86–90, Sept. 2016.

[89] X. Cheng, L. Fang, X. Hong, and L. Yang, "Exploiting mobile big data: Sources, features, and applications," *IEEE Netw.*, vol. 31, no. 1, pp. 72–79, Jan./Feb. 2017.

[90] S. Bi, R. Zhang, Z. Ding, and S. Cui, "Wireless communications in the era of big data," *IEEE Commun. Mag.*, vol. 53, no. 10, pp. 190–199, Oct. 2015.

[91] K. Zheng, Z. Yang, K. Zhang et al., "Big data-driven optimization for mobile networks toward 5G," *IEEE Netw.*, vol. 30, no. 1, pp. 44–51, Jan. 2016.

[92] C.-L. I, Q. Sun, Z. Liu, S. Zhang, and S. Han, "The big-dat-driven intelligent wireless network: Architecture, use cases, solutions, and future trends," *IEEE Veh. Technol. Mag.*, vol. 12, no. 4, pp. 20–29, Dec., 2017.

[93] FuTURE Forum 5G SIG, Whitepaper, "Wireless big data for smart 5G," Nov. 2017, available: http://www.future-forum.org/dl/171114/whitepaper2017.rar.

[94] Y. Li, B. Cao, and C. Wang, "Handover schemes in heterogeneous LTE networks: Challenges and opportunities," *IEEE Wireless Commun.*, vol. 23, no. 2, pp. 112–117, Feb. 2016.

[95] C.-L. I, S. Han, Z. Xu et al., "New paradigm of 5G wireless Internet," *IEEE J. on Sel. Areas in Commun.*, vol. 34, no. 3, pp. 474–482, Mar. 2016.

[96] A. Ray, S. Deb, and P. Monogioudis, "Localization of LTE measurement records with missing information," *IEEE INFOCOM 2016*.

[97] J. Tadrous, A. Eryilmaz, and H. E. Gamal, "Proactive content download and user demand shaping for data networks," *IEEE/ACM Transactions on Networking*, vol. 23, no. 6, pp. 1917–1930, 2015.

[98] E. Baştuğ, M. Bennis, M. Kountouris, and M. Debbah, "Cache-enabled small cell networks: Modeling and tradeoffs," *EURASIP Journal on Wireless Communications and Networking*, vol. 2015, no. 1, pp. 1–11, 2015.

[99] D. Liu and C. Yang, "Cache-enabled heterogeneous cellular networks: Comparison and tradeoffs," IEEE ICC, 2016, accepted.

[100] D. Liu and C. Yang, "Energy efficiency of downlink networks with caching at base stations," Accepted, IEEE Journal on Selected Areas in Communications, 2015. http://arxiv .org/abs/1505.06615.

[101] H. Li and G. Ascheid, "Long-term window scheduling in multiuser OFDM systems based on large scale fading maps," IEEE SPAWC 2012 *IEEE SPAWC 2012*.

[102] Z. Lu and G. De Veciana, "Optimizing stored video delivery for mobile networks: The value of knowing the future," IEEE INFOCOM 2013 *IEEE INFOCOM 2013*.

[103] J. Tadrous, A. Eryilmaz, and H. El Gamal, "Proactive resource allocation: Harnessing the diversity and multicast gains," *IEEE Trans. Information Theory*, vol. 59, no. 8, pp. 4833–4854, 2013.

[104] C. Yao, B. Chen, and C. Yang, "Energy Saving pushing based on personal interest and context information," Accepted, *Proc. IEEE Vehicular Technology Conference 2016*.

[105] V. A. Siris and D. Kalyvas, "Enhancing mobile data offloading with mobility prediction and prefetching," ACM MobiArch, 2012.

[106] S.-Y. Lien, K.-C. Chen, and Y.-C. Liang, "Ultra-low latency ubiquitous connections in heterogeneous cloud radio access networks," *IEEE Wireless Commun.*, vol. 22, no. 3, pp. 22–31, Jun. 2015.

[107] Y. Huang, J. Tan, and Y.-C. Liang, "Wireless big data: Transforming heterogeneous networks to smart networks," *Journal of Communications and Information Networks*, vol. 2, no. 1, pp. 19–32, Mar. 2017.

[108] Y. Sun, G. Feng, S. Qin, S. Sun, and L. Zhang, "User behavior aware cell association in heterogeneous cellular networks," *Proc. of IEEE Wireless Communications and Networking Conference (WCNC)*, San Francisco, Mar. 2017.

[109] 3GPP TS 36.331, "Radio resource control (RRC); Protocol specification (Release 14)."

[110] N. Sinclair, D. Harle, I. A. Glover, J. Irvine, and R. C. Atkinson, "Parameter optimization for LTE handover using an advanced SOM algorithm," *Proc. IEEE Vehicular Technology Conference 2013*.

[111] W. Stevens, "TCP slow start, congestion avoidance, fast retransmit, and fast recovery algorithms." Request for comment: 2001, 1997

[112] S. Bajeja and A. Gosai, "Performance evaluation of traditional TCP variants in wireless multihop networks," *3rd Int. Conf. on Comput. for Sustainable Global Development (INDIACom)*, New Delhi, India, pp. 3517–3522, 2016.

4 Energy-Efficient Signaling Design and Resource Management

The fundamental issues of energy-efficient design have been discussed in Chapter 2, from the perspectives of information theory and optimization theory. Those ideas and methods mainly focus on single-cell network scenarios where energy-efficient design can be coordinated at a single base station (BS). These approaches merely aim at link-level energy efficiency (EE) by optimizing radio resource allocation. However, the operation of a BS contributes to 60–80% of the overall network energy consumption and is the main source of energy usage. Therefore, energy-saving of BSs plays an important role in green cellular networks.

Beyond energy-efficient radio resource optimization at the link level, in this chapter we discuss green communication approaches from the perspective of network architecture. Currently, most cellular networks consist of homogenous BSs, which can cover an area with a radius from 100m to 10km. The energy consumption of these macro-cell BSs is usually very high since high-transmit power is required to maintain large network coverage. Moreover, most of them are equipped with cooling systems, consuming a large amount of energy. Thus, the dynamic management of the operation status of BSs, i.e., adaptive sleep control, is an essential issue for saving network energy consumption.

On the other hand, a heterogenous cellular structure with different kinds of densely deployed BSs will be adopted by future cellular networks. Such heterogenous networks (HetNets) with small-cell BSs can greatly reduce the energy consumption of both network and user devices. First, the transmit power of BSs can be considerably reduced due to the smaller network coverage size. Furthermore, compared with macro BSs, the operation of small-cell BSs is much more energy efficient without the cooling system. The HetNet architecture can also reduce the energy consumption of mobile devices due to the shortened distance between user devices and serving BSs.

According to the radio access technology (RAT) used, HetNets can be classified into two types: single-RAT HetNet and multi-RAT HetNet. Single-RAT HetNet consists of BSs with the same radio access protocol, such as LTE BSs. On the other hand, multi-RAT HetNet consists of BSs with different radio access protocols, e.g., LTE BSs and WiFi access points. A single-RAT HetNet has a much easier network control, whereas intercell interference coordination is much more complicated due to frequency reuse among cells. On the other hand, the network control of multi-RAT HetNet is somewhat difficult due to various types of access points. However, intercell interference coordination could be much easier, since different RATs may transmit at different frequency bands.

In the rest of this chapter, we will discuss green communication techniques from the network-layer perspective. In particular, we will introduce several sleep control and cell zooming strategies for BSs in Section 4.1. In Section 4.2, we will present a joint downlink and uplink energy-efficient resource allocation algorithm. Sections 4.3 and 4.4 discuss the energy-efficient design issues in homogenous and heterogeneous networks, respectively.

4.1   Sleeping Strategy and Cell Zooming

4.1.1   Dynamic Base Station Sleep Control

Cellular network operators have been continuously seeking ways to increase EE in all components of cellular networks, including mobile devices, BSs, and core (backhaul) networks. There has been a tremendous amount of work on mobile device EE with the objective of prolonging battery life. Similarly, green operation of the Internet has been considered, and some of the techniques can be extended to the cellular backhaul networks. However, as mentioned before, the key source of energy usage in cellular networks is the operation of BS equipment, which contributes to almost 60–80% of total energy consumption. Therefore, the energy-efficient operation of cellular BSs is the key challenge to implementing the so-called green cellular network.

The traditional network architecture is designed based on the assumption that user requests may happen anytime and anyplace. Therefore, to guarantee cell coverage and provide appropriate services for potential requests, most existing cellular networks have been designed to keep the transmit power always on, which is clearly not energy efficient, since the user requests occur only sometimes and somewhere in practice.

Energy-efficient design of BSs has been considered in all stages of cellular networks, including hardware design and manufacture, deployment, and operation. However, there is also room for significant improvement in cellular operation. In fact, the BS consumes more than 90% of its peak energy while experiencing little or even no activity. As a result, turning off some radio transceivers at BSs with low traffic load can save some energy consumption but is still not sufficient for green cellular operation. To obtain significant energy savings, BS sleeping, a carefully coordinated dynamic approach, has been developed and involves the operation of shutting entire BSs and transferring the corresponding load to neighboring cells during periods of low utilization.

BS sleeping has attracted more and more attention in recent years [1–3], and there have been many ways to facilitate its implementation. Early works mainly focused on static BS sleep according to deterministic traffic patterns over time, without considering randomness and spatial variation. However, for the path-loss-dominant cellular network, a dynamic clustering algorithm considering BS collaboration is more effective. A common principle is to dynamically adjust the work modes (active or sleeping) of BSs according to the traffic variation with respect to certain blocking probability requirements. Moreover, BSs should hold their current working modes for at least a given interval to prevent frequent mode switching.

The most challenging issue of dynamic BS sleep control is to maintain the cell coverage. Among all methods, the most direct one is to increase the coverage area of the nearby BSs. There are two alternative ways to realize this: increasing the transmission power or utilizing lower-frequency bands with better penetration capacity under the same transmission power constraint. Besides, appropriate design of multi-hop relay and coordinated multipoint transmission will also be clearly useful to ensure that the dynamic shutting down of BSs will not leave any coverage hole. More detailed granularity of control and more differentiated heterogeneous cell sizes should also be carefully designed.

4.1.2   Cell Zooming for Green Cellular Networks

Different from the dynamic BS sleep control in the previous sections, in this section, we introduce a dynamic cell zooming strategy to facilitate the implementation of green cellular networks [4].

The cell sizes and the transmit power on the control channel of traditional cellular BSs are usually fixed. The cell size of a BS can be regarded as the area where the power of the received control signal from the BS is above a given threshold. The cell size of each BS is usually predetermined according to the estimated traffic load when the network is established. In the past voice traffic dominated the cellular traffic load, and thus fixed cell size could achieve satisfactory network performance with simple network management. However, as wireless data traffic has grown very rapidly in the past few years, traffic loads are significantly dynamic among different cells. Both spatial and temporal traffic fluctuations can be observed. As discussed before, dynamic BS sleep control can be applied to save network energy consumption. For example, in the nighttime, some BSs located in office areas can be switched off due to the relatively light traffic load. In this situation, the cell size of each BS should be adaptively adjusted according to the traffic dynamics. This phenomenon is known as cell zooming, which can not only balance the traffic load but also reduce network energy consumption.

There have been many ways to facilitate the implementation of cell zooming. The simplest one is to adjust the physical parameters of BSs, such as transmit power control and antenna tilt adaption. Increasing the transmit power can zoom out the cell size, while adjusting the antenna tilt can change the coverage area. Another effective means is BS cooperation where neighboring BSs can utilize the coordinated multipoint (CoMP) transmission to form a cluster. Besides these two methods, leveraging the BS relaying technique or D2D communications is also an effective method to realize the cell zooming [5, 6].

Figure 4.1 shows an example of cell zooming for green cellular networks in which the BS in the center cell decides to sleep due to the low traffic load. In this situation, the nearby BS will trigger the cell zooming operation. Among them, BS1 and BS4 increase their transmit powers to enlarge cell coverage, BS2 and BS3 leverage the relay technique to cover some areas of the sleeping cell, while BS5 and BS6 utilize the CoMP transmission to expand the network coverage.

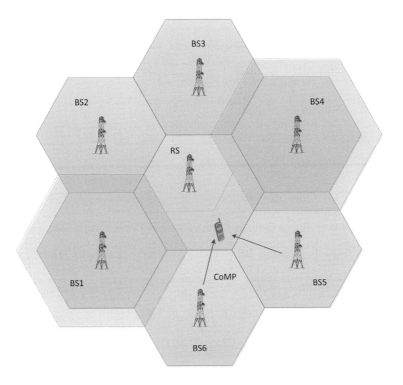

Figure 4.1 An illustration of cell zooming for green cellular networks.

4.1.3   Soft-Defined Network Architecture for Green Cellular Networks

In traditional cellular systems, the physical infrastructure consists of geographically distributed hardware subsystems including the BSs. These subsystems are often dedicated for specific tasks and present network functions together by communicating with each other via predefined network protocols. Although traditional cellular systems have sufficient capabilities to serve traffic with traditional moderate QoS requirements, there are several severe problems that must be addressed in order to reach the demand of next-generation communications.

- Flexibility. The proprietary subsystems and interfacing protocols make it hard to update the traditional cellular system's functions or add services. It would be energy-consuming for the widely-deployed subsystems.
- Efficiency. The distributed system layout prohibits cooperation technologies like CoMP. As a result, the network performance cannot be improved via intercell and inter-network cooperation. Dynamic BS sleeping control and cell zooming cannot be implemented either.
- Resource utilization. Geographically distributed subsystems rely on local physical resources to function. Thus, resources must be overprovisioned for peak load, leading to a low resource utilization efficiency.

Thus, it is proposed that 5G networks should be constructed with centralized physical resource placement and become increasingly software-defined. In general, the emerging cellular system architectures should be developed from the following three aspects [7, 8].

- Separation of the air interface. Aiming at flexible and efficient control of small cells for throughput boosting, energy-saving, and BS sleeping, the separation of signaling and data should be featured in the new air interface architecture of cellular networks. Therefore, the data traffic service is relative to demand, whereas the control plane is always "on," to guarantee basic coverage.
- Base station functions virtualization. By providing programmable BS functions and reconfigurable radio elements, flexible and efficient signal processing can be implemented to reconstruct the frame components from the control and traffic layers of the air interface. In this way, control-traffic-decoupled air interface can be realized.
- Soft-defined networking. By separating the control and data planes, the SDN architecture enables centralized optimization of data transmission. In addition, the control-data separation in SDN can be extended to wireless access layer by the control-traffic-decoupled air interface.

4.2   Joint Optimization of Uplink and Downlink Energy Efficiency

There are generally two different trends for energy-efficient design in cellular networks. One is energy-efficient design for downlink transmission, the other is energy-efficient design for uplink transmission. The former mainly saves power consumption of BSs from the perspective of cellular network operators, and the latter mainly saves power consumption of mobile devices from the perspective of users. Most previous studies have only focused on one aspect of energy-efficient design, either uplink [9–11] or downlink [11–15]. Those energy-efficient algorithms proposed in the existing literature may cause unbalance of uplink EE and downlink EE. In this section, we will present a framework to simultaneously optimize both EEs by joint downlink and uplink resource allocation.

To do so, we consider a cellular network with carrier aggregation (CA), which allows users to aggregate different sub-bands to achieve a larger data rate. We also consider a time division duplex (TDD) operation, which enables dynamic allocation of uplink and downlink resource to guarantee the performance of both BS and user devices. In a TDD CA system, the uplink to downlink resource ratio on each sub-band can be optimized. Existing literature has investigated this problem from the perspective of throughput enhancement and network load balance while the EE issue has not been studied yet [16–19].

In what follows, we will first briefly introduce the system model, and then formulate the EE optimization problem. Then, a joint uplink and downlink resource allocation

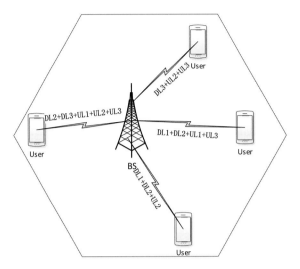

DL1: Downlink bandwidth from band 1.   UL1: Uplink bandwidth from band 1.
DL2: Downlink bandwidth from band 2.   UL2: Uplink bandwidth from band 2.
DL3: Downlink bandwidth from band 3.   UL3: Uplink bandwidth from band 3.

Figure 4.2 The network model of the TDD CA system.

algorithm will be developed. Finally, numerical simulation results are provided to validate the effectiveness of the proposed algorithm.

4.2.1 System Model and Problem Formulation

As depicted in Fig. 4.2, we consider a TDD CA system with M users, $\mathcal{M} = \{1, 2, \cdots, M\}$, and N available sub-bands, $\mathcal{N} = \{1, 2, \cdots, N\}$. Each user can dynamically transmit on several sub-bands according to the user's channel power gains and data rate requirement. Moreover, the uplink to downlink resource ratio on each sub-band can also be dynamically chosen to balance the uplink and downlink performance. The channel between user m and the BS on the n-th sub-band follows the i.i.d. Rayleigh fading model, as

$$g_{m,n} = \Gamma_n h_{m,n} d_m^{-\alpha_n}. \tag{4.1}$$

In (4.1), d_m denotes the distance between user m and the BS, α_n is the path loss exponent of the sub-band n, Γ_n is the path loss constant of the sub-band n, and $h_{m,n}$ is the random variable reflecting the small-scale fading and the shadowing effect. We further assume that uplink and downlink transmissions have the same channel power gain due to the channel reciprocity of TDD.

To simplify our analysis, we only consider a single-cell TDD system where intercell interference can be ignored. Then, we can express the signal-to-noise ratio (SNR) for user m on sub-band n as

$$\text{SNR}^x_{m,n} = \frac{p^x_{m,n} g_{m,n}}{b^x_{m,n} N_0}, \ \forall m,n, \tag{4.2}$$

where x can be UL or DL (standing for uplink or downlink, respectively), N_0 is the power spectral density of the additive white Gaussian noise (AWGN), $p^x_{m,n} \geq 0$ denotes the transmit power allocated to user m on sub-band n, and $b^x_{m,n} \geq 0$ denotes the bandwidth allocation. According to the Shannon formulation, the transmit data rate for user m on sub-band n can be expressed as

$$R^x_{m,n} = b^x_{m,n} \log_2\left(1 + SNR^x_{m,n}\right), \ x = \text{'UL' or 'DL'}, \ \forall m,n. \tag{4.3}$$

Since we need to optimize the EE of both downlink and uplink, we shall calculate the overall data rate of both the BS and user devices. The overall transmitted data rate of the BS can be expressed as

$$R_0 = \sum_{n=1}^{N}\sum_{m=1}^{M} b^{\text{DL}}_{m,n} \log_2\left(1 + \frac{g_{m,n} p^{\text{DL}}_{m,n}}{b^{\text{DL}}_{m,n} N_0}\right), \tag{4.4}$$

and the transmitted data rate of user m can be expressed as

$$R_m = \sum_{n=1}^{N} b^{\text{UL}}_{m,n} \log_2\left(1 + \frac{g_{m,n} p^{\text{UL}}_{m,n}}{b^{\text{UL}}_{m,n} N_0}\right). \tag{4.5}$$

We shall also model the power consumption of the BS and user devices. The power consumption mainly includes three parts, which can be expressed as

$$P_0 = \sum_{n=1}^{N}\sum_{m=1}^{M} w_n p^{\text{DL}}_{m,n} + P^{\text{DL,I}} + \sum_{n=1}^{N} P^{\text{DL,D}}_n \sum_{m=1}^{M}\left(b^{\text{DL}}_{m,n} + b^{\text{UL}}_{m,n}\right), \tag{4.6}$$

where the first item is the radio frequency (RF) power consumption (w_n denotes the inverse of the power amplifier efficiency on sub-band n), the second item is the fixed circuit power consumption, and the third item is the circuit power consumption related to the total occupied bandwidth ($P^{\text{DL,D}}_n$ denotes the power consumption per unit bandwidth).

The total power consumption for user m can be modeled in a similar way as

$$P_m = \sum_{n=1}^{N} v_n p^{\text{UL}}_{m,n} + P^{\text{UL,I}} + \sum_{n=1}^{N}\left(b^{\text{DL}}_{m,n} + b^{\text{UL}}_{m,n}\right) P^{\text{UL,D}}_n. \tag{4.7}$$

This power consumption also contains three parts, with v_n representing the inverse of the power amplifier efficiency at the user device, $P^{\text{UL,I}}$ being the fixed circuit power consumption, and $P^{\text{UL,D}}_n$ being the bandwidth-related power consumption.

Now we are ready to express the EEs for the downlink and the uplink as

$$\chi_0 = \frac{R_0}{P_0}, \tag{4.8}$$

and

$$\chi_m = \frac{R_m}{P_m}, m = 1, 2, \cdots, M, \tag{4.9}$$

respectively.

The objective of this section is to develop a joint downlink and uplink resource allocation algorithm to maximize the EE of both the BS and user devices. Therefore, the optimization problem can be formulated as

$$\max_{\mathbf{P}, \mathbf{B}} \sum_{m=0}^{M} \eta_m \chi_m, \tag{4.10a}$$

subject to

$$\sum_{n=1}^{N} p_{m,n}^{\mathrm{UL}} \leq p_m^{\mathrm{UL, max}}, \forall m, \tag{4.10b}$$

$$\sum_{m=1}^{M} \sum_{n=1}^{N} p_{m,n}^{\mathrm{DL}} \leq p^{\mathrm{DL, max}}, \tag{4.10c}$$

$$\sum_{m=1}^{M} \left(b_{m,n}^{\mathrm{DL}} + b_{m,n}^{\mathrm{UL}} \right) \leq W_n, \forall n, \tag{4.10d}$$

$$\sum_{n=1}^{N} b_{m,n}^{\mathrm{UL}} \log_2 \left(1 + \frac{g_{m,n} p_{m,n}^{\mathrm{UL}}}{b_{m,n}^{\mathrm{UL}} N_0} \right) \geq R_m^{\mathrm{UL, min}}, \forall m, \tag{4.10e}$$

$$\sum_{n=1}^{N} b_{m,n}^{\mathrm{DL}} \log_2 \left(1 + \frac{g_{m,n} p_{m,n}^{\mathrm{DL}}}{b_{m,n}^{\mathrm{DL}} N_0} \right) \geq R_m^{\mathrm{DL, min}}, \forall m, \tag{4.10f}$$

$$b_{m,n}^{\mathrm{DL}} \geq 0, b_{m,n}^{\mathrm{UL}} \geq 0, p_{m,n}^{\mathrm{DL}} \geq 0, p_{m,n}^{\mathrm{UL}} \geq 0, \forall m, n. \tag{4.10g}$$

Here, η_0 is the weight value for downlink EE and η_m ($m \in \{1, 2, \cdots M\}$) is the weight value for uplink EE of user m. They are all predetermined by the network operator to characterize the relative importance of network energy consumption and user energy consumption. $\mathbf{P}_{M \times 2N} = [\ \mathbf{p}^{\mathrm{DL}}, \mathbf{p}^{\mathrm{UL}}\ ]$ and $\mathbf{B}_{M \times 2N} = [\ \mathbf{b}^{\mathrm{DL}}, \mathbf{b}^{\mathrm{UL}}\ ]$, where $\mathbf{p}^{\mathrm{DL}}$, $\mathbf{p}^{\mathrm{UL}}$, $\mathbf{b}^{\mathrm{DL}}$, and $\mathbf{b}^{\mathrm{UL}}$ are all $M \times N$ matrices collecting the elements of $p_{m,n}^{\mathrm{DL}}$, $p_{m,n}^{\mathrm{UL}}$, $b_{m,n}^{\mathrm{DL}}$, and $b_{m,n}^{\mathrm{UL}}$, respectively. The constraints in (4.10b) and (4.10c) restrict the maximum transmit power for each user and the BS, respectively. The constraint in (4.10d) indicates the total uplink and downlink bandwidth on each sub-band. The constraints in (4.10e) and (4.10f) guarantee the minimum uplink and downlink data rates for user m, respectively.

4.2.2   Joint Uplink and Downlink Resource Allocation

The problem in (4.10a) is obviously a sum-of-ratios optimization problem, i.e., maximizing the summation of several fractional functions. As described in the Appendix in Chapter 2, this problem can be effectively solved by the sum-of-ratios algorithm by transforming it into an equivalent convex optimization problem, as shown in the following theorem.

THEOREM 4.1   *If* $(\mathbf{P}^*, \mathbf{B}^*)$ *is the optimal solution to* (4.10), *then there exist* $\boldsymbol{u}^* = \left(u_0^*, u_1^*, \cdots, u_M^*\right)$ *and* $\boldsymbol{\beta}^* = \left(\beta_0^*, \beta_1^*, \cdots, \beta_M^*\right)$ *such that* $(\mathbf{P}^*, \mathbf{B}^*)$ *is a solution to the following problem under the constraints from* (4.10b) *to* (4.10g) *for* $\boldsymbol{u} = \boldsymbol{u}^*$ *and* $\boldsymbol{\beta} = \boldsymbol{\beta}^*$

$$\max_{\mathbf{P},\mathbf{B}} \sum_{m=0}^{M} u_m \left(\eta_m R_m - \beta_m P_m\right). \tag{4.11}$$

Moreover, $(\mathbf{P}^*, \mathbf{B}^*)$ *also satisfies the following system of equations for* $\boldsymbol{u} = \boldsymbol{u}^*$ *and* $\boldsymbol{\beta} = \boldsymbol{\beta}^*$:

$$u_m = \frac{1}{P_m}, m = 0, 1, \cdots, M, \tag{4.12}$$

$$\beta_m = \eta_m \chi_m, m = 0, 1, \cdots, M. \tag{4.13}$$

From the above theorem, the fractional objective function has been changed into a subtractive form as $\sum_{m=0}^{M} u_m \left(\eta_m R_m - \beta_m P_m\right)$. Thus, we can achieve the global optimal solution to the original problem by equivalently solving this problem. We now prove that the equivalent problem is jointly concave on $(\mathbf{P}, \mathbf{B})$.

Define $U(\mathbf{P}, \mathbf{B}) = \sum_{m=0}^{M} u_m \left(\eta_m R_m - \beta_m P_m\right)$ and $\mathbf{H}\left(U_{m,n}(\mathbf{P}, \mathbf{B})\right)$ as the Hessian of $U(\mathbf{P}, \mathbf{B})$ on the n-th sub-band and the m-th user, as

$$\mathbf{H}\left(U_{m,n}(\mathbf{P}, \mathbf{B})\right) = \begin{bmatrix} -\Lambda_1^{\mathrm{DL}} & \Lambda_1^{\mathrm{DL}} & 0 & 0 \\ \Lambda_1^{\mathrm{DL}} & -\Lambda_2^{\mathrm{DL}} & 0 & 0 \\ 0 & 0 & -\Lambda_1^{\mathrm{UL}} & \Lambda_1^{\mathrm{UL}} \\ 0 & 0 & \Lambda_1^{\mathrm{UL}} & -\Lambda_2^{\mathrm{UL}} \end{bmatrix}, \tag{4.14}$$

where

$$\Lambda_1^x = u_0 \eta_0 \cdot \frac{g_{m,n}^2 b_{m,n}^x}{\left(b_{m,n}^x N_0 + g_{m,n} p_{m,n}^x\right)^2 \ln 2},$$

and

$$\Lambda_2^x = u_m \eta_m \cdot \frac{\left(g_{m,n} p_{m,n}^x\right)^2}{\left(b_{m,n}^x N_0 + g_{m,n} p_{m,n}^x\right)^2 \ln 2}.$$

The four eigenvalues of the Hessian matrix are $\phi_1 = \phi_2 = 0$, $\phi_3 = -\left(\Lambda_1^{\mathrm{DL}} + \Lambda_2^{\mathrm{DL}}\right) \leq 0$, and $\phi_4 = -\left(\Lambda_1^{\mathrm{UL}} + \Lambda_2^{\mathrm{UL}}\right) \leq 0$. Since all its eigenvalues are nonpositive, the Hessian matrix is negative semi-definite. Therefore, we can conclude that $U_{m,n}(\mathbf{P}, \mathbf{B})$ is jointly concave in $(\mathbf{P}, \mathbf{B})$, and $U(\mathbf{P}, \mathbf{B})$ is also jointly concave in $(\mathbf{P}, \mathbf{B})$ since linear operation preserves concavity.

In the expressions just discussed, we have proved that the objective function is a concave one. It is also rather obvious that the constraints also comprise a convex set. Thus, the problem is a convex one and standard convex tools, such as Lagrangian duality method, can be applied to solve it. The detailed steps of the convex optimization approach can be found in classical textbooks [20] and are therefore omitted.

In summary, the initial optimization problem in (4.10a) can be solved by two steps: the first one finds the optimal bandwidth and power allocation in (4.11) for given $(\boldsymbol{\beta}, \boldsymbol{u})$, and the second one finds the optimal $(\boldsymbol{\beta}^*, \boldsymbol{u}^*)$ that satisfies (4.12) and (4.13).

For the second step, we can develop a modified Newton method to solve it. We first introduce the following theorem, which can be proved in a similar way as Theorem 3.1 in [21].

THEOREM 4.2 *Let* $\psi_m(\boldsymbol{\beta}, \boldsymbol{u}) = -\eta_m R_m + \beta_m P_m$, $\psi_{M+1+m}(\boldsymbol{\beta}, \boldsymbol{u}) = -1 + u_m P_m$, $m = 0, 1, \cdots, M$, *and* $\boldsymbol{\psi}(\boldsymbol{\beta}, \boldsymbol{u}) = [\psi_0(\boldsymbol{\beta}, \boldsymbol{u}), \psi_1(\boldsymbol{\beta}, \boldsymbol{u}), \cdots, \psi_{2M+1}(\boldsymbol{\beta}, \boldsymbol{u})]$. *Then, the optimal* $(\boldsymbol{\beta}^*, \boldsymbol{u}^*)$ *is achieved if and only if*

$$\boldsymbol{\psi}(\boldsymbol{\beta}, \boldsymbol{u}) = \mathbf{0}. \tag{4.15}$$

Moreover, the optimal $(\boldsymbol{\beta}^*, \boldsymbol{u}^*)$ *is unique for problem (4.11).*

Furthermore, the modified Newton (MN) method can be used to solve the equation in (4.15), as expressed in the follows

$$\boldsymbol{\beta}^{k+1} = \boldsymbol{\beta}^k + \lambda_k \mathbf{q}^k, \ \boldsymbol{u}^{k+1} = \boldsymbol{u}^k + \lambda_k \mathbf{q}^k, \tag{4.16}$$

$$\mathbf{q}^k = -\left[\boldsymbol{\psi}'\left(\boldsymbol{\beta}^k, \boldsymbol{u}^k\right)\right]^{-1} \boldsymbol{\psi}\left(\boldsymbol{\beta}^k, \boldsymbol{u}^k\right), \tag{4.17}$$

where λ_k is the greatest ξ^i that satisfies

$$\left\|\boldsymbol{\psi}\left(\boldsymbol{\beta}^k + \xi^i \mathbf{q}^k, \boldsymbol{u}^k + \xi^i \mathbf{q}^k\right)\right\| \le \left(1 - \epsilon \xi^i\right) \left\|\boldsymbol{\psi}\left(\boldsymbol{\beta}^k, \boldsymbol{u}^k\right)\right\|, \tag{4.18}$$

and $i \in \{0, 1, 2, \cdots\}$, $\xi \in (0, 1)$, and $\epsilon \in (0, 1)$. Specifically, $\boldsymbol{\beta}^{k+1}$ and $\boldsymbol{u}^{k+1}$ can also be expressed component-wise as

$$\beta_m^{k+1} = (1 - \lambda_k) \beta_m^k + \lambda_k \frac{\eta_m R_m^k}{P_m^k}, \forall m, \tag{4.19}$$

$$u_m^{k+1} = (1 - \lambda_k) u_m^k + \lambda_k \frac{1}{P_m^k}, \forall m. \tag{4.20}$$

In summary, the algorithm includes two loops: the inner loop is the optimal bandwidth and power allocation and the outer loop is the sum-of-ratios optimization that finds the optimal $(\boldsymbol{\beta}, \boldsymbol{u})$. As mentioned in Chapter 2, the sum-of-ratios algorithm is a global-optimal solution. Moreover, the inner loop is a standard convex problem and also its global optimum solution can be achieved. Therefore, our proposed algorithm can achieve the global optimal solution to the original problem in (4.10). Regarding

Table 4.1 The global optimal solution for joint downlink and uplink resource allocation.

Algorithm 7 The global optimal solution for joint downlink and uplink resource allocation.

1: Initialize the maximum tolerance Δ and the maximum number of iterations $I_{\max}$.

2: Choose $\xi \in (0,1)$, $\epsilon \in (0,1)$, and $(\mathbf{P}^0, \mathbf{B}^0) \in \mathbf{dom}\ EE$. Let

$$\beta_0^0 = \eta_0 \frac{\sum_{n=1}^{N}\sum_{m=1}^{M} b_{m,n}^{\mathrm{DL}0} \log_2 \left(1 + \frac{g_{m,n} p_{m,n}^{\mathrm{DL}0}}{b_{m,n}^{\mathrm{DL}0} N_0}\right)}{\sum_{n=1}^{N}\sum_{m=1}^{M} w_n p_{m,n}^{\mathrm{DL}0} + P^{\mathrm{DL,I}} + \sum_{n=1}^{N} P_n^{\mathrm{DL,D}} \sum_{m=1}^{M} \left(b_{m,n}^{\mathrm{DL}0} + b_{m,n}^{\mathrm{UL}0}\right)},$$

$$\beta_m^0 = \eta_m \frac{\sum_{n=1}^{N} b_{m,n}^{\mathrm{UL}0} \log_2 \left(1 + \frac{g_{m,n} p_{m,n}^{\mathrm{UL}0}}{b_{m,n}^{\mathrm{UL}0} N_0}\right)}{\sum_{n=1}^{N} v_n p_{m,n}^{\mathrm{UL}0} + P^{\mathrm{UL,I}} + \sum_{n=1}^{N} \left(b_{m,n}^{\mathrm{DL}0} + b_{m,n}^{\mathrm{UL}0}\right) P_n^{\mathrm{UL,D}}}, \forall m,$$

$$u_0^0 = \left(\sum_{n=1}^{N}\sum_{m=1}^{M} w_n p_{m,n}^{\mathrm{DL}0} + P^{\mathrm{DL,I}} + \sum_{n=1}^{N} P_n^{\mathrm{DL,D}} \sum_{m=1}^{M} \left(b_{m,n}^{\mathrm{DL}0} + b_{m,n}^{\mathrm{UL}0}\right)\right)^{-1},$$

$$u_m^0 = \left(\sum_{n=1}^{N} v_n p_{m,n}^{\mathrm{UL}0} + P^{\mathrm{UL,I}} + \sum_{n=1}^{N} \left(b_{m,n}^{\mathrm{DL}0} + b_{m,n}^{\mathrm{UL}0}\right) P_n^{\mathrm{UL,D}}\right)^{-1}, \forall m.$$

3: Initialize the iteration index $k = 0$. Denote $\boldsymbol{\beta}^k = \left(\beta_0^k, \cdots, \beta_M^k\right)$, $\boldsymbol{u}^k = \left(u_0^k, \cdots, u_M^k\right)$.

4: Find the optimal $\left(\mathbf{P}^k, \mathbf{B}^k\right)$ for a given $\left(\boldsymbol{\beta}^k, \boldsymbol{u}^k\right)$ by convex optimization tool.

5: **if** $\psi\left(\boldsymbol{\beta}^k, \boldsymbol{u}^k\right) \preceq \boldsymbol{\Delta}$ **then**

6:    $\left(\mathbf{P}^k, \mathbf{B}^k\right)$ is the optimal solution and stop the algorithm.

7: **else**

8:    Denote i_k as the smallest i satisfying (4.18).

9:    Let $\lambda_k = \xi^{i_k}$, update $\boldsymbol{\beta}^{k+1}$, $\boldsymbol{u}^{k+1}$ and $\mathbf{q}^k$ by (4.16), (4.17).

10:    $k = k + 1$.

11:    **if** $k < I_{\max}$ **then**

12:       go to step 4.

13:    **else**

14:       go to step 16.

15:    **end if**

16: **end if**

---

the computational complexity, the sum-of-ratios optimization algorithm converges in superlinear/quadratic rate and the subgradient method of convex optimization converges in linear rate [22]. Therefore, the computational complexity of the proposed algorithm is low and acceptable for practical implementation. The detailed procedures of the proposed algorithm are illustrated in Table 4.1.

Table 4.2 System Parameters

Parameters	Settings
Noise	-174 dBm/Hz
User noise figure	9 dB
Base station noise figure	5 dB
Base station antenna gain	15 dBi
User antenna gain	0 dBi
Wavelength of sub-band 1, μ_1	0.375 m
Path loss of sub-band 1, α_1	3
Bandwidth of sub-band 1, W_1	4 MHz
Wavelength of sub-band 2, μ_2	0.12 m
Path loss of sub-band 2, α_2	4
Bandwidth of sub-band 2, W_2	8 MHz
$P^{DL,I}$, $P^{UL,I}$	46 dBm, 24.8 dBm
$P_1^{DL,D}$, $P_1^{UL,D}$	-23 dBm/Hz, -40 dBm/Hz
$P_2^{DL,D}$, $P_2^{UL,D}$	-26 dBm/Hz, -43 dBm/Hz
$P^{DL,max}$, $P_m^{UL,max}$, $\forall m$	46 dBm, 24 dBm
Δ, Δ'	$\left[10^{-4}, \cdots, 10^{-4}\right]$, 10^{-4}
I_{max}, I'_{max}	5, 100
$R_m^{DL,min}$, $R_m^{UL,min}$, $\forall m$	10 Mbit/s
w_n, v_n, $n = 1, 2$	1, 1
(ξ, ϵ)	(0.5, 0.5)

4.2.3 Numerical Results

In what follows, we shall provide numerical results to verify the proposed joint downlink and uplink energy-efficient resource allocation algorithm. In the simulation, we consider a single-cell network with a radius of 250 m. There are two sub-bands, sub-band 1 and sub-band 2, with different central frequencies of 800 MHz and 2.5 GHz, respectively. Four users are located in the network with distances of 50 m, 100 m, 150 m, and 200 m from the BS. The channel between the BS and users follows i.i.d Rayleigh model. Other major simulation parameters are given in Table 4.2 unless otherwise stated.

We first test the convergence speed of the proposed algorithm, as presented in Fig. 4.3. In this test, we set the same weight for all users, i.e., $\eta_m = \eta$, $m = 1,2,3,4$. We can observe from the figure that the sum-of-ratios optimization converges only after about 4–5 iterations, demonstrating a fast convergence speed of our proposal.

Figure 4.4 shows the bandwidth allocation on each sub-band with different weights of uplink EEs. In this test, we set the same weight to each uplink EE, i.e., $\eta_m = \eta$, $m = 1,2,3,4$. We also assume that sub-band 1 is more energy-efficient than sub-band 2 since sub-band 1 has a lower center frequency. From Fig. 4.4c, more bandwidth will be allocated to the uplink if more importance is put up to users than the BS. We can also observe that more bandwidth on sub-band 1 will be allocated to users with the increase of the weight value, η, to achieve higher uplink EE. Meanwhile, to

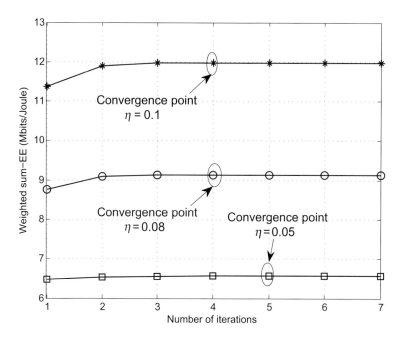

Figure 4.3 The convergence of the sum-of-ratios algorithm. © 2015 IEEE. Reprinted, with permission, from Yu, G., 2015, 'Joint Downlink and Uplink Resource Allocation for Energy-Efficient Carrier Aggregation', IEEE Transactions on Wireless Communications, Vol. 14, No. 6, pp. 3207–3218.

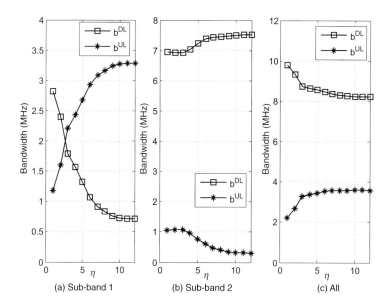

Figure 4.4 The bandwidth allocation for different weights of uplink EE. © 2015 IEEE. Reprinted, with permission, from Yu, G., 2015, 'Joint Downlink and Uplink Resource Allocation for Energy-Efficient Carrier Aggregation', *IEEE Transactions on Wireless Communications*, Vol. 14, No. 6, pp. 3207–3218.

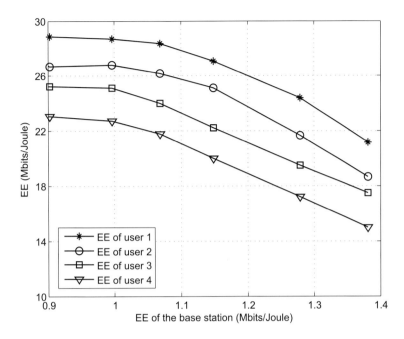

Figure 4.5 EE tradeoff between downlink and uplink. © 2015 IEEE. Reprinted, with permission, from Yu, G., 2015, 'Joint Downlink and Uplink Resource Allocation for Energy-Efficient Carrier Aggregation', *IEEE Transactions on Wireless Communications*, Vol. 14, No. 6, pp. 3207–3218.

guarantee the minimum data rate requirement, the downlink bandwidth on sub-band 2 will increase. Due to the same reason, the downlink bandwidth allocation will eventually remain at its minimum value as η goes large.

In Fig. 4.5, we assume that each user has the same weight, i.e., $\eta_m = \eta, m = 1, 2, 3, 4$, and adjust η from 0.005 to 0.03. From the figure, a clear EE tradeoff between downlink and uplink can be observed. To further show the EE tradeoff among users, we fix $\eta_m = 0.05, m = 2, 3, 4$ while changing η_1 from 0.005 to 0.05, and the results are depicted in Fig. 4.6. From the figure, the EEs of other users, i.e., $\chi_m, m = 2, 3, 4$, decreases with the EE of user 1. Therefore, we can conclude that the proposed algorithm can achieve a flexible EE tradeoff between downlink and uplink as well as among different users.

4.3 Energy-Efficient Resource Allocation in Homogeneous Networks

In Chapter 2, we introduced several works on the energy-efficient design in single-cell networks, such as fundamental EE–SE tradeoff, EE design in OFDMA systems, and EE design in non-orthogonal systems. In this chapter, we will discuss the energy-efficient design in multicell systems where intercell interference poses a major challenge for EE design. We will first investigate multicell homogeneous networks with the same kind of LTE BSs. For such a scenario, there have been several algorithms to improve the overall system EE, which is defined as the ratio between the overall transmit data rate and the

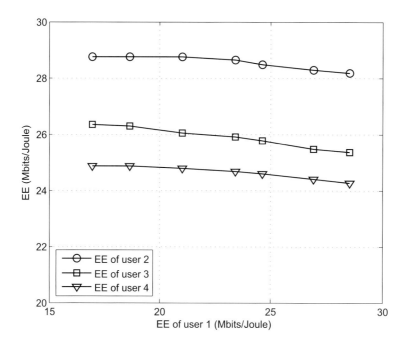

Figure 4.6 EE tradeoff among users. © 2015 IEEE. Reprinted, with permission, from Yu, G., 2015, 'Joint Downlink and Uplink Resource Allocation for Energy-Efficient Carrier Aggregation', *IEEE Transactions on Wireless Communications*, Vol. 14, No. 6, pp. 3207–3218.

overall energy consumption. In this section, we will utilize a more general EE objective, that is, to maximize the individual EE of each BS.

The motivation of this work is to maximize the individual EE of each BS, which has a more practical meaning than the overall EE maximization, as explained in the following. Currently, there exist some BSs powered by renewable external energy (e.g., solar and wind) in addition to those traditional BSs powered by the electricity grid. In such a situation, the EEs of different BSs may have different levels of importance. For example, to prolong the battery lifetime, BSs powered by external energy source should be more energy-efficient than those powered by the electricity grid. Therefore, it is more meaningful to consider the individual EE of each BS rather than the overall system EE. Nevertheless, the overall system EE maximization could be regarded as a special case of individual EE maximization.

In a multicell homogeneous network, the multi-stream aggregation (MSA) technique can be applied to improve the user data rate by aggregating data from different channels belonging to different BSs [23, 24]. This technique can greatly help improve the user experience of cell-edge users that can easily gather data from several nearby BSs. However, there are also many challenges related with the MSA technique. Among others, how to associate users with proper BSs, and how to allocate channel and power resource on each channel, are two critical challenges worth investigating. Several existing works have explored the user association and resource allocation algorithm in MSA systems from the perspective of system throughput maximization [25, 26]. However, there is no work considering the EE issue in such systems.

In the following subsection, we will first describe the system model and formulate the problem before introducing an effective algorithm to solve the problem. After that, we will provide numerical results to test our algorithm.

4.3.1   System Model and Problem Formulation

As depicted in Fig. 4.7, a multicell homogeneous system with K BSs and M users is investigated. The whole system has N orthogonal channels, and each has a bandwidth of W Hz. It is assumed that each user can receive data from several different BSs by the MSA technique.

Assuming user m is associated with the k-th BS on the n-th channel, we can express its received signal-to-interference-plus-noise ratio (SINR) as

$$\gamma_{mk}^n = \frac{p_{mk}^n h_{mk}^n}{WN_0 + \sum\limits_{i=1, i \neq m}^{M} \sum\limits_{j=1, j \neq k}^{K} p_{ij}^n h_{mj}^n}, \tag{4.21}$$

where h_{mk}^n denotes the channel power gain of the m-th user on the n-th channel at the k-th BS, p_{mk}^n is the transmit power on that channel, and N_0 is the power density of Gaussian noise. Based on this, the data rate of user m, that is associated with the k-th

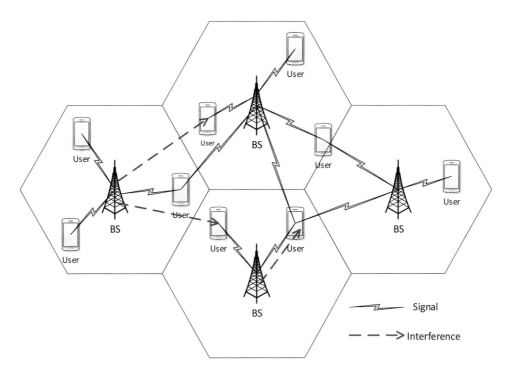

Figure 4.7 The network model for a homogenous cellular system with MSA.

BS on the n-th channel, can be written as

$$R_{mk}^n = W \log \left(1 + \frac{p_{mk}^n h_{mk}^n}{WN_0 + \displaystyle\sum_{i=1, i \neq m}^{M} \sum_{j=1, j \neq k}^{K} p_{ij}^n h_{mj}^n} \right). \tag{4.22}$$

To calculate the EE of each BS, we should first figure out the overall data rate transmitted by the k-th BS as

$$R_k = \sum_{m=1}^{M} \sum_{n=1}^{N} \rho_{mk}^n R_{mk}^n. \tag{4.23}$$

In the above equation, ρ_{mk}^n denotes the binary user association and channel allocation indicator. That is, $\rho_{mk}^n = 1$ if user m is associated with the k-th BS, and the n-th channel is allocated to that user. Otherwise, $\rho_{mk}^n = 0$.

Similarly, the total transmit power of the k-th BS can be expressed as

$$P_k = \sum_{m=1}^{M} \sum_{n=1}^{N} \rho_{mk}^n p_{mk}^n. \tag{4.24}$$

Then, based on the definition of EE, we can express the EE of the k-th BS as

$$\varepsilon_k = \frac{R_k}{\omega_k \cdot P_k + P_k^c}, \tag{4.25}$$

where ω_k and P_k^c denote the inverse of the power amplifier efficiency and the fixed circuit power consumption of the k-th BS, respectively.

As we have mentioned, we want to maximize the EE among different BSs with the constraints of user data rate requirement and the backhaul limit of each BS. Therefore, to indicate the levels of importance, we shall assign different weights to the EEs of different BSs. Let us denote η_k ($0 \leq \eta_k \leq 1$) as the weight factor for the k-th BS, which is a fixed value determined by the network management utility. Let $\boldsymbol{\Omega}$ and $\boldsymbol{P}$ denote the resource and power allocation matrices with a dimensional of $M \times N \times K$, respectively. Then, we can mathematically formulate the optimization problem as

$$\max_{\{\boldsymbol{\Omega}, \boldsymbol{P}\}} \sum_{k=1}^{K} \eta_k \varepsilon_k, \tag{4.26}$$

subject to

$$\rho_{mk}^n \in \{0, 1\} \text{ and } \sum_{m=1}^{M} \rho_{mk}^n \leq 1, \forall k, n, \tag{4.26a}$$

$$\sum_{n=1}^{N} \sum_{k=1}^{K} \rho_{mk}^n R_{mk}^n \geq R_m^{\min}, \forall m, \tag{4.26b}$$

$$\sum_{m=1}^{M}\sum_{n=1}^{N} \rho_{mk}^{n} p_{mk}^{n} \leq P_{k}^{\max}, \forall k, \tag{4.26c}$$

$$\sum_{m=1}^{M}\sum_{n=1}^{N} \rho_{mk}^{n} R_{mk}^{n} \leq R_{k}^{\max}, \forall k, \tag{4.26d}$$

$$p_{mk}^{n} \geq 0, \forall m, n, k. \tag{4.26e}$$

The above constraints include the orthogonality of channel allocation in each BS (4.26a), the minimum data rate requirement of each user (4.26b), the maximum transmit power limit (4.26c), and the backhaul capacity constraints of each BS (4.26d).

4.3.2 Problem Analysis and the Sub-Optimal Algorithm

We can observe that the problem in (4.26) is a sum-of-ratios optimization, since its objective function is the summation of several fractional functions, which is very complicated. Furthermore, this problem is also a combinatorial one due to the binary channel allocation matrix, $\boldsymbol{\Omega}$. Therefore, the problem is NP-hard and there exists no direct solution to solve it. To tackle it, we shall again leverage the sum-of-ratios optimization theory to convert it into a more tractable one, as presented in the following expressions.

First, the rate expression in (4.23) can be rewritten as

$$R_k = \sum_{m=1}^{M}\sum_{n=1}^{N} \rho_{mk}^{n} W \log\left(1 + \frac{p_{mk}^{n} h_{mk}^{n}}{WN_0 + \sum_{i=1, i \neq m}^{M}\sum_{j=1, j \neq k}^{K} p_{ij}^{n} h_{mj}^{n}}\right) \tag{4.27}$$

$$= \sum_{m=1}^{M}\sum_{n=1}^{N} W \log\left(1 + \frac{\rho_{mk}^{n} p_{mk}^{n} h_{mk}^{n}}{WN_0 + \sum_{i=1, i \neq m}^{M}\sum_{j=1, j \neq k}^{K} \rho_{mk}^{n} p_{ij}^{n} h_{mj}^{n}}\right), \forall k.$$

Then, we can simply prove that $p_{mk}^{n} = 0$ if $\rho_{mk}^{n} = 0$. This is very clear that no power will be allocated if channel n of BS k is not allocated to user m. Based on this, the constraint in (4.23) is equivalent to

$$R_k = \sum_{m=1}^{M}\sum_{n=1}^{N} W \log\left(1 + \frac{\rho_{mk}^{n} p_{mk}^{n} h_{mk}^{n}}{WN_0 + \sum_{i=1, i \neq m}^{M}\sum_{j=1, j \neq k}^{K} \rho_{ij}^{n} p_{ij}^{n} h_{mj}^{n}}\right), \forall k. \tag{4.28}$$

To deal with the combinatorial variable of ρ_{mk}^n, we can relax it into a continuous one, denoted as $\tilde{\rho}_{mk}^n$. Then, the initial problem in (4.26) can be converted into

$$\max_{\mathbf{P}} \sum_{k=1}^{K} \eta_k \tilde{\varepsilon}_k, \tag{4.29}$$

subject to

$$\sum_{n=1}^{N} \sum_{k=1}^{K} \tilde{R}_{mk}^n \geq R_m^{\min}, \forall m, \tag{4.29a}$$

$$\sum_{m=1}^{M} \sum_{n=1}^{N} \tilde{p}_{mk}^n \leq P_k^{\max}, \forall k, \tag{4.29b}$$

$$\sum_{m=1}^{M} \sum_{n=1}^{N} \tilde{R}_{mk}^n \leq R_k^{\max}, \forall k, \tag{4.29c}$$

$$\tilde{p}_{mk}^n \geq 0, \forall m, n, k, \tag{4.29d}$$

where

$$\tilde{p}_{mk}^n = \tilde{\rho}_{mk}^n p_{mk}^n,$$

$$\tilde{R}_{mk}^n = W\log\left(1 + \frac{\tilde{p}_{mk}^n h_{mk}^n}{WN_0 + \sum_{i=1, i \neq m}^{M} \sum_{j=1, j \neq k}^{K} \sum_{j=1, j \neq k} \tilde{p}_{ij}^n h_{mj}^n}\right),$$

$$\tilde{R}_k = \sum_{m=1}^{M} \sum_{n=1}^{N} \tilde{R}_{mk}^n,$$

$$\tilde{P}_k = \sum_{m=1}^{M} \sum_{n=1}^{N} \tilde{p}_{mk}^n,$$

and

$$\tilde{\varepsilon}_k = \frac{\tilde{R}_k}{\omega_k \cdot \tilde{P}_k + P_k^c}.$$

Since the constraint in (4.26a) has been relaxed, the optimal solution to (4.29) is naturally larger than the optimal solution to the initial problem in (4.26). However, the modified problem is unfortunately non-concave, since the objective function is fractional, which is not concave. We shall utilize the sum-of-ratios optimization tool to transform the problem into a more tractable and equivalent one, as stated in the following theorem. The proof of this theorem can be found in the Appendix of Chapter 2.

THEOREM 4.3   *If $\tilde{\mathbf{P}}^*$ is the optimal solution to (4.29), then there exist $\mathbf{u}^* = \left(u_1^*, u_2^*, \cdots, u_K^*\right)$ and $\boldsymbol{\tau}^* = \left(\tau_1^*, \tau_2^*, \cdots, \tau_K^*\right)$ such that $\tilde{\mathbf{P}}^*$ is a solution to the following problem under the constraints from (4.29a) to (4.29d) for $\mathbf{u} = \mathbf{u}^*$ and $\boldsymbol{\tau} = \boldsymbol{\tau}^*$*

$$\max_{\tilde{\mathbf{P}}} \; \sum_{k=1}^{K} u_k \left(\eta_k \tilde{R}_k - \tau_k(\omega_k \tilde{P}_k + P_k^c)\right). \tag{4.30}$$

Moreover, $\tilde{\mathbf{P}}^*$ also satisfies the following system of equations for $\mathbf{u} = \mathbf{u}^*$ and $\boldsymbol{\tau} = \boldsymbol{\tau}^*$:

$$u_k = \frac{1}{\omega_k \tilde{P}_k + P_k^c}, k = 1, 2, \cdots, K. \tag{4.31}$$

$$\tau_k = \eta_k \tilde{\varepsilon}_k, k = 1, 2, \cdots, K. \tag{4.32}$$

According to the above theorem and the sum-of-ratios optimization theory (Section A2.1.2), we shall now solve the equivalent optimization problem in the sum-of-ratios subtractive form, i.e., $\max_{\tilde{\mathbf{P}}} \; \sum_{k=1}^{K} u_k \left(\eta_k \tilde{R}_k - \tau_k(\omega_k \tilde{P}_k + P_k^c)\right)$, for this case. We first analyze the property of the objective function in (4.30), which can be rewritten as

$$\tilde{R}_k = \sum_{m=1}^{M} \sum_{n=1}^{N} W \log \left(WN_0 + \tilde{p}_{mk}^n h_{mk}^n + \sum_{i=1, i\neq m}^{M} \sum_{j=1, j\neq k}^{K} \tilde{p}_{ij}^n h_{mj}^n\right)$$
$$- \sum_{m=1}^{M} \sum_{n=1}^{N} W \log \left(WN_0 + \sum_{i=1, i\neq m}^{M} \sum_{j=1, j\neq k}^{K} \tilde{p}_{ij}^n h_{mj}^n\right). \tag{4.33}$$

Obviously, it is a summation of several difference-of-convex (d.c.) functions. Therefore, it is non-concave. To further solve the problem, we can utilize the successive convex approximation (SCA) approach to convert the objective function into the concave form. The main idea of the SCA method is to approximate the d.c. optimization problem into a series of convex optimization problems. The detailed approaches are described in the follows.

We can first express the lower bound of $\tilde{R}_k$ as

$$\tilde{R}_k = \sum_{m=1}^{M} \sum_{n=1}^{N} W \log \left(1 + \tilde{\gamma}_{mk}^n\right) \geq \sum_{m=1}^{M} \sum_{n=1}^{N} W \left(\alpha_{mk}^n \log(\tilde{\gamma}_{mk}^n) + \beta_{mk}^n\right), \tag{4.34}$$

which is tight when

$$\alpha_{mk}^n = \frac{\tilde{\gamma}_{mk}^n}{1 + \tilde{\gamma}_{mk}^n}, \tag{4.35}$$

$$\beta_{mk}^n = \log \left(1 + \tilde{\gamma}_{mk}^n\right) - \frac{\tilde{\gamma}_{mk}^n}{1 + \tilde{\gamma}_{mk}^n} \log(\tilde{\gamma}_{mk}^n). \tag{4.36}$$

Based on this, the objective function in (4.30) can be approximated into

$$\sum_{k=1}^{K} u_k \left(\eta_k \sum_{m=1}^{M} \sum_{n=1}^{N} W \left(\alpha_{mk}^n \log(\tilde{\gamma}_{mk}^n) + \beta_{mk}^n\right) - \tau_k(\omega_k \sum_{m=1}^{M} \sum_{n=1}^{N} \tilde{p}_{mk}^n + P_k^c)\right), \tag{4.37}$$

for given approximation coefficients $\boldsymbol{\alpha} \triangleq \{\alpha_{mk}^n\}$ and $\boldsymbol{\beta} \triangleq \{\beta_{mk}^n\}$. We then define the auxiliary variable $\hat{p}_{mk}^n = \log(\tilde{p}_{mk}^n)$. Therefore, the initial problem in (4.29) can be eventually transformed into the following equivalent problems

$$\max_{\hat{\mathbf{P}}} \sum_{k=1}^{K} u_k \left[\eta_k \sum_{m=1}^{M} \sum_{n=1}^{N} W\left(\alpha_{mk}^n \log(\hat{\gamma}_{mk}^n) + \beta_{mk}^n\right) - \tau_k \left(\omega_k \sum_{m=1}^{M} \sum_{n=1}^{N} \exp\left(\hat{p}_{mk}^n\right) + P_k^c \right) \right], \tag{4.38}$$

subject to

$$\sum_{n=1}^{N} \sum_{k=1}^{K} W\left(\alpha_{mk}^n \log(\hat{\gamma}_{mk}^n) + \beta_{mk}^n\right) \geq R_m^{\min}, \forall m, \tag{4.38a}$$

$$\sum_{m=1}^{M} \sum_{n=1}^{N} \exp\left(\hat{p}_{mk}^n\right) \leq P_k^{\max}, \forall k, \tag{4.38b}$$

$$\sum_{m=1}^{M} \sum_{n=1}^{N} W\left(\alpha_{mk}^n \log(\hat{\gamma}_{mk}^n) + \beta_{mk}^n\right) \leq R_k^{\max}, \forall k, \tag{4.38c}$$

where $\log(\hat{\gamma}_{mk}^n) = \hat{p}_{mk}^n + \log(h_{mk}^n) - \log\left(WN_0 + \sum_{i=1, i\neq m}^{M} \sum_{j=1, j\neq k}^{K} \exp\left(\hat{p}_{ij}^n\right) h_{mj}^n\right)$.

Now, the problem in (4.38) is eventually a convex problem due to the convexity of log-sum-exp functions. Thus, classical convex optimization approaches can be utilized to solve it. After the problem in (4.38) is solved, we still need the following operations. First, the optimal power should be transformed back according to $\tilde{p}_{mk}^n = \exp(\hat{p}_{mk}^n)$. Then, the bound in (4.37) should be iteratively updated until reaching a tighten one. Finally, the optimal $(\boldsymbol{u}^*, \boldsymbol{\tau}^*)$ should be updated according to the sum-of-ratios algorithm, i.e., iteratively updating $\boldsymbol{\alpha}$ and $\boldsymbol{\beta}$ until both (4.31) and (4.32) are satisfied.

In summary, the original problem can be solved by three nested iterative loops, as depicted in Fig. 4.8. The innermost loop is the optimal power allocation based on the standard Lagrangian solution. The intermediate loop is the SCA step that finds the appropriate $\left\{\alpha_{m,k}^n\right\}$ and $\left\{\beta_{m,k}^n\right\}$ to approximate $\tilde{R}_k$. The outermost loop is the sum-of-ratios algorithm, finding the optimal $(\boldsymbol{\tau}, \mathbf{u})$ that satisfies (4.31) and (4.32). This step can be realized by the Newton updating method as described earlier. Since all loops converge, the convergence of the whole algorithm can be guaranteed.

4.3.3 Numerical Results

In what follows, we shall provide numerical results to demonstrate the effectiveness of the proposed suboptimal algorithm. We adopt a simple scenario with two BSs and four users, where each BS covers a circle area with a radius of 500 m. There are four channels in each BS and users are randomly located in the network. An independent Rayleigh channel model is used to model the channel fading between the BSs and the users. The other major simulation parameters are summarized in Table 4.3.

We first test the convergence performance of the proposed algorithm. Figures 4.9(a) and 4.9(b) show the convergence rate of the sum-of-ratios algorithm (the outermost loop) and the SCA algorithm (the intermediate loop), respectively. From the figure, both

Table 4.3 Simulation Parameters

Parameters	Settings
Cellular radius	500 m
Noise	−174 dBm/Hz
Path loss exp.	3.5
$P_k^c, \forall k$	40 W
$R_k^{\max}, \forall k$	50 Mbit/s
$w_k, \forall k$	1
W	1 MHz
$P_k^{\max}$	2.6 W

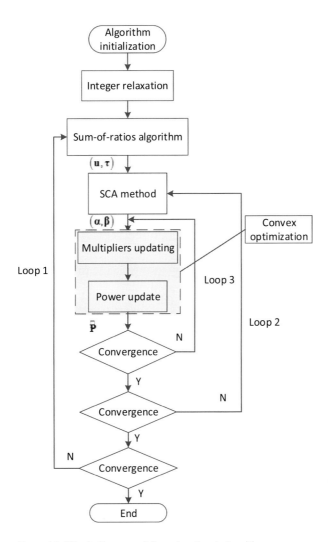

Figure 4.8 Block diagram of the suboptimal algorithm.

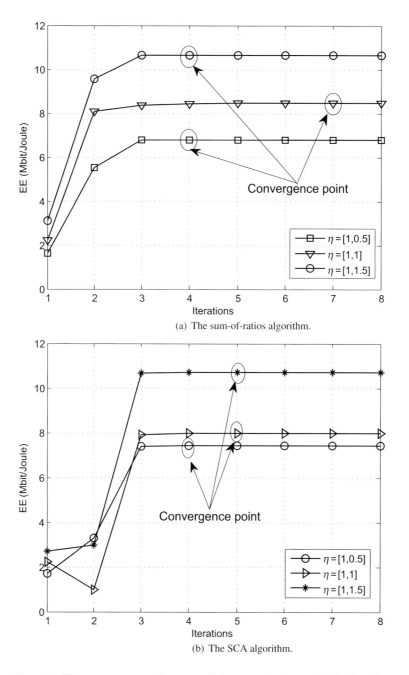

(a) The sum-of-ratios algorithm.

(b) The SCA algorithm.

Figure 4.9 The convergence performance of the sum-of-ratios and SCA algorithms. (a) The sum-of-ratios algorithm. (b) The SCA algorithm. © 2016 IEEE. Reprinted, with permission, from Chen, Q., 2016, 'Energy-Efficient User Association and Resource Allocation for Multistream Carrier Aggregation', *IEEE Transactions of Vehicular Technology*, vol. 65, no. 8, pp. 6366–6376.

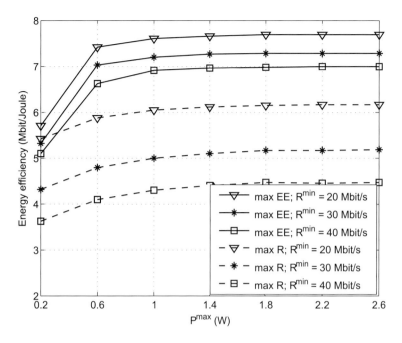

Figure 4.10 EE comparison between the EE-optimal algorithm and the rate-optimal algorithm. © 2016 IEEE. Reprinted, with permission, from Chen, Q., 2016, 'Energy-Efficient User Association and Resource Allocation for Multistream Carrier Aggregation', *IEEE Transactions of Vehicular Technology*, vol. 65, no. 8, pp. 6366–6376.

algorithms only require a few iterations to reach the optimal solutions, demonstrating the quick convergence of our proposed algorithms.

We then compare the proposed EE-optimal algorithm with the rate-optimal algorithm that aims to maximize the overall system throughput. Note that the rate-optimal algorithm can be designed by simply replacing the objective function with the overall system data rate. It can be also achieved by the SCA algorithm with the standard convex optimization in a similar way. Figures 4.10 and 4.11 illustrate the EE performance and data rate performance of our proposed algorithm, respectively, as compared with the rate-optimal algorithm. As depicted in the figures, although the rate-optimal algorithm can maximize the overall system throughput, the EE-optimal algorithm can achieve a better EE with a small loss of data rate. It shows that our proposal can indeed achieve a higher EE and higher system throughput simultaneously.

4.4 Energy-Efficient Resource Allocation in Heterogenous Networks

So far, we have discussed the energy-efficient resource allocation for the homogenous cellular system with multiple macro-cell BSs. With the dramatic increase of mobile data traffic, the cellular network structure is undergoing major changes. In addition to those macro-cell BSs, various kinds of low-power access points have been deployed to

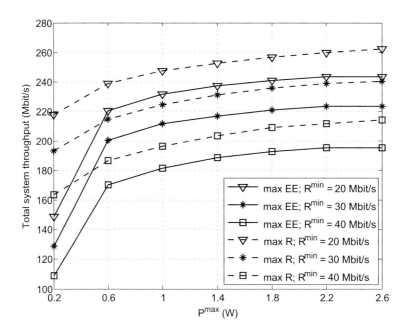

Figure 4.11 System throughput comparison between the EE-optimal algorithm and the rate-optimal algorithm. © 2016 IEEE. Reprinted, with permission, from Chen, Q., 2016, 'Energy-Efficient User Association and Resource Allocation for Multistream Carrier Aggregation', *IEEE Transactions of Vehicular Technology*, vol. 65, no. 8, pp. 6366–6376.

improve network capacity, which is referred to as heterogeneous wireless networks [27]. Besides the improvement on the area spectral efficiency because of dense frequency reuse, the EE can also be enhanced due to the low-power transmitter. There are two kinds of heterogeneous networks (HetNets), namely single-RAT HetNets and multi-RAT HetNets. The access points of the former one belong to the same RAT, e.g., LTE macro-cell, pico-cell, and femto-cell. On the other hand, the latter one consists of different access points with different radio access technologies, such as the coexistence of LTE BSs and WiFi access points. The major characteristic of the multi-RAT heterogenous network (HetNet) is that different access points are usually deployed by different operators and work on different frequency bands. This brings new communication freedom into multi-RAT mobile devices to dynamically and optimally select the most suitable network for better performances. Moreover, users can also utilize the multi-homing technology to further improve the throughput by aggregating data from different networks. Therefore, the multi-RAT HetNet has aroused considerable research interest in recent years.

Undoubtedly, the EE of both single-RAT and multi-RAT HetNets is also an important issue worth investigation [28]. Early studies show that, by offloading cellular traffic to the WiFi network, both system capacity and EE can be significantly improved [29–31]. The works in [32, 33] proposed partial spectrum reuse and adaptive BS sleeping control strategies to improve the EE of two-tier heterogeneous networks, respectively. In [34, 35], energy-efficient coordinated beamforming and precoding schemes have been proposed to maximize the system-level EE of MIMO heterogeneous networks. The large-scale user behavior and traffic dynamics can also be leveraged to design energy-efficient communication protocols [36, 37]. Moreover, energy-efficient optimal access point selection [38] and resource allocation algorithms [39] for multi-RAT heterogeneous networks have been investigated.

In this section, we will introduce energy-efficient joint bandwidth and power allocation for multi-RAT HetNets to maximize the EE of uplink users. In contrast to existing works that generally maximize the system-level EE, we aim at maximizing the EE of each individual user. This objective is much more general than the overall EE and has the merit of providing a better understanding of EE tradeoff. The problem of maximizing the overall EE is generally a fractional programming problem. However, maximizing all uplink EEs is much more different and new methods should be involved. Therefore, we introduce a multi-objective optimization tool to model the individual EE maximization problem. To deal with it, we first propose a novel concept of utopia EE for each user, defined as the maximum EE that a particular user can achieve. After that, we develop an effective multi-objective resource allocation algorithm based on the weighted Tchebycheff method.

In what follows, we will first introduce the multi-RAT HetNet and formulate the optimization problem. Then, we will describe the concept of Utopia EE and devise the novel algorithm to solve the problem. Finally, numerical results will be provided to demonstrate the performance of the proposed algorithm.

4.4.1   System Model and Problem Formulation

The system model of multi-RAT HetNets is depicted in Fig. 4.12. The system has M ($M \in \mathcal{M} = \{1, 2, \ldots, M\}$) different types of access points (APs), each equipped with

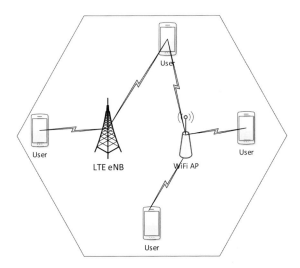

Figure 4.12 System model for multi-RAT heterogenous networks.

different RAT. For example, there are three different access points, namely, LTE BS, LTE-A small-cell access point, and WiFi access point. Different APs work on different frequency bands and thus inter-RAT interference does not exist. There are N ($N \in \mathcal{N} = \{1, 2, \ldots, N\}$) users randomly distributed in the network. It is assumed that each user can utilize the multi-homing technology to opportunistically choose several APs. This can be implemented by splitting the data stream into multiple sub-streams, and each sub-stream is transmitted via one RAT. For simplicity, both APs and users are assumed to be equipped with one antenna, while our work can be easily extended to the multi-antenna scenario with few modification.

The overall available bandwidth of the m-th RAT is denoted as W_m, which can be allocated to all users associated with this RAT. Moreover, we have the channel power gain model between user n and RAT m as

$$g_{m,n} = h_{m,n} d_{m,n}^{-\alpha_m},$$

which is a function of the distance between user n and RAT m ($d_{m,n}$), the small-scale channel fading coefficient ($h_{m,n}$), and the path-loss exponent α_m. To perform the bandwidth and power allocation, it is assumed that all channel state information (CSI) is accurate and known at the centralized scheduler.

Let $b_{m,n}$ be the bandwidth allocated for user n in RAT m. Then, the achievable data rate for user n in RAT m can be modeled as

$$R_{m,n} = b_{m,n} \log_2(1 + SNR_{m,n}), \forall m, n, \tag{4.39}$$

where $SNR_{m,n}$ is the uplink SNR that can be expressed as a function of power allocation $p_{m,n}$ and channel power gain $g_{m,n}$, as

$$SNR_{m,n} = \frac{p_{m,n} g_{m,n}}{b_{m,n} N_0}, \forall m, n. \tag{4.40}$$

Then, the overall achievable data rate for user n can be written as

$$R_n = \sum_{m=1}^{M} b_{m,n} \log_2 \left(1 + \frac{p_{m,n} g_{m,n}}{b_{m,n} N_0} \right), \forall n. \tag{4.41}$$

Similar to the previous analysis, the power consumption for each user can be modelled as

$$P_n = \sum_{m=1}^{M} \omega_{m,n} p_{m,n} + P^{\mathrm{I}} + \sum_{m=1}^{M} b_{m,n} P_m^{\mathrm{D}}, \forall n. \tag{4.42}$$

Again, the power consumption model consists of three items: the transmit power with $\omega_{m,n}$ being the inverse of the power amplifier efficiency; the fixed power consumption; and the power consumption as a function of the occupied bandwidth, where P_m^{D} denotes the fixed power consumption if RAT m is used. We shall note that this model is more general than that in [39] in which the last item is not considered.

According to (4.42), the EE of user n can be expressed as

$$\eta_n = \frac{R_n}{P_n}, \forall n. \tag{4.43}$$

As mentioned earlier, we aim at maximizing the EE for each individual user by joint bandwidth and power allocation. Therefore, the optimization problem can be formulated as

$$\max_{\mathbf{P}, \mathbf{B}} \{\eta_1, \eta_2, \cdots, \eta_N\}, \tag{4.44}$$

subject to

$$\sum_{n=1}^{N} b_{m,n} \leq W_m, \forall m, \tag{4.44a}$$

$$\sum_{m=1}^{M} p_{m,n} \leq P_n^{\max}, \forall n, \tag{4.44b}$$

$$\sum_{m=1}^{M} R_{m,n} \geq R_n^{\min}, \forall n, \tag{4.44c}$$

$$b_{m,n} \geq 0, \ p_{m,n} \geq 0, \forall m, n. \tag{4.44d}$$

In the above problem, the optimization variables are $\mathbf{P} = \{p_{m,n}\}_{M \times N}$ and $\mathbf{B} = \{b_{m,n}\}_{M \times N}$, the constraint in (4.44a) limits the overall available bandwidth in each RAT, the constraint in (4.44b) is the maximum allowable transmit power of each user, and the constraint in (4.44c) guarantees the minimum uplink data rate requirement of users. Here, the problem is assumed to be always feasible. This can be achieved by proper admission control strategies.

Equation (4.44) and its four parts are a multi-objective optimization problem [40], which aims to maximize several different objective functions. Since EEs of all users cannot be maximized simultaneously, those objective functions are generally conflicting with each other. Therefore, the optimal solutions are in general Pareto-optimal, which indicates that one cannot improve the EE of one user without decreasing the EE of other users. Different from a single-objective optimization problem, there may exist many Pareto-optimal solutions to a multi-objective optimization problem in general. For our case, Pareto-optimal EE can be defined as follows.

Pareto optimal EE: *An EE vector* $\eta^* = \{\eta_1^*, \eta_2^*, \cdots, \eta_N^*\}$ *is Pareto-optimal if and only if there does not exist another EE vector* η *such that* $\eta_n \geq \eta_n^*$ *for all users, but* $\eta_i > \eta_i^*$ *for at least one user.*

4.4.2 The Multi-Objective Energy-Efficient Algorithm

Now we will try to develop a bandwidth and power allocation algorithm to achieve Pareto-optimal EE. Before solving it, we first introduce the concept of Utopia EE for each individual user, defined as.

Utopia EE: *The Utopia EE for each user can be defined as the maximal EE it can achieve, i.e., $\eta_n^o = \max\limits_{\eta \in \mathcal{F}} \{\eta_n\}$, where $\mathcal{F}$ denotes the feasible set of bandwidth and power allocation strategies.*

Based on this definition, we can formulate the following optimization problem to find the Utopia EE for user n

$$\max_{\mathbf{P},\mathbf{B}} \ \eta_n, \tag{4.45}$$

subject to (4.44a)–(4.44d).

Clearly, this problem is a standard convex fractional programming, and therefore the classical Dinkelbach algorithm described in Chapter 2 can be utilized to solve it. The detailed procedures are omitted. After Utopia EE is solved for each user, we can convert the multiple-objective optimization problem into a single-objective one by utilizing the weighted Tchebycheff method as described in Chapter 2. The objective function of the single-objective optimization problem can be written as

$$\min_{\mathbf{P},\mathbf{B}} \max_{n} \ \left\{ \phi_n (\eta_n^o - \eta_n) \right\}, \tag{4.46}$$

where $\boldsymbol{\phi} = \{\phi_1, \cdots, \phi_N\}$ is any weight vector with all positive elements. By changing the weight vector, one can achieve all Pareto-optimal solutions from the above formulation. The equivalence between (4.46) and (4.44) is given in the following theorem.

THEOREM 4.4   *Let an EE vector η be a Pareto optimal solution to problem (4.44). Then, there must exist a positive weight vector $\boldsymbol{\phi} = \{\phi_1, \cdots, \phi_N\}$, such that η is the solution to problem (4.46).*

The relationship between Pareto-optimal EE and Utopia EE for a two-user case is depicted in Fig. 4.13. The shadowing area in this figure illustrates the feasible EE region for the problem in (4.44), and the upper bound of the region is the Pareto-optimal EE set. Moreover, the points of η_1^o and η_2^o are the two Utopia EE for user 1 and user 2, respectively. The point (η_1^*, η_2^*) is the Pareto optimal EE corresponding to the same weight vector, i.e., $\phi_1 = \phi_2$. From the figure, point (η_1^*, η_2^*) can achieve the minimum of $\max\{\eta_1^o - \eta_1^*, \eta_2^o - \eta_2^*\}$, as demonstrated in Theorem 4.4. Other Pareto-optimal EE points can be also achieved by setting different ϕ_i, for $i = 1, 2$, in a similar way. Indeed, the weight value ϕ_n reflects the different levels of importance for each user, which can be predetermined by the network operator according to practical applications. For example, one can assign a larger weight to those users with lower battery power or higher priority. In this way, the EEs for these users will be much closer to their maximum/Utopia EEs, as can be derived in the above Theorem 4.4.

Now we will develop an iterative algorithm to solve the problem in (4.46). First, inserting (4.43) into (4.46), the objective function can be converted into

$$\min_{\mathbf{P},\mathbf{B}} \max_{n} \ \left\{ \phi_n \left(\frac{\eta_n^o P_n - R_n}{P_n} \right) \right\}. \tag{4.47}$$

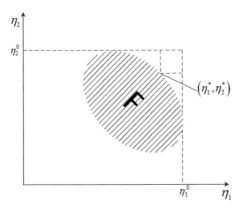

Figure 4.13 The illustration of Utopia EE and Pareto-optimal EE.

Now this problem becomes a generalized fractional programming (GFP), aiming at maximizing the minimum of several fractional functions [41, 42]. We can further prove that the objective function is quasiconvex and can be transformed into an equivalent but more tractable one, as shown in the following theorem.

THEOREM 4.5 *The objective function in (4.47) is quasiconvex and equivalent to*

$$\max_{\mathbf{y} \in \mathcal{Y}} \min_{\mathbf{P}, \mathbf{B}} f(\mathbf{y}, \mathbf{B}, \mathbf{P}) = \frac{\sum_{n=1}^{N} y_n \phi_n(\eta_n^o P_n - R_n)}{\sum_{n=1}^{N} y_n P_n}, \tag{4.48}$$

where $\mathcal{Y} \triangleq \{(y_1, \cdots, y_N) | y_n \geq 0, \forall n, \sum_{n=1}^{N} y_n = 1\}$.

Proof First, we will show that $R_{m,n}$ is concave over $b_{m,n}$ and $p_{m,n}$. Define $x \triangleq b_{m,n}, y \triangleq \frac{p_{m,n} g_{m,n}}{N_0}$, and $f(x, y) = -x \log_2 \left(1 + \frac{y}{x}\right)$.

The Hessian of $f(x, y)$ is

$$\mathbf{H} = \begin{bmatrix} \dfrac{y^2/x}{(x+y)^2} & -\dfrac{y}{(x+y)^2} \\ -\dfrac{y}{(x+y)^2} & \dfrac{x}{(x+y)^2} \end{bmatrix},$$

which is positive semi-defined, since its eigenvalues are

$$\lambda_1 = 0,$$

$$\lambda_2 = \frac{x^2 + y^2}{x^3 + 2x^2 y + x y^2} \geq 0.$$

Therefore, $R_{m,n} = -f(x, y)$ is concave and R_n is also concave since it is a linear combination of $R_{m,n}$.

To prove the quasiconvexity of (4.47), we write the sublevel set of $-\dfrac{R_n}{P_n}$ as

$$\tau_\alpha = \left\{ b_{m,n} \geq 0, p_{m,n} \geq 0, \forall m, n \left| -\frac{R_n}{P_n} \leq \alpha \right. \right\},$$

which is equal to

$$\tau_\alpha = \left\{ b_{m,n} \geqslant 0, p_{m,n} \geqslant 0, \forall m, n \mid -\alpha P_n - R_n \leq 0 \right\} .$$

We see that τ_α is convex due to the convexity of $-\alpha P_n - R_n$, which leads to the quasi-convexity of $\eta_n^o - \frac{R_n}{P_n}$. According to [20], $\max_n \left\{ \phi_n \left(\frac{\eta_n^o P_n - R_n}{P_n} \right) \right\}$ is also quasiconvex.

Then, due to the fact that a quasiconvex function will attain its maximum in a vertex of a convex polyhedron [43], we have

$$\max_n \left\{ \phi_n \left(\frac{\eta_n^o P_n - R_n}{P_n} \right) \right\} = \max_{\mathbf{y} \in \mathcal{Y}} \frac{\sum_{n=1}^N y_n \phi_n (\eta_n^o P_n - R_n)}{\sum_{n=1}^N y_n P_n} .$$

We will now show that $f(\mathbf{y}, \mathbf{B}, \mathbf{P})$ is quasiconvex over $(\mathbf{B}, \mathbf{P})$ and quasilinear over $\mathbf{y}$. For a given $\mathbf{y}$, the sublevel set of $f(\mathbf{y}, \mathbf{B}, \mathbf{P})$ can be denoted as

$$\tau_\alpha = \left\{ b_{m,n} \geqslant 0, p_{m,n} \geqslant 0, \forall m, n \mid f(\mathbf{y}, \mathbf{B}, \mathbf{P}) \leq \alpha \right\} .$$

Furthermore, we can rewrite $f(\mathbf{y}, \mathbf{B}, \mathbf{P}) \leq \alpha$ as

$$\sum_{n=1}^N y_n (\phi_n (\eta_n^o P_n - R_n) - \alpha P_n) \leq 0 .$$

Therefore, it can be easily observed that τ_α is convex since $\sum_{n=1}^N y_n (\phi_n (\eta_n^o P_n - R_n) - \alpha P_n)$ is convex over $(\mathbf{B}, \mathbf{P})$, which leads to the quasiconvexity of $f(\mathbf{y}, \mathbf{B}, \mathbf{P})$ over $(\mathbf{B}, \mathbf{P})$.

Again, for a given $(\mathbf{B}, \mathbf{P})$, the sublevel set of $f(\mathbf{y}, \mathbf{B}, \mathbf{P})$ is

$$S_\alpha = \left\{ y_n \geqslant 0, \forall n \mid f_y(\mathbf{y}) \leq \alpha \right\}, \tag{4.49}$$

which equals to

$$S_\alpha = \left\{ y_n \geqslant 0, \forall n \mid \sum_{n=1}^N y_n (\phi_n (\eta_n^o P_n - R_n) - \alpha P_n) \leq 0 \right\} .$$

Due to that $\sum_{n=1}^N y_n (\phi_n (\eta_n^o P_n - R_n) - \alpha P_n)$ is an affine function of $\mathbf{y}$, which is both convex and concave, $f(\mathbf{y}, \mathbf{B}, \mathbf{P})$ is quasilinear over $\mathbf{y}$.

Then, according to Sion's mini-max theorem [44], (4.47) can be finally converted into

$$\max_{\mathbf{y} \in \mathcal{Y}} \min_{\mathbf{P}, \mathbf{B}} f(\mathbf{y}, \mathbf{B}, \mathbf{P}).$$

This ends the proof. □

According to Theorem 4.6, we can obtain the optimal solution to the problem in (4.46) by iteratively solving the following two subproblems: a) finding the optimal

$\{\mathbf{B}^*, \mathbf{P}^*\}$ for a given $\mathbf{y}$; b) finding the optimal $\mathbf{y}$. Mathematically, they can be respectively formulated as

$$\eta(\mathbf{y}) = \min_{\mathbf{P},\mathbf{B}} \ f(\mathbf{y}, \mathbf{B}, \mathbf{P}),$$

and

$$\eta^* = \max_{\mathbf{y} \in \mathcal{Y}} \ \eta(\mathbf{y}).$$

We can use the Dinkelbach algorithm [45] to solve the above problem where the optimal η^* should satisfy $U^* = 0$. However, it is very hard to directly find the optimal η^* to satisfy this condition. In what follows, we shall present an effective algorithm to solve the above two subproblems, as presented in the following theorem.

THEOREM 4.6 *Define*

$$U(\mathbf{y}, \alpha) = \sum_{n=1}^{N} y_n (\phi_n(\eta_n^o P_n - R_n) - \alpha P_n),$$

and let $\mathbf{y}^{(k)}, k = 0, 1, \cdots$, be a sequence updated by the following equation for any initial $\mathbf{y}^{(0)}$:

$$\mathbf{y}^{(k+1)} = \arg \max_{\mathbf{y} \in \mathcal{Y}} \min_{\mathbf{P},\mathbf{B}} U(\mathbf{y}, \eta(\mathbf{y}^{(k)})).$$

Then we have
 (a) For the first subproblem, $\eta(\mathbf{y})$ is achieved when

$$\min_{\mathbf{P},\mathbf{B}} U(\mathbf{y}, \eta(\mathbf{y})) = 0.$$

 (b) For the second subproblem, we have $\eta(\mathbf{y}^{(k+1)}) \geq \eta(\mathbf{y}^{(k)})$ and the optimal solution $\eta^* = \eta(\mathbf{y}^{(k)})$ is achieved when $\eta(\mathbf{y}^{(k+1)}) = \eta(\mathbf{y}^{(k)})$. In this case,

$$U^* \triangleq \max_{\mathbf{y} \in \mathcal{Y}} \min_{\mathbf{P},\mathbf{B}} U(\mathbf{y}, \eta^*) = 0.$$

Proof The first subproblem is a standard concave fractional programming problem, thus (a) can be easily proved, based on the parametric algorithm in [45] by introducing the parameter α and the function $U(\mathbf{y}, \alpha)$.
 For the second part, since

$$\eta(\mathbf{y}^{(k+1)}) = \min_{\mathbf{P},\mathbf{B}} f(\mathbf{y}^{(k+1)}, \mathbf{B}, \mathbf{P}),$$

we have

$$\min_{\mathbf{P},\mathbf{B}} U(\mathbf{y}^{(k+1)}, \eta(\mathbf{y}^{(k+1)})) = 0. \tag{4.50}$$

On the other hand, according to the definition of $\mathbf{y}^{(k+1)}$,

$$\min_{\mathbf{P},\mathbf{B}} U(\mathbf{y}^{(k+1)}, \eta(\mathbf{y}^{(k)})) \geq \min_{\mathbf{P},\mathbf{B}} U(\mathbf{y}^{(k)}, \eta(\mathbf{y}^{(k)})) = 0. \tag{4.51}$$

Table 4.4 The algorithm to achieve Pareto-optimal EE

Algorithm 8 The algorithm to achieve the Pareto optimal EE

1: Initialize $\mathbf{y}^{(0)} \in \mathcal{Y}$, η, $k = 0$, ϵ_1, and ϵ_2.

2: Find $\{\mathbf{P}^*, \mathbf{B}^*\} = \arg\min_{\mathbf{P}, \mathbf{B}} U(\mathbf{y}^{(\mathbf{k})}, \eta)$.

3: **If** $\left| \sum_{n=1}^{N} y_n^{(k)}(\phi_n(\eta_n^o P_n^* - R_n^*) - \eta P_n^*) \right| < \epsilon_1$, **then**

4: $\quad$ Set $\eta(\mathbf{y}^{(\mathbf{k})}) = \eta$.

5: $\quad$ **goto** step (10).

6: **Else**

7: $\quad$ Update $\eta = \dfrac{\sum_{n=1}^{N} y_n^{(k)} \phi_n (\eta_n^o P_n^* - R_n^*)}{\sum_{n=1}^{N} y_n^{(k)} P_n^*}$.

8: $\quad$ **goto** step (2).

9: **end**

10: Update $\mathbf{y}^{(k+1)} = \arg\max_{\mathbf{y} \in \mathcal{Y}} \min_{\mathbf{P}, \mathbf{B}} U(\mathbf{y}, \eta(\mathbf{y}^{(k)}))$.

11: **If** $\eta(\mathbf{y}^{(k+1)}) - \eta(\mathbf{y}^{(k)}) < \epsilon_2$, **then**

12: $\quad$ Set $\eta^* = \eta(\mathbf{y}^{(k)})$.

13: $\quad$ **exit**.

14: **else**

15: $\quad$ Update $k = k + 1$.

16: $\quad$ **goto** step (2).

17: **end**

Then we can prove that $\eta(\mathbf{y}^{(k+1)}) \geq \eta(\mathbf{y}^{(k)})$ by following (4.50) and (4.51), and the fact that $U(\mathbf{y}, \alpha)$ decreases with α.

If the equality is achieved in (4.51), it means

$$U^* \triangleq \max_{\mathbf{y} \in \mathcal{Y}} \min_{\mathbf{P}, \mathbf{B}} U(\mathbf{y}, \eta(\mathbf{y}^{(k)})) = 0.$$

This yields the global solution of $\eta^* = \eta(\mathbf{y}^{(k)})$ according to [41]. This ends the proof.
$\qquad\square$

According to Theorem 4.6, we can now develop an iterative algorithm to solve the problem, as summarized in Table 4.4.

As shown in step 2 of the algorithm, the objective function of the first subproblem can be written as

$$\max_{\mathbf{P}, \mathbf{B}} \left\{ \sum_{n=1}^{N} y_n \left(\phi_n R_n - (\phi_n \eta_n^o - \alpha) P_n \right) \right\}. \tag{4.52}$$

We can easily prove that this objective function is jointly concave on $(\mathbf{P}, \mathbf{B})$. Therefore, we can utilize the classic convex optimization tools to solve it, such as the subgradient method and the interior point method.

In step 10 of the algorithm, the second subproblem can be expressed as

$$\max_{\mathbf{y}\in\mathcal{Y}} \min_{\mathbf{P},\mathbf{B}} \sum_{n=1}^{N} y_n(\phi_n(\eta_n^o P_n - R_n) - \alpha P_n).$$

This problem is a bit difficult to solve. However, since a quasi-convex function attains its maximum on a vertex of a convex polyhedron [43], we can rewrite the objective function as

$$\min_{\mathbf{P},\mathbf{B}} \max_{n} \left\{ \phi_n(\eta_n^o P_n - R_n) - \alpha P_n \right\}.$$

Moreover, according to the parametric optimization theory, we can further convert the problem into

$$\min_{\mathbf{P},\mathbf{B}} \tau,$$

subject to (4.44a)–(4.44d), and

$$\phi_n(\eta_n^o P_n - R_n) - \alpha P_n \leq \tau, \forall n.$$

Now the problem has been converted into a convex one, thus standard convex optimization methods can be utilized to effectively solve it.

4.4.3 Numerical Results

In this section, we will provide simulation results to show the performance of the proposed multi-objective energy-efficient algorithm. A multi-RAT HetNet consisting of two RATs is considered. The two networks have bandwidths of 1 MHz and 2 MHz, respectively. The channel model between the AP and users follows the i.i.d Rayleigh fading model. We assume that each uplink user has the same maximum transmit power of 1 W and the same data rate requirement of $R_{\min}$. Without loss of generality, the power amplifier efficiency is set as 100% for all users. Other major parameters are listed in Table 4.5 unless otherwise stated. We consider a simple symmetric scenario to validate the effectiveness of our proposal where the distances between each user and its access point are the same.

Table 4.5 System parameters

Parameters	Settings
Noise	-174 dBm/Hz
P^{I}	0.1 W
$P_1^{\mathrm{D}}, P_2^{\mathrm{D}}$	10 mW/MHz, 5 mW/MHz
W_1, W_2	2 MHz, 1 MHz
$P_n^{\max}, \forall n$	1 W
Δ	10^{-4}
ϵ_1, ϵ_2	$10^{-4}, 0.1$
$\omega_{m,n}, \forall m,n$	1
Γ_1, Γ_2	1
α_1, α_2	3, 4

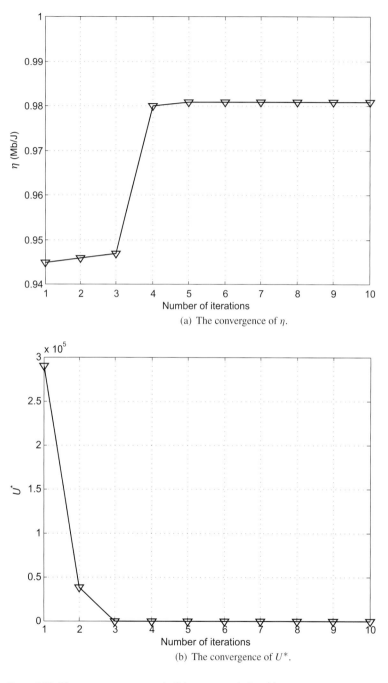

(a) The convergence of η.

(b) The convergence of U^*.

Figure 4.14 The convergence speed of the proposed algorithm.

First, we examine the convergence rate of the proposed algorithm in Fig. 4.14. From the figure, $\eta(y^{(k)})$ converges to its maximum value while $U^{(k)}$ converges to 0 after only 4–6 iterations, which confirms our analysis. Moreover, since the first subproblem is a standard convex fractional programming problem, it converges fast. Therefore,

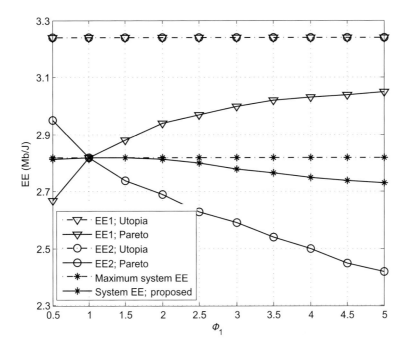

Figure 4.15 EE versus ϕ_1 in the symmetric scenario, $\phi_2 = 1$, $R_{\min} = 2$ Mbps. © 2015 IEEE. Reprinted, with permission, from Yu, G., 2015, 'Muilt-Objective Energy Efficient Resource Allocation for Multi-RAT Heterogeneous Networks', *IEEE Journal on Selected Areas in Communications*, vol. 33, no. 10, pp. 2118–2127.

the convergence speed of our proposed algorithm is fast and suitable for practical implementation.

Figure 4.15 shows the EE with different weights for user 1, ϕ_1, while the weight of user 2 is fixed to 1, i.e., $\phi_2 = 1$, and $R_{\min} = 2$Mbps. From the figure, the Utopia EE exactly serves as an upper bound of the EE for both users according to its definition. Moreover, with the increase of the weight, ϕ_1, the EE of user 1 increases gradually and finally reaches its maximum/Utopia EE. Meanwhile, the EE of user 2 decreases with ϕ_1, since more resource will be allocated to user 1 as its weight value increases, which demonstrates that an EE tradeoff between the two users can be achieved by the proposed algorithm. In the figure, we also compare the overall system EE of the proposed algorithm with the maximum overall system EE, which can be achieved by the classical Dinkelbach algorithm. The results in this figure show that, when $\phi_1 = \phi_2 = 1$, both users can achieve the same EE, which is also the maximum system EE. When the two weights are different, there exists a gap between the EE of our algorithm and the maximum system EE. Fortunately, this gap is rather small, which indicates that our algorithm can achieve a near-optimal EE performance.

We further test the EE performance with different user data rate requirements. In Fig. 4.16, the EE for users with different data rate requirements, $R_{\min}$, is depicted, where $\phi_1 : \phi_2 = 1 : 1$. The results in this figure show that the EE decreases with $R_{\min}$, which can be readily explained. When $R_{\min}$ is smaller, a larger EE can be achieved due to more freedom in the resource allocation algorithm. However, when $R_{\min}$ goes larger,

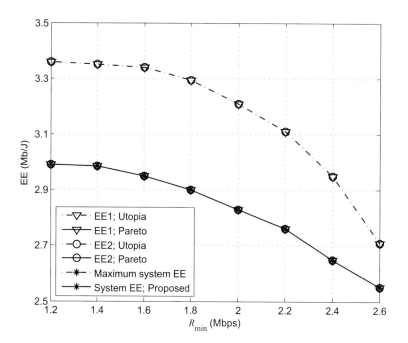

Figure 4.16 EE versus R_{min}. $\phi_1 : \phi_2 = 1 : 1$. © 2015 IEEE. Reprinted, with permission, from Yu, G., 2015, 'Muilt-Objective Energy Efficient Resource Allocation for Multi-RAT Heterogeneous Networks', *IEEE Journal on Selected Areas in Communications*, vol. 33, no. 10, pp. 2118–2127.

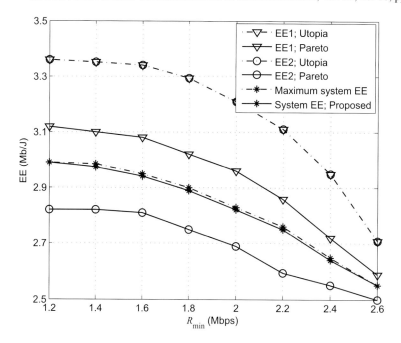

Figure 4.17 EE versus R_{min}. $\phi_1 : \phi_2 = 2 : 1$. © 2015 IEEE. Reprinted, with permission, from Yu, G., 2015, 'Muilt-Objective Energy Efficient Resource Allocation for Multi-RAT Heterogeneous Networks', *IEEE Journal on Selected Areas in Communications*, vol. 33, no. 10, pp. 2118–2127.

more resources will be allocated to increase the data rate, and in this case the EE will be inevitably decreased. We can also observe from the figure that the proposed algorithm can achieve the maximum system EE when $\phi_1 = \phi_2$, which is similar to the results in Fig. 4.15.

In Fig. 4.17, we further illustrate the results when $\phi_1 : \phi_2 = 2 : 1$. In this case, although both users can achieve the same Utopia EE, user 1 will achieve a larger EE than user 2 due to its relatively larger weight. We can also see from the results that the gap between Utopia EE and the Pareto-optimal EE for user 2 is almost twice larger than that of user 1. This is resulting from the weighted Tchebycheff algorithm, which minimizes the maximum gap between Utopia and Pareto-optimal EEs. Furthermore, in the case of $\phi_1 : \phi_2 = 2 : 1$, our algorithm cannot achieve maximum system EE, as has been explained previously.

References

[1] Z. Niu, "TANGO: Traffic-aware network planning and green operation," *IEEE Wireless Commun.*, vol. 18, no. 5, pp. 25–29, Oct. 2011.

[2] J. Wu, S. Zhou, and Z. Niu, "Traffic-aware BS sleeping control and power matching for energy-delay tradeoffs in green cellular networks," *IEEE Trans. Wireless Commun.*, vol. 12, no. 8, pp. 4196–4209, Aug. 2013.

[3] E. Oh, B. Krishnamachari, X. Liu, and Z. Niu, "Toward dynamic energy-efficient operation of cellular network infrastructure," *IEEE Commun. Mag.*, vol. 49, no. 6, pp. 56–61, Jun. 2011.

[4] Z. Niu, Y. Wu, J. Gong, and Z. Yang, "Cell zooming for cost-efficient green cellular networks," *IEEE Commun. Mag.*, vol. 48, no. 11, pp. 74–79, Nov. 2011.

[5] D. Feng, L. Lu, Y. Wu, G. Li, and S. Li, "Device-to-device communications underlaying cellular networks," *IEEE Trans. Commun.*, vol. 61, no. 8, pp. 3541–3551, Aug. 2013.

[6] G. Yu, L. Xu, D. Feng et al., "Joint mode selection and resource allocation for device-to-device communications," *IEEE Trans. Commun.*, vol. 62, no. 11, pp. 3814–3824, Nov. 2014.

[7] J. Liu, T. Zhao, S. Zhou, Y. Cheng, and Z. Niu, "CONCERT: A cloud-based architecture for next-generation cellular systems," *IEEE Wireless Commun.*, vol. 21, no. 6, pp. 14–22, Dec. 2014.

[8] S. Zhou, T. Zhao, Z. Niu, and S. Zhou, "Software-defined hyper-cellular architecture for green and elastic wireless access ," *IEEE Commun. Mag.*, vol. 54, no. 1, pp. 12–19, Jan. 2016.

[9] C. Xiong, G. Y. Li, S. Zhang, Y. Chen, and S. Xu, "Energy- and spectral-efficiency tradeoff in downlink OFDMA networks," *IEEE Trans. Wireless Commun.*, vol. 10, no. 11, pp. 3874–3886, Nov. 2011.

[10] D. W. K. Ng, E. S. Lo, and R. Schober, "Energy-efficient resource allocation in multi-cell OFDMA systems with limited backhaul capacity," *IEEE Trans. Wireless Commun.*, vol. 11, no. 10, pp. 3618–3631, Oct. 2012.

[11] C. Xiong, G. Y. Li, S. Zhang, Y. Chen, and S. Xu, "Energy-efficient resource allocation in OFDMA networks," *IEEE Trans. Commun.*, vol. 60, no. 12, pp. 3767–3778, Dec. 2012.

[12] G. Miao, N. Himayat. G. Y. Li, and S. Talwar, "Low-complexity energy-efficient scheduling for uplink OFDMA," *IEEE Trans. Commun.*, vol. 60, no. 1, pp. 112–120, Jan. 2012.

[13] O. Onireti, F. Heliot, and M. A. Imran, "On the energy efficiency-spectral efficiency trade-off in the uplink of CoMP system," *IEEE Trans. Wireless Commun.*, vol. 11, no. 2, pp. 556–561, Dec. 2012.

[14] G. Miao, "Energy-efficient uplink multi-user MIMO," *IEEE Trans. Wireless Commun.*, vol. 12, no. 5, pp. 2302–2313, May 2013.

[15] S. Buzzi, G. Colavolpe, D. Saturnino, and A. Zappone, "Potential games for energy-efficient power control and subcarrier allocation in uplink multicell OFDMA systems," *IEEE J. Sel. Topics Signal Process.*, vol. 6, no. 2, pp. 89–103, Nov. 2012.

[16] Z. Shen, A. Khoryaev, E. Eriksson et al., "Dynamic uplink-downlink configuration and interference management in TD-LTE," *IEEE Commun. Mag.*, vol. 50, no. 11, pp. 51–59, Nov. 2012.

[17] R. Ratasuk, A. Ghosh, W. Xiao, et al., "TDD design for UMTS long-term evolution," in *Proc. IEEE PIMRC 2008*, Cannes, France, pp. 1–5, Sept. 2008.

[18] 3GPP TS 36.828, "Evolved universal terrestrial radio access (E-UTRA); Further enhancements to LTE time division duplex (TDD) for downlink-uplink (DL-UL) interference management and traffic adaptation (Release 11)."

[19] H. Holma, S. Hekkinen, O. A. Lehtinen, and A. Toskala, "Interference considerations for the time division duplix node of the UMTS terrestrial radio access," *IEEE J. Sel. Areas Commun.*, vol. 18, no. 8, pp. 1386–1393, Aug. 2000.

[20] S. Boyd and L. Vandenberghe, *Convex Optimization*, Cambridge University Press, 2004.

[21] Y.-C. Jong, "An efficient global optimization algorithm for nonlinear sum-of-ratios problem," May 2012. www.optimization-online.org/DB_FILE/2012/08/3586.pdf.

[22] S. Boyd, L. Xiao, and A. Mutapcic, "Subgradient methods," lecture notes, EE392o: Optimization Projects, Stanford University, Autumn Quarter 2003–2004.

[23] "MSA: A key technology for the evolution of future wireless networks," Huawei Whitepaper, Jun 2013. www.huawei.com/mediafiles/CORPORATE/PDF/Magazine/communicate/70/HW_267909.pdf.

[24] I. F. Akyildiz, D. M. Gutierrez-Estevez, R. Balakrishnan, and E. Chavarria-Reyes, "LTE-Advanced and the evolution to Beyond 4G (B4G) systems," *Physical Commun.*, vol. 10, no. 1, pp. 31–60, Mar. 2014.

[25] C. Kim, R. Ford, and S. Rangan, "Joint interference and user association optimization in cellular wireless networks," in *Proc. IEEE Asilomar Conf. on Signals, Systems and Computers*, Pacific Grove, CA, Nov. 2014, pp. 511–515.

[26] Q. Ye, B. Rong, Y. Chen et al., "User association for load balancing in heterogeneous cellular networks," *IEEE Trans. Wireless Commun.*, vol. 12, no. 6, pp. 2706–2716, Jun. 2013.

[27] A. Damnjanovic, J. Montojo, Y. Wei, et. al, "A survey on 3GPP heterogeneous networks," *IEEE Wireless Commun.*, vol. 18, no. 3, pp. 10–21, Jun. 2011.

[28] S. S. Yong, T. Q. S. Quek, M. Kountouris, and H. Shin, "Energy efficient heterogeneous cellular network," *IEEE J. Sel. Areas Commun.*, vol. 31, no. 5, pp. 840–850, May 2013.

[29] S. Navaratnarajah, A. Saeed, M. Dianati, and M. A. Imran, "Energy efficiency in heterogeneous wireless access networks," *IEEE Wireless Commun.*, vol. 20, no. 5, pp. 37–43, Oct. 2013.

[30] Y. Choi, H. Kim, S. Han, and Y. Han, "Joint resource allocation for parallel multi-radio access in heterogeneous wireless retworks," *IEEE Trans. Wireless Commun.*, vol. 9, no. 11, pp. 3324–3329, Nov. 2010.

[31] S. Singh, H. S. Dhillon, and J. G. Andrews, "Offloading in heterogeneous networks: Modeling, analysis and design insights," *IEEE Trans. Wireless Commun.*, vol. 12, no. 5, pp. 2484–2497, May 2013.

[32] D. Cao, S. Zhou, and Z. Niu, "Improving the energy efficiency of two-tier heterogeneous cellular networks through partial spectrum reuse," *IEEE Trans. Wireless Commun.*, vol. 12, no. 8, pp. 4129–4141, Aug. 2013.

[33] D. Cao, S. Zhou, and Z. Niu, "Optimal combination of BS densities for energy-efficient two-tier heterogeneous cellular networks," *IEEE Trans. Wireless Commun.*, vol. 12, no. 9, pp. 4350–4362, Sept. 2013.

[34] S. He, Y. Huang, H. Wang, S. Jin, and L. Yang, "Leakage-aware energy-efficient beamforming for heterogeneous multicell multiuser systems," *IEEE J. Sel. Areas Commun.*, vol. 32, no. 6, pp. 1268–1281, Jul. 2014.

[35] Z. Xu, C. Yang, G. Y. Li, Y. Liu, and S. Xu, "Energy-efficient CoMP precoding in heterogeneous networks," *IEEE Trans. Signal Process.*, vol. 62, no. 4, pp. 1005–1017, Feb. 2014.

[36] Y. Huang, X. Zhang, J. Zhang et al., "Energy efficient design in heterogeneous cellular networks based on large-scale user behavior constraints," *IEEE Trans. Wireless Commun.*, vol. 13, no. 9, pp. 4746–4757, Sept. 2014.

[37] X. Ma, M. Sheng, and Y. Zhang, "Green communications with network cooperation: A concurrent transmission approach," *IEEE Commun. Lett.*, vol. 16, no. 12, pp. 1952–1955, Dec. 2012.

[38] G. Lim and L. J. Cimini, "Energy-efficient cooperative relaying in heterogeneous radio access networks," *IEEE Wireless Commun. Lett.*, vol. 1, no. 5, pp. 476–479, Oct. 2012.

[39] G. Lim, C. Xiong, L. J. Cimini, and G. Y. Li, "Energy-efficient resource allocation for OFDMA-based multi-RAT networks," *IEEE Trans. Wireless Commun.* vol. 13, no. 5, pp. 2696–2705, May 2014.

[40] R. T. Marler and J. S. Arora, "Survey of multi-objective optimization methods for engineering," *Struct. Multidiscip. O.*, vol. 26, no. 6, pp. 369–395, Apr. 2004.

[41] A. I. Barros, J. B. G. Frenk, S. Schaible, and S. Zhang, "A new algorithm for generalized fractional programs," *Math. Program.*, vol. 72, no. 2, pp. 147–175, Feb. 1996.

[42] J. P. Crouzeix and J. A. Ferland, "Algorithms for generalized fractional programming," *Math. Program.*, vol. 52, no. 2, pp. 191–207, Oct. 1991.

[43] M. Avriel, W. E. Diewert, S. Schaible, and I. Zang. "Generalized concavity," *Mathematical Concepts and Methods in Science and Engineering*, vol. 36, Plenum Press, 1988.

[44] M. Sion, "On general minimax theorems," *Pacific J. Math.*, vol. 8, no. 1, pp. 171–176, Mar. 1958.

[45] W. Dinkelbach, "On nonlinear fractional programming," *Manage. Sci.*, vol. 13, pp. 492–498, Mar. 1967.

5 Software-Defined Air Interface (SDAI) for a Greener Network

Air interface is the most fundamental aspect of the physical layer, and has been continuously evolving in each wireless communication generation. The paradigm of air interface design in the previous four generations took peak data rate and system capacity as the dominant objectives and adopted a one-size-fits-all approach. As a result, a global optimized or trade-off air interface design, being not necessarily optimal for each individual application scenario, was adopted by previous standards. As for 5G, air interface design is expected to expand and support diverse use case scenarios and applications that will continue beyond the current 4G standards. Three typical use case scenarios for 5G have been identified: enhanced mobile broadband (eMBB), ultra-reliable low-latency communication (URLLC), and massive machine type communication (mMTC). eMBB aims to provide high data rate mobile broadband services; URLLC is designed for applications that have stringent latency and reliability requirements; and mMTC is the basis for ultra dense connectivity in internet of things (IoT). The consensus has been made in the community that 5G new radio (NR) should be more agile, efficient, and even able to update itself on demand. A unified air interface design to accommodate different applications and services shall become the major consideration in 5G NR and beyond.

This chapter focuses on the soft and green design of 5G NR air interface. It starts with an introduction of SDAI's framework, which is proposed by the CMCC. Then the wireless propagation channels are discussed, which serve as the foundation for designing a flexible air interface. Then the design considerations of frame structure, multiple input multiple output (MIMO), waveforms, multiple access (MA) scheme, full duplex, and signaling/control/protocol are elaborated.

5.1 SDAI Framework

SDAI extends the concept of "soft design" to the air interface in communication networks [1, 2]. Instead of a global optimized air interface, which is the trade-off among many factors, SDAI is highly motivated to meet the massive connections and diverse demands by reconfiguration and combination of multiple physical-layer building blocks, including frame structure, duplex mode, waveforms, MA scheme, modulation and coding, a MIMO transmission scheme, etc., as shown in Fig. 5.1. To enable SDAI, each potential building block could be predefined, then suitable building blocks are selected

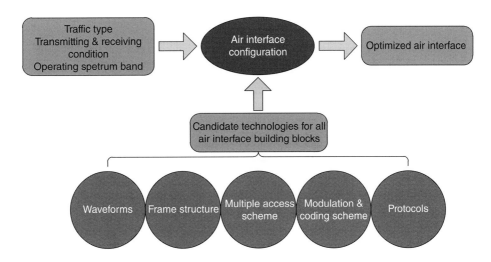

Figure 5.1 Building blocks of SDAI.

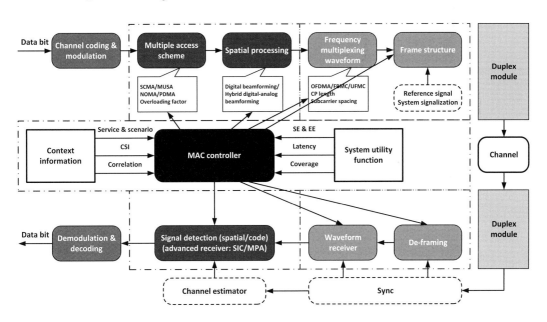

Figure 5.2 An illustrative structure of SDAI.

and properly configured and combined according to varying service requirements and network/UE capabilities to obtain the optimized technical solutions. It is expected to greatly improve the efficiency of wireless resources, reduce network deployment costs, and effectively cope with conceivable new scenarios and services. SDAI is expected to be a key driver of a green and soft RAN in 5G.

As shown in Fig. 5.2, the core of SDAI structure is an intelligent central controller. The controller is able to receive, sense, even predict the context information of the

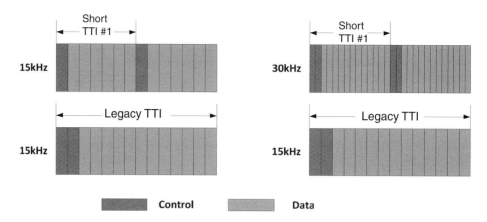

Figure 5.3 An illustrative example of short TTI.

network, including service types and requirements, traffic volume, channel state information, etc. Given the goal of optimizing specific system utility functions like spectrum efficiency, energy efficiency (EE), latency, coverage, and user QoE, the controller could select and configure the building blocks.

For example, a stringent user plane latency requirement of 1ms has been set for URLLC service in 5G standardization. Legacy frame structure and transmission time interval (TTI) length in LTE cannot meet the latency demand. A possible solution is to define a shorter TTI length in the air interface. A short TTI either contains fewer OFDM symbols than a legacy TTI, or contains same number of OFDM symbols but with a larger subcarrier spacing. The duration of a short TTI could be freely configured via the number of OFDM symbols and/or subcarrier spacing, as illustrated in Fig. 5.3. For constant URLLC traffic, the controller can configure legacy TTI transmission and short TTI transmission on separate frequency bands. Then filter-based OFDM waveform could be applied to mitigate the cross-numerology interference. For bursty URLLC traffic, the controller can configure both the legacy and short TTI transmissions on the same band to improve spectrum efficiency. The location of the short TTI transmission could be dynamically configured within the TTI duration.

Since there are diverse deployment scenarios and use cases emerging, or soon to emerge in future 5G networks, it is very important and beneficial for operators to deploy one network to support all use scenarios and use cases. To implement this, it is critical to adopt one unified and flexible air interface framework like SDAI to meet the diverse requirements of diversified usage scenarios. Following the E2E network slicing concept discussed in previous chapters, for each scenario, the air interface can be configured based on a physical layer (PHY) slicing, and corresponding layer 2 and layer 3 slicing. Each air interface slicing is tailored for specific service requirements and network/UE capabilities, while the coexistence of multiple slicing in one carrier needs to be well studied in a radio access network (RAN). In detail, each air interface slice may include the following aspects:

- Frame structure: Different DL and UL configurations and numerologies, variable subframe lengths to support different use cases. Flexible frame structure design is elaborated in Section 5.3.
- MIMO: Various single-user and multiple-user MIMO modes, with digital beamforming, analog beamforming, or hybrid analog and digital beamforming structures. Flexible MIMO design will be elaborated in Section 5.4.
- Waveforms: Various waveforms, including OFDM (orthogonal frequency division multiplexing), f-OFDM (filter-based OFDM), UFMC (universal filter multiple carrier), FBMC (filer bank multiple carrier), GFDM (generalized frequency division multiplexing), and OTFS (orthogonal time frequency spacing). Flexible waveform design will be elaborated in Section 5.5.
- MA schemes: Various orthogonal schemes and non-orthogonal schemes, such as MUSA (multiuser shared access), NoMA (non-orthogonal multiple access), SCMA (sparse code multiple access), and PDMA (pattern division multiple access). Flexible MA schemes will be elaborated in Section 5.6.
- Duplex: TDD, FDD, flexible duplex, full duplex, which will be elaborated in Section 5.7.
- Signaling: Layer 2 and layer 3 signaling associated with MAC, RLC, PDCP and RRC procedures, and possibly with cross layer interaction between RAN and the application layer. This will be covered in Section 5.8.
- Coding and modulation: various coding and modulation schemes, e.g., LDPC codes or polar codes.

5.2   Wireless Propagation in 5G Use Cases

The vision for 5G has been fueled by the development of two core PHY technologies that fundamentally set 5G NR apart from previous radio access technologies (RAT), namely, millimeter Wave (mmWave) and massive multi-input multi-output (MIMO) for below 6GHz frequencies. They both put forth a different paradigm that breaks with many current understandings of signal processing, device manufacturing, and network design, but most importantly the wireless propagation.

The physical layer technologies involved in air interface are generally built on the understanding of the fundamental of wireless propogation. In this section, we take a focused look upon the wireless propagation, i.e., its importance, its modeling methodology, existing channel models for cellular communications, and finally some challenging future research directions in 5G.

5.2.1   The Importance of Propagation Channels

The aim of this section is to briefly describe the fundamentals of radio wave propagation between the transmitter and the receiver. The main channel characteristics include path loss, shadow fading, and small-scale fading, as well as channel variations in time, frequency, and angle. The propagation channels dominate the actual performance of any practical system, since physical law, described by Shannon's capacity equation, dictates the amount of information that can be carried through the wireless media. To evaluate the performance of a wireless system, one could rely on software simulation, which is probably the most cost-effective and time-saving evaluation method. It offers a mathematically tractable process and can be used to predict trends and average performance reasonably and accurately. Wireless channel models are generally needed for these simulations. Standardized channel models are furthermore essential to enable fair comparisons of different system proposals. Channel behaviors are well understood at traditional cellular frequency bands. However, in the realm of 5G, we see high-frequency bands (i.e., mmWave bands with wide bandwidth), and high mobility scenarios (e.g., 500km/hr for high-speed railway [HSR]) are considered for mobile access. Yet in both areas there is very limited research done for channel modeling. This hinges on the system design for the new technologies and new use cases.

5.2.2   Channel Modeling Principle and Fundamentals

Channel modeling should always be based on the understanding of the physics of the propagation. It needs to be an accurate representation of what the transmitter or the receiver "sees." Moreover, its mathematical formulation should be intuitive. Unlike propagation channels, channel models are dependent on the system in which it operates. This means it should only be as complex as necessary, neglecting any effects that won't affect the system performance.

Channel models used to be only in two dimension, however it is increasingly obvious that the elevation domain and the azimuth domain both have huge impacts on radio propagation, hence wireless performance. There are four basic propagation mechanisms: reflection, diffraction, scattering, and transmission (penetration), as illustrated in Fig. 5.4.

Propagation in the cellular and WLANs scenarios is more complex. In many cases, the receiver is placed in non-line-of-sight (NLOS) of the base station (BS) or access point (AP). The transmitted signal usually reflects off surfaces, diffracts around object edges, or transmits through obstacles, following different paths before reaching the receiver. This results in the multipath effect, where reflections, diffractions, and transmissions all attenuate the signal power. Each multipath signal can be abstracted as a vector, which comprises both magnitude and phase. After arriving at the receiver, vector summation occurs, leading to small-scale fading caused by the moving terminals and Doppler, which leads to frequency dispersion and time-selective fading.

5.2.3   Channel Modeling Methods in Cellular Systems

The state-of-the-art channel models can be classified as physical and analytical channel models, as shown in Fig. 5.5. The physical channel models concern the physical propagation environment, thus modeling signal parameters such as AOA, AOD, complex power, and time of flight. They can be further divided into deterministic channel models (e.g., ray-tracing and measurement-based models), geometry-based

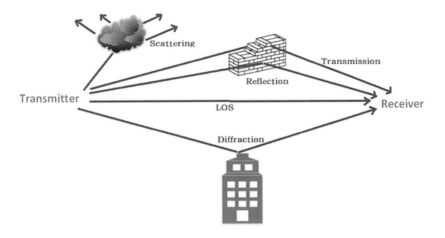

Figure 5.4 Wireless propagation.

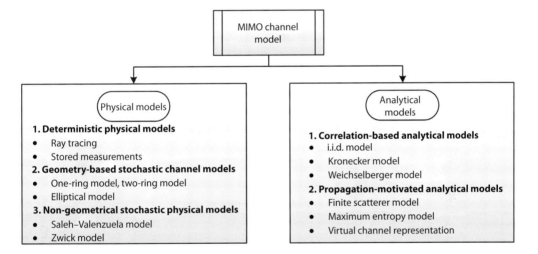

Figure 5.5 Traditional channel model classification.

(e.g., one/two-ring and elliptical models), and non-geometry stochastic channel models (e.g., Saleh-Valenzuela and Zwick models). Physical channel models can be in 2D or 3D.

The analytical channel models are derived from the statistical characteristics of the channels, which are obtained from mathematical representations such as the channel impulse response (CIR) between the transmitter and the receiver. Fitted models are often derived from the interpolation of measurement data points for a specific parameter. The analytical channel models can be divided into correlation-based stochastic models (CBSM) and propagation-based models (e.g., virtual channel representation [VCR]). Many of the current standardized wireless models can be put into these two categories, or a combination of them [3]. There is much great literature on the traditional channel modeling methodology for cellular systems, for instance [4–7]. Therefore, we will only focus on the new and challenging channel model research emerging in 5G.

5.2.4 New and Exciting Challenges in Channel Modeling

Massive MIMO Channels

Massive MIMO, is a disruptive 5G technology where the number of BS antennas grows to the hundreds, and aggressive spatial multiplexing supports tens of users with the same frequency and time radio resources [8]. Massive MIMO in the sub-6GHz band relies on the favorable propagation condition to deliver the superior spectral and energy efficiency. There is also the channel hardening effect, where the large number of BS antennas effectively average out the frequency selectiveness of users' channels due to spatial diversity [9, 10]. It is nevertheless very difficult to model the nonstationary phenomenon across the large antenna array elements and spherical wave effects when the UE is close enough to the BS and no longer in the far field of the array. The spatial, temporal, and angular correlations of the channel thus depart from the traditional understanding of MIMO channels [11].

There are already many research efforts in developing a massive MIMO channel model, for instance, the extension of the COST 2100 channel model [12], the extension of 3GPP-SCM [13], and the WINNER+ model [14], as well as mathematical models based on correlation or mutual coupling or geometry. They all have their accomplishment compared to traditional smaller MIMO models. However, none is without ambiguous assumptions or simply over-prediction of theoretical performance. Channel measurement campaigns are still needed to understand the necessary characteristics for an accurate model and its parameters.

Massive MIMO is not a stand-alone technology, and it is envisioned to be applied in other 5G applications, for instance, mmWave communication and distributed cellular deployment, as well as IoT scenarios. This further necessitates the understanding of the impact of large antenna arrays/RF chains on channel response seen by the BS and the UE before incorporating with other technologies.

mmWave Channels

Channel measurement is a necessary exercise for wireless researchers to understand the characteristics of a new spectrum. The mmWave band that is currently under consideration for mobile communications has a range from 30GHz to 100GHz. Since 2011, there had been many channel measurement campaigns conducted at 28, 38, 60, 72, and 73GHz. Channel modeling in this regard is critical for evaluating wireless technology in a timely and cost-effective manner. Considering many existing channel models and their evolution, such as the COST 2100 channel model, COST IC 1004 [15], ETSI model [16], and 3GPP TR 38.900 [17], it is always preferable to adapt and reuse the existing model structure. However, it is no longer enough to model the mmWave channel the same way as the microwave channel. Setting aside the obvious weather and oxgen absorption phenomenon, both large-scale and small-scale parameters need to be measured and analyzed. There has been continuous effort on modeling the large-scale spatial and angular parameters of the measured channels (e.g., path loss, shadow fading, delay spread, and angular power spectrum in azimuth/elevation domain) for standardization purposes in urban macro, urban micro, and indoor scenarios [18]. Due to lack of measurement data, small-scale parameterization has not been validated extensively. Therefore, the current channel models such as the 3GPP 3D model, the QuaDRiGa model [19], the IEEE 802.11ad model [20], the MiWEBA model [21], the METIS model [22] and the mmMAGIC model [23] are not adequate to predict actual system performance.

Furthermore, there are unique mmWave channel characteristics that are yet to be specified and accurately modeled. For example, temporal channel statistics (e.g., the birth and death of multipath components [MPCs]) should be measured and investigated; the small-scale dynamics, e.g., diffuse scattering, significantly impact system performance and should be truthfully modeled; an intra-cluster model on PDP and power angular spectrum (PAS) allows for more accurate correlation modeling between MPCs; spatial consistency characteristics at such high frequency determine how the MIMO channel works; the blockage model of clusters is based on AOAs and attenuations, and its ability to represent the dynamic reality is highly desired. A first-attempt statistical channel model was presented in [24], taking the spatial, angular, and temporal statistics into consideration and consolidating them into a time cluster-spatial lobe (TCSL) approach.

It is increasingly obvious to us that, given the knowledge we have now about wireless propagation, there may not be a "one-size-fits-all" kind of channel model for such a vast range of spectrum and different deployment scenarios, as well as different levels of accuracy requirement.

Device-to-Device Channel

Mass connectivity has been a major selling point for 5G, where anyone and anything can be connected through the internet to facilitate everything from day-to-day convenience to high productivity. This requires device-to-device (D2D) communication, where the outdoor interaction with the environment is very different than traditional cellular device-to-infrastructure (D2I) channels (the indoor difference is less pronounced). The most obvious differences between D2D and D2I are the immediate surroundings of the transmitter, and the receivers being similar and correlated, and that both ends of the link could be mobile, i.e., dual mobility. There are also many more deployment scenarios than typically identified by standard bodies such as 3GPP; to name a few, indoor offices and shopping malls, as well as urban roads and highways. In each scenario, the link types, i.e., outdoor to outdoor (O2O), outdoor to indoor (O2I), indoor to indoor (I2I), or vehicle-to-vehicle (V2V), should also be taken into consideration when a more realistic channel representation is desired. Currently, there a three major established D2D channel models: WINNER, 3GPP, and COST 2100. However, the WINNER model derives the LSPs from deterministic maps, and the small-scale parameters do not have spatial and temporal consistency, leading to inaccurate MIMO channels; 3GPP D2D channels [25] are based on measurements, incorporating dual-mobility and Doppler effects, but making unrealistic assumptions about angular statistics such as uniform AOAs; COST 2100 is a GBSCM (geometry-based stochastic channel model) with cluster distribution based on real measurements and the visibility region feature enables moving clusters, yet it is not designed for D2D where the BS is not mobile.

In this section, we take V2V for illustrative purposes as a typical example of D2D channels. V2V considers both peer-to-peer 802.11p and future 5G-based communication as the data bearer. The final desired channel model should be general, easy to use, and able to transit smoothly among different scenarios.

Taking the V2V scenario at an intersection as an example, this wireless propagation parameter involves path loss, delay dispersion, Doppler spread, temporal variation, etc. Path loss is determined by the distances from TX/RX to side buildings, the width of the road, the distance between TX and RX, as well as the intersection spacing $[PL(d_r, d_t, w_r, x_t, i_s)]$. Delay spread can be modeled as random variables with a fitted distribution and a mean RMS value dependent on the deployment environment. Measurements have shown that the typical value is between 100 and 400ns. Temporal variation depends on the dynamics of the end nodes, since both ends of the link can move, sometimes at a fast speed, and the shadowing objects and scatters can move as well. The nonstationarities in the channel cannot be simply ignored, as what is done in 3GPP models; one of the less complicated solutions is GBSCM with randomly placed scatters or a tapped delay line model with a birth and death process and continuously changing delay. The final challenge is the ability to simulate multiple D2D links with the correct correlation/interaction between them. This is critical to the prediction of system performance under interference [26].

High-Speed Railway (HSR) Channels

High-speed railway is another much-discussed application that is considering 5G as a candidate technology for both safety-related critical data transmission and in-car passenger communication. Traditionally, GSM-R is used for trains at a low data rate of around 200kbps, but mainly for safety-related data. Currently, LTE-R and 5G-R are being actively researched and developed for next-generation railway communications [27]. It may have many new featured technologies, such as a distributed antenna system (DAS), coordinated multipoint (CoMP), or a mobile relay station (MRS). However, the challenges ahead will not be addressed properly if channel models are not accurate enough. There are unique features in HSR scenarios that lead to unique channel characteristics. Besides some challenges inherited from conventional trains such as high penetration losses, limited visibility in tunnels, and the harsh electromagnetic environment, typical hurdles involve fast handover, fast travel through diverse scenarios, large Doppler spreads, and nonstationarities. All the issues mentioned above are heavily dependent on the implementation method, i.e., mobile relay or direct link.

[28] has identified 12 different propagation environments, including viaduct, cutting, station, rural open area, and mountainous terrain, to name a few. Each environment has drastically different values and distributions of channel parameters. At microwave frequency, extensive measurements have been carried out and some practical models for power and fading are in use [29]; standardized models can be categorized as reused rural/urban models, such as a IMT-A channel model or simplified (quasi-LOS) model, neither of which is valid. At mmWave frequency, there is only one paper [30].

There is also a misunderstanding that "fortunately, most scenarios can be LOS, leading to simple channel characteristics." Unfortunately, this is a naive simplification of the problem; due to rich scattering in some environments, the Ricean K factor is surprisingly very small, and even becomes negative in tunnels. Additionally, there is not just a Doppler shift, but a Doppler spread in most situations, even in LOS. Angular spread is also very pronounced across short distances, e.g., 20 degrees of RMS angular spread in 100m in viaduct (assuming the train speed is 500km/hr, the time it takes is 0.72s). Finally, the power delay profile is closely related to the immediate environment; as we can see in [31], the measured PDP has a strong periodicity due to the power poles along the tracks. Moving forward, nonstationary HST channel models could be deterministic or stochastic, or the hybrid of the two, catering to different applications of the channel models. Deterministic models can be pure ray-tracing and a random graph; stochastic models can be GBSCM (e.g., a finite-state Markov channel can effectively capture the dynamic nature of fast fading/time-varying in some environments). There is also the possibility that large-scale fading models and small-scale models are separately developed.

5.2.5   Concluding Remarks

This section began with a brief description of wireless propagation and the fundamental principles in channel modeling, as well as the characterization of different model methods. The main focus herein is identifying the channel modeling challenges and some possible directions/solutions in various 5G scenarios, i.e., massive MIMO, mmWave, D2D, and HST. In summary, channel modeling work is the foundation for the success of 5G, where diversified use cases and requirements demand statistically accurate and easily applicable channel models. Extensive measurements and innovative methods (e.g., leveraging wireless big data) are needed to blaze the challenges ahead. More importantly, given the fact that channel conditions vary significantly in diversified scenarios and extensive measurements and modeling methods are required, the 5G air interface design is strongly motivated to be as flexible and efficient as possible, and capable of configuring the physical-layer and higher-layer building blocks and parameters. In the following subsections, the soft and green design of physical-layer technologies will be elaborated, including frame structure, waveform, MA, MIMO, and duplex.

5.3   Flexible Frame Structure

Frame structure is the basic DL and UL operation framework for wireless communication systems, which specifies where and when the signaling, control, and data should be transmitted. In LTE, frame structure is generally designed with relatively fixed settings that consider the worst scenarios, which simplifies the system design but at the cost of efficiency. As diversified scenarios, such as eMBB, URLLC, mMTC, and the wide range of spectrum defined in 5G, instead of the traditional "one-size-fits-all" design, the frame structure in 5G is envisioned to be more agile and efficient. It should be dynamically configured to adapt to various propagation and application scenarios. On the other hand, facing new challenges posed by 5G, such as the extremely high mobility (up to

500km/hr), ultra-low latency (less than 0.5ms for URLLC), and massive connectivity, the frame structure is expected to be comprehensively designed to fulfill these stringent demands.

In this section, we begin with the frame structure design principles in SDAI for 5G. Then, key features of the flexible frame structure, e.g., scalable numerology, service multiplexing, configurable subframe/scheduling unit, flexible reference signal/scheduling, and HARQ timing, are discussed. Finally, the standardization progress of frame structure in 3GPP is also summarized, along with the future research directions on full duplex frame structure.

5.3.1   Frame Structure Design Principles

In order to address the great challenges in 5G and facilitate SDAI, the frame structure should be more flexible compared with that of 4G era. In the following, some basic design principles for frame structure are listed:

- Allowing for scalability to address different services requirements;
- Support for efficient multiplexing of different services, e.g., eMBB, URLLC, and mMTC;
- Support for an extremely wide range of physical properties, e.g., very wideband, narrow band, TDD, FDD, sub 6 GHz, and mmWave bands;
- Support for dynamic TDD assignment with efficient interference management;
- Support for self-contained subframe with a single interlace structure (ACK/NACK in the same subframe) and possible multiple interlace structure for forward compatibility;
- Support for tight coupling across aggregated carriers (i.e., supporting carrier aggregation).

With the above-mentioned principles in mind, some key features of the flexible frame structure are illustrated in the following part.

Scalable Numerology

To support a wide range of services, deployment scenarios, and spectrum, the numerologies (including subcarrier spacing, cyclic prefix length, and TTI length) of the frame structure need to be scalable and flexibly configurable.

For eMBB services, LTE numerology of 15kHz subcarrier spacing works well below 6 GHz. For mMTC narrow-band services, a narrower subcarrier spacing, e.g., 3.75KHz, is desired for higher capacity when considering the same coverage. For URLLC service, a larger subcarrier spacing, e.g., 60KHz, is preferred for latency reduction. Besides the service aspects, for numerology design, deployment-related attributes (such as carrier frequency, channel characteristics, inter-site distance, UE speeds, and possible transmission schemes) also should be taken into account. In addition, implementation cost and complexity are also crucial factors for the numerology design. With much broader bandwidth defined for 5G, the required FFT size and frequency domain signal processing with RE- or RB-level granularity may reach a point where

Table 5.1 Numerology for diverse services, deployments and spectrum.

Motivation	Scenarios	Subcarrier spacing	CP length	TTI length
Diverse service	eMBB	>=15kHz	Depends on scenarios	Depends on scenarios
	mMTC	<=15kHz	Longer CP	Longer TTI
	URLLC	FFS	Depends on scenarios	Shorter TTI
Diverse deployment	Low to medium UE speed	15kHz	Depends on services	Depends on services
	High UE speed	>=60kHz	Depends on services	Depends on services
Diverse spectrum	Sub-6GHz	Depends on scenarios	Depends on scenarios	Depends on scenarios
	Above 6GHz	Larger carrier spacing	Shorter CP	Shorter TTI

the implementation cost and complexity become unacceptable if LTE legacy subcarrier spacing is utilized, especially for the mmWave bands. Targeting a unified air interface for diversified services and deployment scenarios with all potential available spectrum, the scalable subcarrier spacing and TTI length could be the starting point.

In Table 5.1, some examples on the numerology design for diverse services, deployments, and spectrum are shown [32].

Service Multiplexing

To have an efficient utilization of scarce air-interface resources, multiplexing of transmissions with different latency and/or reliability requirements for eMBB/URLLC/mMTC should be supported in 5G. When UEs are scheduled in different networks or frequency bands, no attention needs to be paid for services multiplexing. But, for cases in which UEs with different services are scheduled in the same band and same network, flexible frame structure with variable numerologies are suggested to be considered and applied, so as to satisfy the requirements of different services simultaneously.

Different services may require different numerologies. The question of how to support multiplexing of different numerologies in the same frame structure needs be addressed carefully. Both time division multiplexing (TDM) and frequency division multiplexing

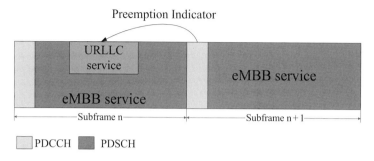

Figure 5.6 Multiplexing of URLLC and eMBB services.

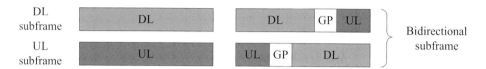

Figure 5.7 Three types of time domain structure.

(FDM) can be considered. For FDM, the interference of both UL to DL and DL to UL when considering the multiplexing of different services needs to be addressed. Applying TDM is in general more beneficial for multiplexing long and short scheduling frames. A typical use case for applying TDM is mission-critical applications. They are usually characterized by a bursty transmission and supported via time domain puncturing. For example, dynamic resource sharing between URLLC and eMBB can be supported by transmitting URLLC traffic in resources scheduled for ongoing eMBB traffic, known as puncturing or preempted transmission, as shown in Fig. 5.6. To avoid the severe performance loss of eMBB in this case, an indicator signal may be necessary to let the eMBB UE know the existence of URLLC transmission.

Configurable Subframe

In order to meet the requirements of different services, three types of subframes are recommended to be supported, i.e., DL subframe, UL subframe, and bidirectional subframe, as illustrated in Fig. 5.7.

Considering the requirement of URLLC on fast response to scheduling, transmission and feedback, bidirectional subframe seems to be a good solution to make self-scheduling and self-feedback possible. A bidirectional subframe contains a DL transmission region (containing DL control, RS and/or data), guard period (GP), and an UL transmission region (containing UL control, RS and/or data). The overhead of GP can be configured flexibly to cater to different scenarios.

Even though a bidirectional subframe provides adequate flexibility for DL and UL scheduling and timing, it is necessary to support a full DL and UL subframe, so as to reduce overhead and ensure coverage, especially in scenarios with low frequency and macro coverage. They are more efficient when deployed on a paired spectrum, and on an unpaired spectrum when latency is not a problem. In addition, it is suggested to keep as many commonalities as possible for operating these subframe types on paired/unpaired spectrums and licensed/unlicensed spectrums.

In summary, three types of subframes should be supported in 5G: DL subframe, UL subframe, and bidirectional subframe. For different use cases, the flexible combination of these types of subframes can be considered.

Configurable Scheduling Unit

Different services are characterized by different features, for example, URLLC traffic may have small packet size but strict delay requirements, while eMBB service may ask for very high data rate with medium latency, and mMTC may put stringent requirements on coverage and connection numbers. It is possible that a common scheduling time unit is defined for all the services, but apparently it is not efficient. To satisfy the delay requirement of URLLC, the time duration of the common time unit should be short enough, and a short time unit will result in limited resource elements in it. For eMBB service with large packet size, several continuous time units may be needed to one UE schedule, with a common short time unit, as URLLC, the control channel overhead, and feedback overhead may increase linearly. Thus, it is more efficient to define a configurable scheduling time unit.

Different from the 1ms scheduling unit defined in LTE, for 5G, mini-slot-, slot-, and slot-aggregation-based scheduling should be supported. Note that mini-slot can be composed of fewer OFDM symbols. A subframe should contain an integral number of mini-slot for timing alignment. Mini-slot-based scheduling can be utilized for traffic with small packet size but strict delay requirements, e.g., URLLC services, while the slot aggregation scheduling unit is suit for the large packet size with stringent requirements on coverage. Data transmitted in the scheduling unit should be self-decodable with its control channel, reference signals, and A/N feedback. With configurable different scheduling time intervals, and with different packet sizes, delay and coverage requirements can be satisfied with high efficiency.

Flexible Reference Signal Design

In general, reference signals (RS) are used for data demodulation, phase tracking, time/frequency tracking, channel state information (CSI) measurement, radio link monitoring, RRM measurement, etc. Various RS signals, such as DMRS, SRS, CSI-RS, and PT-RS are being heatedly discussed in 3GPP NR. Basically, RS design needs to consider the tradeoff between the estimation performance and the overhead. For example, refarding the high-speed train scenario with 500km/hr mobility, denser DMRS is needed to combat the rapidly varying channel for accurate channel estimation. As for the low and medium mobility scenario, much sparse DMRS can be configured to reduce the DMRS overhead.

Flexible Scheduling/HARQ Timing

To support diverse services with different latency requirements and different UE processing capability, configurable scheduling and HARQ timing is expected to be sup-

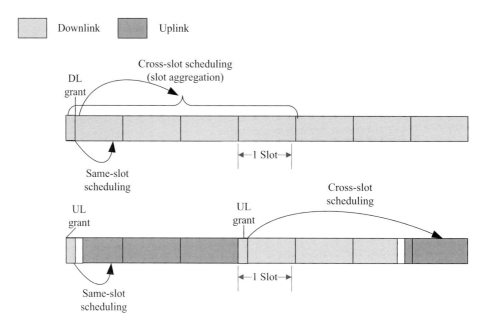

Figure 5.8 An example of configurable scheduling timing.

ported in 5G. Four types of timing relationship need to be considered for the scheduling and HARQ timing.

- K0: Delay between the DL grant and corresponding DL data (PDSCH) reception
- K1: Delay between DL data (PDSCH) reception and the corresponding acknowledgement transmission on UL
- K2: Delay between UL grant reception in DL and the UL data (PUSCH) transmission
- K3: Delay between the ACK/NAK reception in UL and the corresponding retransmission of data (PDSCH) on DL

An example of configurable scheduling timing is shown in Fig. 5.8.

For the DL/UL scheduling, at least the same-slot scheduling and the cross-slot scheduling should be supported. Note that the value of minimum timing depends on many factors such as transmission block size, data RE mapping, the RS location, channel coding and cell range, UE processing capability, etc. Scheduling multiple DL/UL slots from a single DL control occasion enables the reduction of both signaling overhead for NR data transmissions and the interference caused to neighboring cells.

Figure 5.9 illustrates an example of flexible HARQ timing. For low latency service or small packet transmission, self-contained properties can be supported including the short HARQ timing between DL data and the corresponding A/N, UL grant, and the corresponding UL data. For noncritical eMBB service or large packet transmission, more processing time is needed, and accordingly HARQ timing is longer.

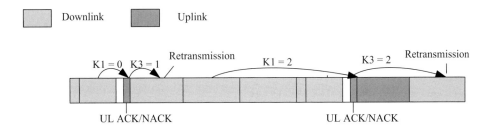

Figure 5.9 An example of flexible HARQ timing.

5.3.2   Progress of Frame Structure in 3GPP 5G NR

Two types of timing units, fixed length and variable length, are defined in state-of-the-art standardization on 5G frame structure. Timing units with fixed length include radio frame and subframe. A radio frame has a 10 ms duration and a subframe has a 1 ms duration. They are chosen to be consistent with LTE to strive for a better LTE-NR coexistence performance in case of co-site deployment. On the other hand, to support diverse services and multiple numerologies, timing units of varying length are also defined, including slot and mini-slot. A slot contains 14 OFDM symbols for normal CP length (nearly 7% CP overhead) and a mini-slot contains X OFDM symbols, where X is less than 14 and greater than 1. Mini-slot mainly targets fast scheduling service and/or delay-critical service, e.g., URLLC. Obviously, the length of a slot depends on the subcarrier spacing and that of a mini-slot depends on both subcarrier spacing and number of OFDM symbols. Note that, for extended CP length (nearly 25% CP overhead), a slot only contains 12 OFDM symbols. Currently, except for 60kHz subcarrier spacing, whether to support extended CP length on other subcarrier spacing is still undecided. The motivation to support 60kHz subcarrier spacing with extended CP length derives from the high-speed train scenario.

Slot format is another aspect in standardization on frame structure that shows flexibility. Now it has been agreed that "A slot can contain all downlink, all uplink, or at least one downlink part and at least one uplink part [54]", which exactly conforms to our principles on frame structure design in SDAI. Moreover, the slot format can be dynamically indicated to UEs through layer 1 signaling, rather than the only semi-static ways in LTE. This further provides adequate flexibility on DL/UL traffic adaption and scheduling.

5.3.3   Concluding Remarks

A flexible frame structure is the basis for green and soft RAN operation, which is capable of flexible configuration of UL and DL slots, numerology, scheduling unit, HARQ timing, RS patterns, etc. This is quite motivated in the 5G era to meet service requirements with diversified KPIs in various usage scenarios. Future research directions of frame structure may include the possible transition to full duplex when full duplex technologies become mature and ready to be implemented in cellular networks.

5.4   Flexible MIMO

MIMO techniques have been widely utilized in 4G LTE systems, where multiple MIMO schemes including diversity, spatial multiplexing, and MU-MIMO are specified [33]. These schemes are implemented in basedband via the digital beamforming structure. In 5G NR, a distinguishing configuration is that digital beamforming, analog beamforming, and hybrid beamforming will all possibly be considered in system deployment [34]. Different structures may be employed at both BS and UE in various scenarios and frequency bands. For example, analog beamforming may be more suitable for indoor scenario in mmWave bands. When more uses need to be supported in spatial domains, hybrid beamforming is motivated, where on top of analog beamforming, digital beam-forming may further help to reduce the inter-user/inter-beam interferences. For lower frequency bands in 5G, traditional digital beamforming may be the most suitable. This necessitates the design of flexible MIMO with a unified beamforming architecture and a unified CSI acquisition/feedback mechanism. In addition, as the antenna number is expected to increase both in BS and UE in 5G communications, the energy efficiency will be an important performance indicator, necessitating energy efficient designs.

In this part, a unified MIMO framework is presented, which includes analog, digital, and hybrid beamforming as special cases. Typical hybrid beamforming structures are also investigated, with various beamforming algorithms surveyed. Furthermore, energy-efficient design considerations of hybrid beamforming structure are presented. Finally, the standardization of hybrid beamforming is discussed.

5.4.1   Unified Framework for MIMO Techniques for 5G

In the last 10 years, the evolution of MIMO techniques tends to employ a large number of antennas (usually hundreds of antennas) at the BS to serve tens of users in the same time and frequency resource, which is commonly referred to as massive MIMO [35]. Massive MIMO is not a straightforward extension of conventional small-scale MIMO. Its systems can acquire "channel hardening," such that the uncorrelated noises and channel vectors for different users are averaged out, and simple linear signal processing procedures can achieve near-optimal performance.

With the severe spectrum shortage in conventional cellular bands, massive MIMO in mmWave bands can potentially help to meet the anticipated demands of mobile traffic in 5G era. There are many challenging issues in the implementation of digital beamforming on mmWave, including complexity, energy consumption, cost, etc. In a practical deployment, hybrid beamforming structures can be an important alternative choice and have been proposed as an enabling technology for 5G cellular communications [36–39].

The main concept of hybrid beamforming is to divide the traditional baseband signal processing into digital and analog domains. A unified framework of hybrid beamform-ing is illustrated in Fig. 5.10, where transmit data from N_s ports are mapped onto a N_t^{RF} transmit and receive unit (TXRU) via digital beamforming, and further mapped onto N_t antennas via analog beamforming. The mapping PA (power amplifier) is determined by the connections between TXRU and antennas, and the phase and amplitude of each RF

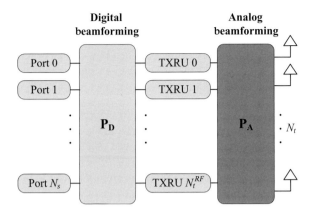

Figure 5.10 A block graph for hybrid beamforming.

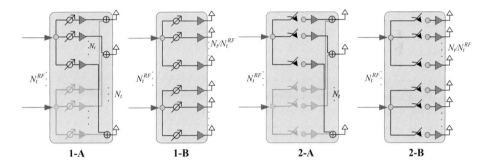

Figure 5.11 Typical structures of the analog part of the hybrid beamforming.

path. This architecture is a generic framework. For example, when $N_t^{RF} = N_t$ and each TXRU is directly linked to each antenna element, this architecture would be a typical digital beamforming structure, which doesn't need analog processing. Based on this generic architecture, various specific realizations of hybrid beamforming architecture and corresponding beamforming schemes can be developed.

Figure 5.11 shows typical structures of the hybrid beamforming. Generally speaking, one category of hybrid beamforming architecture is the full-array architecture (as shown in Fig. 5.11 (1-A) and (2-A), in which each stream of the signal is transmitted via the whole antenna array [40]). Another kind of hybrid beamforming architecture is called the sub-array architecture [41] (as shown in Fig. 5.11 Architecture 1-B and Architecture 2-B), in which each stream of the signal is transmitted using a set of antenna elements instead of the whole antenna array. The key difference between full-array structures and sub-array structures is that the full-array structure totally needs $N_t^{RF} \times N_t$ analog devices (e.g., phase shifters or switches), since each RF chain is connected to all N_t antennas, while the sub-array structure only requires N_t analog devices, since each RF chain is only connected to N_t/N_t^{RF} antennas, leading to this leads to greatly increased complexity for the former. The difference between the structures with phase shifters and those with switches is that the cost in the latter case is much reduced, though there can be some performance loss.

5.4.2   Schemes of Hybrid Beamforming

We consider a typical massive MIMO system with hybrid beamforming structure, where the BS with N_t transmit antennas sends N_s independent data streams to the user with N_r receiving antennas. Furthermore, it could be assumed that the BS and the user have N_t^{RF} and N_r^{RF} chains, respectively, which satisfy $N_s \leq N_t^{RF} \leq N_t$ and $N_s \leq N_r^{RF} \leq N_r$. In the hybrid precoding structure, as shown in Fig. 5.10, the hybrid precoding matrix $\mathbf{P} \in \mathbb{C}^{N_t \times N_s}$ at the BS can be written as the product of two parts: the first part is a low-dimension digital precoding matrix $\mathbf{P}_D \in \mathbb{C}^{N_t^{RF} \times N_s}$; the second part is a high-dimension analog precoding matrix $\mathbf{P}_A \in \mathbb{C}^{N_t \times N_t^{RF}}$, i.e., $\mathbf{P} = \mathbf{P}_A \times \mathbf{P}_D$, where $\mathbb{C}$ is the set of complex numbers. Note that these remarks on precoding matrix can also be applied to the combining matrix. Thus, the transmitted signal vector $\mathbf{x}$ is

$$\mathbf{x} = \mathbf{Ps} = \mathbf{P}_A \mathbf{P}_D \mathbf{s}. \tag{5.1}$$

where $\mathbf{s} \in \mathbb{C}^{N_s \times 1}$ denotes the source signal vector. After receiving the signal vector, the user utilizes a hybrid combining matrix $\mathbf{W}$ to combine the signal vector:

$$\mathbf{y} = \sqrt{\rho} \mathbf{W}^H \mathbf{H} \mathbf{P}_A \mathbf{P}_D \mathbf{s} + \mathbf{W}^H \mathbf{n}. \tag{5.2}$$

where ρ is the average received power, $\mathbf{H} \in \mathbb{C}^{N_r \times N_r}$ denotes the channel matrix between BS and the user, and $\mathbf{n} \in \mathbb{C}^{N_r \times 1}$ is the additive noise vector.

The optimal beamforming matrix could be inferred through an optimization method to achieve certain objectives such as maximizing sum-rate or minimizing power consumption with QoS constraints. Following the pioneering work [42] published in 2014, various hybrid beamforming schemes have been recently proposed to achieve different trade-offs between performance and costs, as surveyed in [43]. In [37], two alternating hybrid beamforming methods are proposed to jointly optimize the analog and digital beamforming matrices to maximize the achievable rate with different practical constraints. The authors in [44] propose to decompose the total achievable rate optimization problem with non-convex constraints into a series of simple subrate optimization problems, each of which only considers one subantenna array.

In the remainder of the subsection, some prior works are surveyed, which consider low-resolution and low-cost ADCs/DACs and phase shifters in the hybrid beamforming algorithm design. A DAC/ADC (digital-to-analog converter or analog-to-digital converter) performs the function of transforming the digital (analog) signal to the corresponding analog (digital) signal.

1.   **Hybrid beamforming with few-bits ADC/DAC**: With bandwidths on the order of a gigahertz in mmWave communication systems, high-resolution ADCs or

DACs become a power consumption bottleneck and cause excessive signal processing. One solution is to employ low resolution one-bit or few-bits ADCs. Considering the low cost of low-resolution ADCs (or DACs), it would be a particularly attractive solution for massive MIMO systems.

Since the channel state information at transmitter (CSIT) and channel state information at receiver (CSIR) needs transmitter beamforming and receiver combining, the CSI estimation for one-bit (or few-bits) ADC is particularly important for massive MIMO hybrid beamforming. For conventional MIMO systems, the closed-form ML estimator of such a SISO channel has been derived in [45]. Yet, for ultra wideband mmWave systems, these works may not be efficient for their frequency-selective characteristics. One method is to transmit the burst reference signal and estimate the CSI of each tap of the channel separately. Another more efficient scheme is to employ the generalized approximate message passing method using the channel correlation information based on the sparsity of the mmWave channel. Since channel estimation errors with one-bit ADCs would decrease with the sparsity of the channel [46], it would be more convenient to use compressive sensing techniques for channel estimation with relatively few CSI measurements (one-bit or few-bit CSI).

As for transmitter beamforming, a detailed capacity analysis of one-bit quantized MIMO systems with available CSIT is provided in [47]. At low and medium SNRs, when CSIT is assumed to be available, simple channel inversion beamforming is shown to be nearly optimal if the channel has full row rank. At high SNRs, a specific beamforming scheme was proposed to achieve system capacity even if the channel matrix is full rank. Hence, this transmitter beamforming technique would eliminate the gap between unquantized and quantized CSI. [48] studied the impact of CSI on the sum capacity of massive MIMO systems with quantized hybrid beamforming where the RF beamformer is selected from a finite size codebook. Considering the sparsity characteristics of the mmWave channel, channel inversion beamforming may be better than eigen-beamforming. This means that, kinds of simplified hybrid beamforming optimizations would be suitable for one-bit ADCs. With the increasing of the number of antenna elements, CSI acquiring (or recovery) would also be more challenging. In [49], a method of hybrid beamforming was proposed to reduce the quantization errors introduced at the analog beamformer part, which would lead to performance degradation.

2. **Hybrid Beamforming with few-bits phase shifters:** Due to limitations in cost and power supply, analog beamforming with a low-resolution phase shifter, instead of pure baseband digital beamforming, tends to be more favorable in mmWave hybrid beamforming systems. In IEEE 802.11ad, beamforming is based on a codebook with a 2-bit phase shifter. However, the use of a low-resolution phase shifter would degrade the link performance, as analyzed in [50]. Moreover, the gain loss brought by the low-precision phase shifter is limited; a 3-bit phase shifter can get a performance close to the ideal one.

5.4.3   EE–SE Analysis of Hybrid Beamforming

The EE and SE analysis of digital and hybrid beamforming has been addressed in many previous works, e.g., [51–53]. Take the sub-array structure as an example, where perfect analog beamforming is assumed within each sub-array with M antennas, which points to one user (there are N users in total). Assuming there is no inter-user interference, i.e., there is proper user scheduling (the BS schedules users with orthogonal channels), then the sum capacity of this structure for N users is:

$$C = W \times N \times \log\left(1 + \frac{M\eta_{PA}P}{W N_0}\right),\tag{5.3}$$

where W is the bandwidth, P is transmit power of each transceiver (the total power of M antenna PAs), η_{PA} is the PA efficiency, and N_0 is the thermal noise density. Without loss of generality, the channel gain is assumed to be the unity. The SE of this structure is:

$$\eta_{SE} = C/W = N \times \log\left(1 + \frac{M\eta_{PA}P}{W N_0}\right).\tag{5.4}$$

Because the accurate power model is nontrivial, the following simple power model is used:

$$P_{total} = NP + P_{static} = NP + N P_0 + P_{common} + NM P_{rf\_circuit},\tag{5.5}$$

where P_{total} is the total power; NP is the RF power of N transceivers; P_{static} is the static power of the BS, including NP_0, which scales with N; P_{common}, which is common for any number of transceivers; and $NMP_{rf\_circuit}$, which scales with NM. The relationship between EE and SE is

$$\eta_{EE} = C/P_{total}$$

$$= \frac{\eta_{SE}}{\left(2^{\frac{\eta_{SE}}{N}} - 1\right) \frac{N_0}{\eta_{PA}} \frac{N}{M} + \frac{N P_0 + P_{common} + NM P_{rf\_circuit}}{W}}.\tag{5.6}$$

Therefore, for a required SE, the hybrid LSAS beamforming should be designed to maximize EE through joint design of N, M, P_0, P_{common}, $P_{rf\_circuit}$, and η_{PA}.

Relationship at Green Points

When we take circuit power into consideration, there is a "green" point on the EE–SE curve where EE is at its maximum and is denoted η_{EE}^* [36]. Here, we discuss two cases for the $N \times M$ sub-array hybrid beamforming structure: 1) $NM = L$ (i.e., the total number of antennas is fixed as L, but N and M are variable), and 2) N and M are independent. For the former case, we allow the first-order derivative of EE over SE to be zero:

$$\eta_{EE}' = \frac{aN^2\left(2^{\frac{\eta_{SE}}{N}} - 1\right) + bN + c - \eta_{SE}aN2^{\frac{\eta_{SE}}{N}}\ln 2}{\left(aN^2\left(2^{\frac{\eta_{SE}}{N}} - 1\right) + bN + c\right)^2} = 0,\tag{5.7}$$

where $a = \frac{N_0}{L\eta_{PA}}, b = \frac{P_0}{W}$, and $c = \frac{P_{common} + L P_{rf\_circuit}}{W}$.

Combining (5.7) with (5.6), the relationship between the maximum EE η_{EE}^* and corresponding SE η_{SE}^* is

$$\eta_{EE}^* = \left(\frac{n_0 N 2^{\frac{\eta_{SE}^*}{N}} \ln 2}{L\eta_{PA}} \right)^{-1}. \tag{5.8}$$

The relationship between η_{EE}^* and η_{SE}^* is further given as

$$\lg\left(\eta_{EE}^*\right) = -\frac{\lg 2}{N}\eta_{SE}^* + \lg\left(\frac{L\eta_{PA}}{n_0 N \ln 2}\right), \tag{5.9}$$

which indicates that $\log \eta_{EE}^*$ scales linearly with η_{SE}^* and has a slope of $-\log 2/N$. Similar to the EE–SE relationship in classic Shannon theory [53], higher η_{SE}^* always leads to lower η_{EE}^*. The relationship between η_{EE}^* and η_{SE}^* is independent of P_0, P_{common}, $P_{rf\_circuit}$, and W, though as can be seen from (5.6), η_{SE}^* and η_{EE}^* are determined on the basis of all the other parameters.

In the case of independent N and M, the relationship is

$$\eta_{EE}^* = \left(\frac{n_0}{\eta_{PA} M} 2^{\frac{\eta_{SE}^*}{N}} \ln 2 \right)^{-1}. \tag{5.10}$$

For each case, there exists only one η_{SE}^* where EE monotonically increases with SE when SE is smaller than η_{SE}^*, and monotonically decreases with SE when SE is larger than η_{SE}^* [36].

It is expected, therefore, that the system operates at the green point. Also, it is important that η_{SE}^* satisfies the system SE requirement, and η_{EE}^* should be high enough. These require careful design of P_0, P_{common}, $P_{rf\_circuit}$, W, η_{PA}, N, and M. For example, when other parameters are given, M or N can be designed to maximize EE.

Optimal M for Maximizing EE for a Given SE, with Independent N and M
It is of practical importance to know how M affects EE for a given SE. If there is one optimal M that results in the highest EE, it is not necessary to implement too many antennas per transceiver. In the following exploration, we derive the optimal M to maximize system EE. Denote the denominator of (5.6) as $f(M)$:

$$f(M) = \left(2^{\frac{\eta_{SE}}{N}} - 1\right) \frac{N_0}{\eta_{PA}} \frac{N}{M} + \frac{N P_0 + P_{common} + NM P_{rf\_circuit}}{W} \tag{5.11}$$

The first- and second-order derivatives of $f(M)$ are

$$f'(M) = \frac{N P_{rf\_circuit}}{W} - \left(2^{\frac{\eta_{SE}}{N}} - 1\right) \frac{N_0}{\eta_{PA}} \frac{N}{M^2} \tag{5.12}$$

and

$$f''(M) = 2\left(2^{\frac{\eta_{SE}}{N}} - 1\right) \frac{N_0}{\eta_{PA}} \frac{N}{M^3} \geq 0 \tag{5.13}$$

Then $f(M)$ is a quasi-convex function of M. The optimal M^* that gives the minimum $f(M)$ is derived by making $f'(M) = 0$:

$$M^* = \sqrt{\frac{W N_0}{\eta_{PA} P_{rf\_circuit}} \left(2^{\frac{\eta_{SE}}{N}} - 1 \right)} \tag{5.14}$$

Because of the definition of η_{EE} in (5.6), EE is a quasi-concave function of M, and the EE is at a maximum when $M = M^*$. When $M < M^*$, EE monotonically increases with M. When $M > M^*$, EE monotonically decreases with M. In practical system design, for a given SE there is one optimal number of antennas per transceiver that results in the highest EE. As in (5.14), the optimal M^* increases with SE and bandwidth, but decreases with PA power efficiency and $P_{rf\_circuit}$. For a given number of tranceivers N, more antennas per transceiver are needed for higher SE. If W increases, the noise power increases correspondingly, and a larger M is needed to achieve the SE. A larger $P_{rf\_circuit}$, however, reduces the optimal M^* because the increased circuit power may reduce EE.

Optimal N for Maximizing EE for a Given SE, with Independent N and M
In order to achieve the performance promised by massive MIMO, a large enough M is required. But the practical implementation of an equal number of transceivers is not trivial with many unresolved issues, including calibration and complexity. When the required SE is predetermined, it's very important to know whether a larger N always brings a better EE. Again, take the denominator of (5.6) as $f(N)$, the first-order derivative of $f(N)$ over N is derived as

$$f'(N) = \frac{M P_{rf\_circuit} + P_0}{W} + \frac{N_0}{\eta_{PA}} \frac{1}{M} \left(\left(2^{\frac{\eta_{SE}}{N}} - 1 \right) - 2^{\frac{\eta_{SE}}{N}} \frac{\eta_{SE}}{N} \ln 2 \right)$$
$$= g\left(\frac{\eta_{SE}}{N} \right) = g(x). \tag{5.15}$$

We have

$$g'(x) = -\frac{N_0}{\eta_{PA}} \frac{1}{M} \left(x 2^x (\ln 2)^2 \right) < 0. \tag{5.16}$$

We also find that $g(0) = \left(M P_{rf\_circuit} + P_0 \right) / W$ and $g(\infty) = -\infty$. This indicates that there exists only one x_0, such that $g(x_0) = 0$. Correspondingly, there exists only one N_0, $N_0 = \eta_{SE}/x_0$. When $N < N_0$, EE is monotonically increasing with N, and when $N \geq N_0$, EE is monotonically decreasing with N.

5.4.4   Standardization

The mmWave band communication technologies have been standardized by multiple international organizations. For example, the IEEE 802.11ad amendment to the 802.11 standard defines a directional communication scheme that takes advantage of beam-forming antenna gain to cope with increased attenuation in the 60 GHz band. Beamforming training is used to determine the appropriate receive and transmit antenna sectors for a pairing of BS and UE. The training procedure is split into two subphases. During the cell-specific beam sweeping, an initial coarse beam (or antenna sector configuration) is determined, which is used in a subsequent optional beam refinement phase. This is a simplified version of hybrid beamforming, in which the analog beamformer of the hybrid beamforming is fixed. Another example is IEEE 802.15.3c, which specifies the physical layer and MAC layer protocols and procedures for indoor wireless personal area networks (WPANs). As multiple antennas are available at both the transmitter and the receiver, codebook-based MIMO beamforming is employed.

The most recent standardization activity of hybrid beamforming technique is in 3GPP 5G NR. It has been agreed that this technology will be deployed in future 5G systems in the conference of 3GPP RAN1 #85 [54]. In this conference, the maximal number of the RF chains for the 5G BS and UE was determined as 32 and 8, and the maximal number of the antenna elements for the 5G BS and UE was 1024 at 70 GHz (128 @ 4 GHz and 256 @ 30 GHz) and 64 at 70 GHz (8 @ 4 GHz and 32 @ 30 GHz), respectively. Besides, the BS hybrid beamforming architecture is likely to employ the Architecture 2-A and the hybrid beamforming architecture of UE tends to be the Architecture 2-B in Fig. 5.11. In the conference of 3GPP RAN1 #86 [55], the DL beam management for hybrid beamforming was agreed upon. Specifically, the UE makes measurements on different BS transmit beams to support beam selection of BS and UE. In the conference discussion of 3GPP RAN1 #86b [56], the key problem of hybrid beamforming was to determine the main process of the beamforming procedure.

In order to guarantee the success probability of data transmission, it is necessary to choose proper TRP/UE beams, which means both analog beamforming and digital beamforming should be implemented properly for the BS and UE. The detailed DL CSI

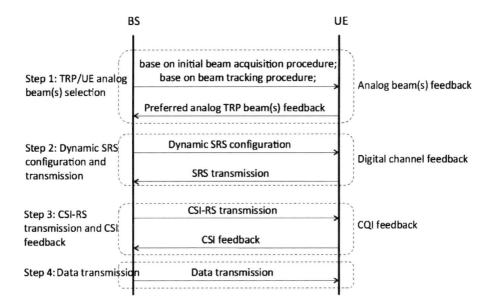

Figure 5.12 DL CSI acquisition framework for NR massive MIMO.

acquisition framework of NR MIMO, which is illustrated in Fig. 5.12, could be divided into the following 4 steps:

1. **TRP/UE analog beam(s) selection**

   First of all, the TRP/UE analog beam(s) should be selected properly to provide appropriate analog beamforming gain. Concretely, analog beam(s) could be selected based on initial beam acquisition procedure and/or beam tracking procedure. Here, we discuss these two analog beam selection schemes:

   - Scheme 1: Analog beam(s) are selected based on an initial beam acquisition procedure. For example, the TRP transmits synchronization signals and/or system information with the beam sweeping method, and a terminal would calculate and compare the power of different beams to identify its preferred TRP beam(s). Then the preferred TRP beam(s) information could be indicated to the network during or after the random access procedure explicitly or implicitly.
   - Scheme 2: Analog beam(s) are selected based on a beam tracking procedure. A beam tracking procedure is needed, since the preferred TRP/UE beam(s) may change when the channel condition between the TRP and UE changes. In order to facilitate the beam tracking procedure, a kind of beam selection RS needs to be designed. The beams carried by beam selection RS may be the same as or different from that carried by synchronization signals in the initial beam acquisition procedure. From the perspective of specification, these two kinds of beams, which applied in the initial beam acquisition procedure and beam tracking procedure separately, can be independent.

   It should be noted that the UE may need to feed back more than one preferred TRP beams in order to support SU-MIMO and MU-MIMO flexibly.

2. **Dynamic SRS configuration and transmission**

   After the base station receives the information of a UE's preferred analog beam(s), it could dynamically configure the UE to transmit UL SRS on some specific time/frequency/code resources. Then the base station will adjust its analog beam on these time/frequency/code resources to receive the UE's SRS. The applied TRP analog beams on these resources are transparent to the UE, and they could be flexibly adjusted not only according to the preferred analog beams fed back in step 1 but also according to the MIMO schemes (e.g., SU-MIMO or MU-MIMO) and/or multi-user pairing schemes. Both periodic and aperiodic SRS transmission could be further investigated.

3. **CSI-RS transmission and CSI feedback**

   For TDD systems, the BS can derive the DL channel information after step 2, and then the proper digital beamforming and rank information can be calculated. However, in order to determine the proper MCS, the BS needs to get the UE's interference information, or CQI report. One method is that the BS can apply the beamforming matrix derived from the SRS to UE-specific beamformed CSI-RS, and then the UE measures the beamformed CSI-RS to derive the CQI and/or PMI information.

4. **Data transmission**

After the analog TRP beam(s), digital beamforming matrix, rank information, and MCS are determined, the BS can transmit traffic data to the UE.

Note that this CSI acquisition framework for NR massive MIMO applies to all the scenarios, no matter what beamforming structure and RS pattern is utilized.

5.4.5 Summary

The diverse use cases and scenarios of 5G motivate software-defined and flexibly configured air interface. It has been a consensus that MIMO technique will be the cornerstone of 5G air interface. However, different beamforming structures may possibly be considered in the system deployment of 5G, including digital beamforming, analog beamforming, and hybrid beamforming, with various mappings between TXRU and antenna elements. This necessitates the design of flexible MIMO with a unified beamforming architecture and a unified CSI acquisition/feedback mechanism. To this end, this section has so far proposed a flexible MIMO framework for 5G communications. Also, various hybrid beamforming algorithms were surveyed for both full-array and sub-array hybrid beamforming architectures. Furthermore, the energy-efficient design of a sub-array structure was examined, with the optimal number of TXRU and antenna elements analyzed. Finally, the standardization of 3GPP 5G NR on hybrid beamforming was investigated, and a unified CSI acquisition and feedback framework was discussed, which is applicable to various beamforming structures and reference signal designs.

5.5 New Waveform

Driven ultimately by diversified applications, new waveform, as one key enabler of SDAI, is envisioned to be able to support various extreme requirements in the physical layer. Orthogonal frequency division multiplexing (OFDM) has been recognized as an effective waveform for mobile communications, due to its ease of implementation, robustness to multipath fading, and MIMO friendliness. Some shortcomings of the current solution emerge when coming across future requirements, including: not-very-well localized in time and frequency domain, and sensitivity to frequency or timing synchronization error, etc. For new radio (NR), several new waveform schemes attract industry's interests, including:

- Filter bank multi-carrier (FBMC),
- Universal filter multi-carrier (UFMC),
- Generalized frequency division multiplexing (GFDM),
- Filtered-OFDM (f-OFDM) and windowed OFDM (w-OFDM),
- Orthogonal Time Frequency Space (OTFS),
- Variants of DFT-S-OFDM, etc.

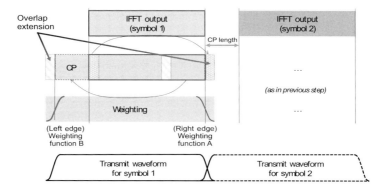

Figure 5.13 WOLA at transmitter with CP-OFDM.

In the following sections, we will briefly introduce the technical principles of each waveform scheme and provide the technical comparison of different solutions. Moreover, the progress of 3GPP 5G NR specification on waveform schemes is discussed, including why two waveforms, DFT-S-OFDM and OFDM, could be flexibly configured as a 5G NR UL waveform scheme (this never happened in previous telecom systems). A unified framework of waveform design is proposed, where multiple waveforms can be flexibly implemented according to the specific use scenario and channel conditions. This is particularly important to achieve green and with respect to green and soft 5G network operations.

5.5.1   w-OFDM/f-OFDM

To offset the poor frequency localization for CP-OFDM, an efficient spectrum-shaping technique of windowing or filtering approach can be utilized. W-OFDM is synthesized by a conventional CP-OFDM waveform, followed by a weighting and overlap-and-add (WOLA) operation. The better-contained frequency response is achieved by adding soft edges to the cyclic extension of the OFDM symbol in the time domain, as shown in Fig. 5.13. When the edges further expand, the overhead is still the same as a CP-OFDM waveform, since adjacent symbols are overlapped in the edge transition region. The shape of the window (or edge) in time domain determines the frequency response of the prototype filter. In general, a raised-cosine edge seems to offer a good compromise with straightforward implementation [57].

In principle, f-OFDM applies a filter with a subband of the CP-OFDM system to reduce the OOB leakage [58]. One example of filter design is the windowed sinc function-based method. To be specific, the transmit filter is computed as the product of the ideal band-pass filter and a time domain mask [59]:

$$f(n) = p_i(n)w(n),$$

where $p_i(n)$ is the ideal band-pass filter covering the allocated bandwidth of the i-th user, and $w(n)$ is a raised-cosine window with duration T_w. The window has smooth transitions to zero on its both ends so that it avoids abrupt jumps at the beginning and end of the truncated filter. Furthermore, up to half symbol length can be used for T_w. The long filter length of $f(n)$ provides good OOB emission suppression. Different from WOLA, the band-pass filter of f-OFDM is bandwidth dependent. Therefore, the filters need to be constructed based on the tone allocation. Another concern of applying f-OFDM, especially for the TDD band, is the long group delay due to the long filter length.

5.5.2   UFMC

Similar to f-OFDM, UFMC is another spectrum-shaping technique utilizing the filtering approach. The main difference is in how the band-pass filter is constructed [60]. Specifically, a band-pass filter is carefully designed for a fixed bandwidth, e.g., a resource block (RB). The same filter can be universally reused only by shifting the center frequency. That is, when n RBs are assigned to the transmitter, n parallel IFFT and filtering operations have to be computed. Instead of CP, a guard interval (GI) filled with zeros is introduced between the symbols to prevent ISI due to filter delay. The filter length is set to be the same as the GI duration (usually not long as the filter in f-OFDM). Since GI is introduced instead of CP, the cyclic convolution property is not preserved in UFMC. Therefore, the receiver structure is not as simple as the one in CP-OFDM. Specifically, doubled-sized FFT is used at the receiver, but only the even tones of the doubled-sized FFT outputs are used for the detection, which increases complexity and latency of the decoder. The detailed operation of the transmitter and receiver could be found in Fig. 5.14.

5.5.3   FBMC

FBMC has drawn much interest due to its excellent spectral containment [61]. It is achieved by optimizing the shape of the prototype filter $p(n)$ through oversampled coefficients on the granularity of each carrier. The modulator and demodulator are conceptually illustrated in Fig. 5.15.

In FBMC, a prototype filter satisfying generalized Nyquist constraints is used for both signal synthesis at the transmitter and signal analysis at the receiver. Figure 5.16 shows a prototype filter with the oversampling factor $K = 4$, where K is denoted as

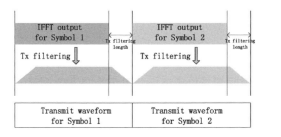
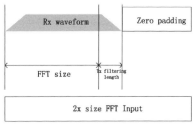

Figure 5.14 UFMC processing at the transmitter and receiver.

Table 5.2 Frequency domain coefficients of prototype filter.

H_{-3}	H_{-2}	H_{-1}	H_0	H_1	H_2	H_3
0.235147	$\sqrt{2}/2$	0.97196	1	0.97196	$\sqrt{2}/2$	0.235147

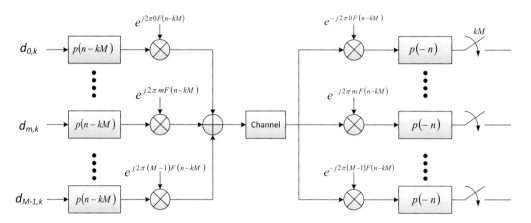

Figure 5.15 Modulator/demodulator of filter bank multi-carrier.

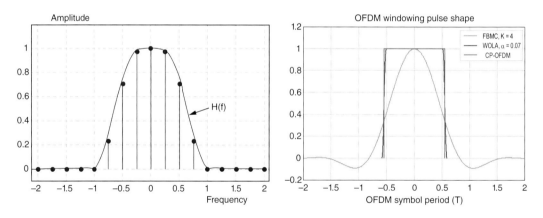

Figure 5.16 Illustration of the prototype filter (left/right: frequency/time domain response).

the FBMC overlapping factor. And the interval between adjacent coefficients is $1/4\triangle_f$, where $\triangle_f = 1/T$ is the sub-channel spacing. Further, because of the oversampled frequency coefficients, the prototype filter spans multiple symbol periods T, as shown in the right figure in Fig. 5.15.

The specified nonzero coefficients are summarized in Table 5.1. It can be verified that the selected frequency coefficients satisfy the Nyquist property.

Note that there are several limitations for practical implementation apart from the complexity issue. Special modulations such as OQAM may be necessary to avoid inter-channel interference introduced by the prototype. In the case of multipath channels, the orthogonality statement at demodulator is no longer valid, since there is no CP protection in this waveform, and channel convolution is not precisely cyclic. Another potential limitation in applying FBMC is the deployment with MIMO when exploiting more degrees of freedom. In addition, reference signals may be less flexible than those in OFDM and hard to enable efficient channel estimation techniques [62].

5.5.4 GFDM

In GFDM, the prototype filter for each sub carrier is also specifically chosen to be well-localized in frequency domain to reduce out-of-band emission, similar as in FBMC. The main difference from FBMC is that, in GFDM: 1) multiple OFDM symbols are grouped into a block, with a CP added to the block; 2) within a block, the prototype filter is "cyclic-shift" in time, for different OFDM symbols [63]. A block of GFDM waveform can be expressed as:

$$x(n) = \sum_{k=0}^{K-1} \sum_{m=0}^{M-1} p_{k,m}(n) d_{k,m} \quad for \, n = 0, 1, \cdots, N-1. \tag{5.17}$$

Each block has $N = KM$ samples, which can be decomposed into M sub-symbols. Each M sub-symbol contains K subcarriers. The pulse $p_{(}k,m)(n)$ is the frequency and time-shifted version of the prototype filter $p(n)$, as shown in (5.18). Specifically, the module operation makes $p(n)$ circularly shifted in time by m sub-symbols, and the exponential term shifts the filter in frequency by k subcarriers.

$$p_{k,m}(n) = p[(n-mK) \, mod \, N] \, e^{j 2 \pi k \frac{n}{K}} \tag{5.18}$$

Figure 5.17 shows an example of GFDM resource partitioning with M sub-symbols per block, with time offset T/M between adjacent sub-symbols. The duration of each symbol can be longer than T/M due to the specially engineered prototype filter. Each sub-symbol contains $K = BT/M$ sub-channels, with spacing of M/T (Hz) spacing between adjacent sub-channels.

In order to avoid interference between sub-symbols within a block, the selected prototype filter should have Nyquist property. Further, special modulations such as OQAM may be necessary to avoid interchannel interference as in FBMC. Otherwise, complicated receiver algorithm is needed to handle the interference [64]. In addition, due to the cyclic structure of the block and the use of CP, GFDM improves the capability against with ISI with the penalty of spectral containment property especially when a waveform spanning multiple blocks in time.

5.5.5 OTFS

OTFS, instead of using filtering or windowing technologies for spectral containment, as above, characterizes the Doppler-induced time varying nature of the wireless channel and parameterizes it as a 2D impulse response in the delay-Doppler domain [65]. Specifically, the channel time-frequency response $H[n,m]$ is related to $h(\tau, v)$ via the

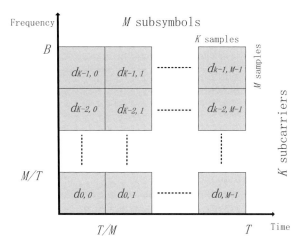

Figure 5.17 Resource partition in GFDM.

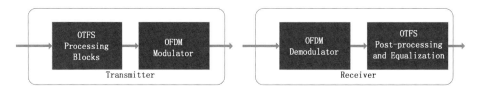

Figure 5.18 OTFS architecture with pre- and post-processing blocks.

transform shown in (5.19), where the delay-Doppler response $h(\tau,\ )$, but not time vary-ing impulse response $h(\tau,t)$, is used for characterization of the channel.

$$H\,[n,m] = \iint^{h} (\tau,\nu)\, e^{j2\pi \nu nT} e^{-j2\pi m\Delta f\tau}\, d\nu d\tau, \qquad (5.19)$$

where τ and ν denote delay and Doppler respectively, m and n are for the frequency bins and time bins, T is the length of the OFDM symbol (plus CP extension). Above equation can be thought of as a 2D Fourier transform version of the delay-Doppler impulse response $h(\tau,)$. More details on the theory analysis of OTFS can be found in [66].

In terms of implementation, 2D OTFS consists of a DFT along the delay/frequency dimension and an IDFT along the Doppler/time dimension. The transformation con-sists of pre- and post-processing blocks in the transmitter and receiver respectively, as depicted in Fig. 5.6. This block diagram is analogous to the blocks used to imple-ment DFT-s-OFDM on top of an underlying OFDM signal chain. The pre- and post-processing blocks could enable QAM modulation in the delay-Doppler domain. In this way, all QAM symbols experience the full diversity of the channel. Further, at a high-mobility scenario, the time invariance property holds for the duration of the TTI. It would make a closed-form transmission mode possible for spectral efficiency (SE) improvement.

No doubt that additional preprocessing blocks introduce higher complexity at both the transmitter and receiver. Another concern is the longer processing latency when the preprocessing block spans a TTI length or multiple OFDM symbol duration. The receiver has to wait until getting the last symbol before it can go on its next step operation, which may be intolerable for the real system.

5.5.6   Variants of DFT-s-OFDM

DFT-s-OFDM, featured by a high-power efficiency due to its low PAPR, is very well known as the UL waveform scheme of LTE. Further, some variants of DFT-s-OFDM, including unique word (UW) DFT-S-OFDM, are designed for superior OOB suppression performance [67].

The transmitter structure for UW-DFT-s-OFDM is illustrated as Fig. 5.19, where a unique word is added prior to the DFT operation. It leads to suppression of out-of-band emission, thanks to the cyclic property obtained from DFT and IDFT operation. Also, unique word at the output of IDFT serves as a guard between two data parts of consecutive OFDM symbols, thus, CP insertion is not necessary. In addition, it allows the possibility of UW-based time domain channel estimation and synchronization design. Note that insertion of UW does not change PAPR of DFT-s-OFDM, as long as the envelope of UW is nearly constant or similar to that of data symbols. Further, if UW is

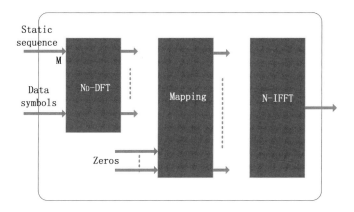

Figure 5.19 Transmitter diagram for UW-DFT-s-OFDM.

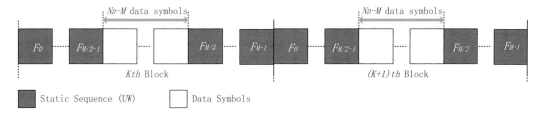

Figure 5.20 Placement of UW in a block.

set to be a zero sequence, i.e., zero-tail (ZT) DFT-S-OFDM, better spectral confinement is expected at the price of PAPR [68].

An example of UW sequence is placed as shown in Fig. 5.20. The number of symbols in UW is denoted by M, and UW is split in half and placed at the head and tail of the DFT-s-OFDM block.

5.5.7   Constant Envelope Waveform

A simple way to achieve high transmit efficiency is to employ a constant envelope waveform, which allows almost any PA to operate at saturation point. Minimum shift-keying (MSK) and Gaussian MSK (GMSK) are the most popular constant envelope waveforms. MSK can equivalently be viewed as offset-QPSK (quadrature phase shift-keying) with sinusoid pulse shaping, which provides efficient modulation and demodulation [69]. Notice that a differential encoder is inserted before the modulator to avoid error propagation at the demodulator. GMSK is a variant of MSK, where a Gaussian-filtered version of the information sequence is applied to an MSK modulator [70]. The Gaussian filter helps to increase the SE of the MSK, with reduced inter-symbol interference. Note that with the introduction of Gaussian filtering, the GMSK signal can no longer be viewed as offset-QPSK. The drawback, however, is the inefficiency from a capacity perspective compared to QAM. But for applications like low data-rate packet transmission in IoT, a constant envelope waveform may be attractive, since it achieves the highest PA efficiency.

5.5.8   Unified Waveform Framework

A common feature of the above new waveforms is that filters are employed to suppress the out of band emission and relax the requirements on time-frequency synchronization. But there are also subtle differences among these waveforms. The filters in UFMC and f-OFDM are implemented at the granularity of each sub-band. The main difference is that f-OFDM uses a longer filter and the signal processing procedure is same as the conventional OFDM in each sub-band for backward compatibility. In contrast, UFMC uses a shorter filter, and the CP of OFDM is replaced with an empty guard period. GFDM can cover CP-OFDM and SC-FDE, which can be regarded as special cases, according to different numbers of subcarriers and sub-symbols in a GFDM block. In addition, the overhead is kept small by adding CP for an entire block that contains multiple sub-symbols. The filter in FBMC is implemented at the granularity of each subcarrier. By a well-designed prototype filter, FBMC can greatly suppress side-lobes of a signal. Moreover, the overhead can also be reduced by removing CP in FBMC as UFMC. At the same time, in order to reduce the interference of adjacent sub-channels and computation complexity, the OQAM modulation and polyphase network is needed in FBMC and GFDM schemes. In addition to the above waveforms, DFT-S-OFDM is also supported by the 3GPP 5G NR as one of the UL waveforms for the improvement of PAPR and UL coverage improvement. In order to better embrace the challenges of fast fading in high-mobility scenarios, e.g., high speed train, OTFS is proposed, which

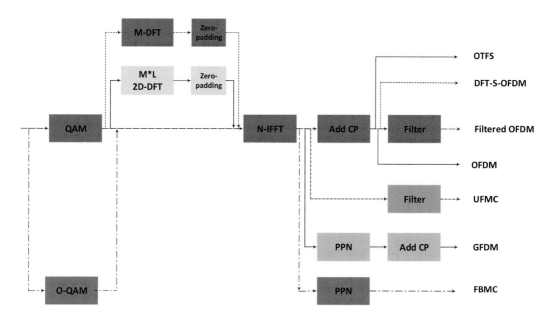

Figure 5.21 A unified framework of waveforms.

transforms the time-varying multipath channel into a time invariant delay-Doppler channel. The transmitter structure of the OTFS is similar to that of the DFT-S-OFDM, except for the 2D-DFT processing.

A unified framework to implement various waveforms is shown in Fig. 5.21, where the waveform can be represented as (5.20),

$$x(t) = \sum_{u \in U} \sum_{k \in K_u} \sum_{n=-\infty}^{+\infty} \sum_{m=1}^{M} s_{k,n}(m) g_{k,m}(t - nT) e^{j2\pi f_k(t-nT)} \otimes h_u(t) \qquad (5.20)$$

where $s_{k,n}(m)$ is the m-th sub-symbol in the n-th transmission symbol and k-th subcarrier. $g_{k,m}(t)$ is the shaping filter in a single symbol. The filter of each user is denoted as $h_u(t)$. And the frequency of subcarrier is denoted as f_k. The symbol duration is T. And the convolution operator is denoted as $\otimes$.

By the unified structure, we can flexibly configure different waveform schemes according to various 5G scenarios on the basis of minimizing the hardware functional module. For instance, if $g_{k,m}(t)$ is a rectangular window and with length T, $M = 1$, $h_u(t) = \delta(t)$, $f_k = k/T_s$ and T_s is the symbol length excluding CP, $x(t)$ is actually the OFDM signal. When $M \neq 1$, $g_{k,m}(t) = g[(t - mK) \bmod (KM)]$, $g(t)$ is the prototype filter and the other parameters set same as the above OFDM parameters, $x(t)$ becomes a GFDM signal.

Based on the above waveform framework, multiple waveforms can be flexibly implemented accordingly to the specific use scenario and channel condition. Note that some waveforms can be implemented without standardization efforts, e.g., f-OFDM, w-OFDM, and UFMC.

5.5.9   Waveform for 5G NR in 3GPP

DL Waveform for NR

For NR DL, OFDM-based waveforms are preferred as the candidate schemes. Specifically, OFDM is identified as the NR DL waveform, while filtering or windowing approach, i.e., f-OFDM or w-OFDM, is used for efficient spectrum-shaping to reduce the in-band and out-of-band emission. It aims to reach 98% spectrum efficiency of the bandwidth in NR while LTE reserves up to 10% of the bandwidth as guard bands to abide by the spectrum mask [71]. It means that a particular filtering or windowing function may be needed, as discussed in the previous subsection. Note that it is an implementation issue for this specific method, from the perspective of 3GPP RAN1's perspective. In the following, the major performance indicators are listed for NR DL waveform [72]. These factors determine the selection of waveform scheme.

- **Spectral efficiency**: To meet extreme data rate requirements for both DL and UL. In general, SE is more important at lower carrier frequencies than at higher frequencies, since the spectrum is not as precious at higher frequencies due to the availability of potentially much larger channel bandwidths.
- **MIMO compatibility**: To enable a straightforward use of MIMO technology. Multiple or massive antenna technology is considered as one of key enablers for data-rate boosting (up to 20Gbps for peak data rate) and coverage improving. Thus, the new waveform should have limited implementation complexity with MIMO integration.
- **Transceiver baseband complexity**: To enable efficient baseband processing at large bandwidths envisioned for NR. Reasonable implementation complexity should be involved for not only the waveform itself, but also its related implementation method too. Signal detection and channel estimation/equalization at the receiver should not have very high complexity. Note that at very high frequencies, the receiver may also have to cope with severe RF impairments.
- **Flexible numerology configuration**: To enable different services (with different numerologies) simultaneously on the same carrier. For example, the uRLLC or synchronization signal may require larger subcarrier spacing for shorter transmission time interval.
- **Frequency localization**: To support the coexistence of different services that are potentially enabled by mixing different numerologies in frequency domain on the same carrier. Further, it is essential to provide minimal loss in SE. Also, efficient asynchronous communication would require a waveform with minimal inter-UE interference leakage, which is achieved by good frequency localization.
- **Time localization**: To efficiently enable (dynamic) TDD and support latency critical applications such as uRLLC. Low latency is very important for all link types.

- **Robustness to synchronization errors**: This important where synchronization is hard to achieve, such as a D2D link.

- **Robustness to channel time-selectivity**: This is important in high speed scenarios. Fast time-varying characteristics would make channel tracking difficult. More reference signals (higher overhead) or additional function blocks (higher complexity) may be needed for robust channel estimation.

- **Robustness to channel frequency-selectivity**: This is always an important measure in multipath channels. Channel frequency selectivity depends on various factors as type of deployment, beamforming technique, and bandwidth. It is important that BSs can cope with frequency selective channels without complicated receiver limitations.

- **Robustness to phase noise**: This is important for all link types especially for a high-frequency device (transmitter/receiver), as phase noise typically increases with carrier frequency. Note that high-quality oscillators mean high cost and may not be affordable.

- **Low cubic metric**: To compensate for the PA's inefficiency. A low cubic metric (or PAPR) is important for power efficient transmissions, and becomes even more important at very high frequencies. Note that small-sized, low-cost BSs are envisioned at high frequencies, therefore, a low cubic metric is also important for DL.

- **Flexibility/scalability**: To support diverse services in wide range of frequencies.

Taking the above factors into account, OFDM shows overwhelming advantages. It reaches the consensus without much debates in 3GPP that OFDM (with filtering or windowing) is identified as a NR DL waveform. Note that for mMTC, high frequency or high mobility, it still retains some possibility for other solutions.

UL Waveform for NR

OFDM-based waveforms are also considered as the candidates for NR UL. Different from NR DL or other communication systems, two schemes, OFDM and DFT-S-OFDM, are both identified as the UL waveform. Specifically, DFT-S-OFDM and OFDM could be flexibly configured by the network. For cell-edge users, DFT-S-OFDM can be configured to improve the coverage. While for users with good channel conditions and multiple antenna ports, OFDM could be used by combining with multi-layered transmissions for data rate boosting. In the following arguments, a general comparison of two waveforms for UL are presented first, and then more detailed technological principles are provided in the next subsections.

Why DFT-S-OFDM is selected for NR UL waveform

1. **PAPR/cubic metric**

   Besides PAPR, a cubic metric is a more accurate measurement metric for the power back-off required for amplifiers [73, 74]. Table 5.3 summarizes the PAPR/CM for OFDM, DFT-S-OFDM, and OFDM with PAPR reduction technology [73].

Table 5.3 PAPR/CM comparison of OFDM, DFT-S-OFDM and OFDM with companding.

	OFDM			DFT-S-OFDM			OFDM with PAPR reduction		
	QPSK	16 QAM	64QAM	QPSK	16 QAM	64QAM	QPSK	16 QAM	64QAM
PAPR(0.1%)	10.614	10.571	10.665	7.446	8.403	8.626	6.432	6.989	7.082
Cubic Metric	3.29	3.31	3.32	1.02	1.80	1.95	2.54	2.58	2.59

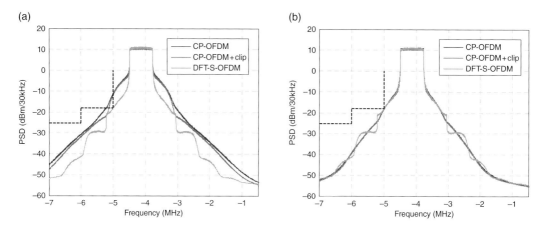

Figure 5.22 OOB emissions performance, where (a): without power back-off; (b): with power back-off.

Note: 4 RBs and subcarrier spacing at 15kHz is assumed.

Both DFT-S-OFDM and OFDM with PAPR reduction technology could give a low PAPR, while DFT-S-OFDM has the lowest cubic metric. That is, in theory, DFT-S-OFDM has the lowest power back-off value, i.e., the maximal output power.

2. **Power back-off**

Figure 5.22 presents the power spectrum density (PSD) for the just-discussed three waveforms without and with power back-off, respectively. The PA model in [75] is considered with post-PA loss as 4dB. Figure 5.22a shows that when using maximal output power [23dBm], PSD of OFDM is out of the emission mask (black dash). Then, by setting different values of power back-off in Table 5.4, all the waveforms can satisfy the requirements of in-band and out-of-band (OOB) emissions (including the reuse of the ACLR and UE emission mask in TS 36.101 for LTE [76]), illustrated in Fig. 5.22b.

Further, Table 5.5 shows the EVM performance, considering both with and without power back-off. With power back-off as described, the EVM fulfills the maximum tolerable limit for QPSK modulation as defined by 3GPP, i.e., 17.5% as in LTE systems [76].

Note: The power back-off values are set the same as Table 5.4.

Table 5.4 Power back-off for in-band and out-of-band emission requirements.

Schemes	Power back-off [dB]	Output power (post-PA loss = 4dB)
DFT-S-OFDM	0	23dBm
OFDM	−2.0	21dBm
OFDM with PAPR reduction	−1.5	21.5dBm

Table 5.5 EVM w/ or w/o power back-off.

Schemes	EVM (w/o power back-off)	EVM (w/ power back-off)
DFT-S-OFDM	7.36%	7.36%
OFDM	22.97%	9.97%
OFDM with PAPR reduction	23.95%	13.5%

Based on the agreed-upon PA model, DFT-S-OFDM could reach the maximal output power [23dBm], and OFDM's maximal output power is reduced to 21.5 dBm. Note that a 1.5dB power gap always exists no matter the transmission scheme or frequency band. Also, the 1.5dB link budget corresponds to a 40m to 50m distance [77] around 4GHz, which is actually a large coverage range. Besides, as no terminal/chip manufacturers would like to use new PAs with larger linear regions, DFT-S-OFDM is identified as the waveform scheme of NR UL.

Why OFDM is selected for NR UL waveform
The BLER vs. SNR curves for different waveforms are presented in Fig. 5.23. It shows that for QPSK, only 0.2dB gain for OFDM is shown. At higher MCS (16QAM/64QAM), OFDM outperforms DFT-S-OFDM greatly, with a gain of 1.5–2dB. As described, DFT-S-OFDM could provide a 1.5dB gain for maximal output power compared with OFDM. But for a cell-center user with high SNR, the gain of DFT-S-OFDM vanishes due to the BLER performance gap for these two waveforms. Furthermore, DFT-S-OFDM may suffer performance loss when no maximal output power is set. For example, if the UE output power is set to be 18dBm, OFDM provides nearly 2dB gain compared with DFT-S-OFDM.

Furthermore, OFDM is preferred due to its friendliness to MIMO. For dynamic TDD, symmetric DL/UL link allows for the possibility of crosslink interference mitigation. Therefore, OFDM is also selected as the waveform scheme of NR UL.

In summary, OFDM used in the UL (and also in side-links) comes with several advantages, as listed below:

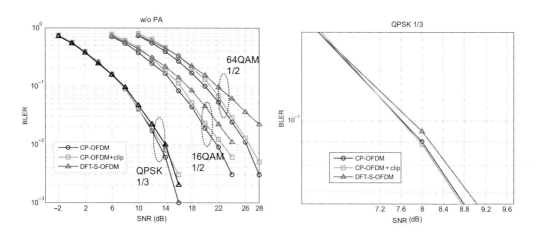

Figure 5.23 BLER performance for OFDM, DFT-S-OFDM, and OFDM with companding.

- It opens up for a more flexible UL scheduling. Having the same transmission scheme in both UL and DL makes the whole system design symmetrical. Further, it simplifies the overall system design by reducing the need for specific baseband receivers for respective link.
- It helps to facilitate the UL MIMO feature in NR. OFDM has shown significant advantages when considering multilayer transmission and hence, OFDM is preferred for UL MIMO use cases.

The agreement is that OFDM is also identified as the waveform scheme of NR UL. In a practical sense, the network can decide and communicate to the UE which CP-OFDM- and DFT-S-OFDM-based waveform to use. Further more, a common framework is targeted in designing CP-OFDM- and DFT-S-OFDM-based waveforms.

5.5.10   Summary

We focused on the discussion of new waveforms in 5G NR in this section. Several new waveform schemes were provided, with technical principles of each scheme and corresponding advantages and disadvantages. Then, the standardization progress of waveform schemes in 3GPP were discussed. For NR DL, OFDM with filtering or windowing was identified as a DL scheme. For NR UL, two waveforms, DFT-S-OFDM and OFDM, could be flexibly configured by the network. The technical principles behind the decision are presented. For mMTC/high frequency/high mobility, there is still some possibility for other solutions. Under the framework of SDAI and the framework of waveform design, various schemes, including those specified in the 5G NR standard and UE-transparent schemes, can be flexibly configured for different scenarios and applications.

5.6 Flexible Multiple Access Schemes

The current wireless communications systems have predominantly adopted orthogonal multiple access (OMA) schemes, where users are allocated with orthogonal radio resources in time, frequency, or space domain. Existing OMA schemes are able to efficiently eliminate multiuser interferences and thus allow relatively simple transceiver implementations. However, to the multiuser case, it is shown that OMA schemes achieve lower capacity than non-orthogonal schemes in the DL broadcast channel (BC) and the UL multiple access channel (MAC). Such inefficiency of OMA schemes is exacerbated in the UL scenario. Utilizing the channel based on existing OMA schemes may lead to a severe waste of radio resources, or even fail to work in massive connectivity scenarios, such as IoT applications.

The design of the 5G radio network is aiming for higher capacity, larger connectivity, and lower latency [78], which should provide better user experience for eMBB, mMTC, and URLLC services. The mMTC application scenario target to support a massive number of devices simultaneously while the URLLC scenario enables mission-critical transmissions with ultra-high reliability and ultra-low latency. Toward these goals, non-orthogonal multiple access (NoMA) opens the horizon for a new angle of thinking. As has been predicted by multiuser information theory, system capacity can be greatly improved by NoMA transmission compared with that of OMA transmission. So it is very suitable for the uplink massive number of simultaneous users. Besides, as a collision resolution method, NoMA combined with grant-free transmission can also improve the reliability and reduce the latency. And due to their non-orthogonal nature, the requirement of precise channel feedback and scheduling for multiuser multiplexing is thus reduced, or even removed in some scenarios. So, since NoMA-based, grant-free transmission schemes have potential advantages over Orthogonal MA in the aspect of collision resolution or robustness in a resource-limited scenario, these schemes may be solutions to future applications that have very stringent latency requirements, e.g., URLLC, etc.

In all, considering the above diverse requirements of different scenarios, the flexible OMA and NoMA schemes should be introduced in the future 5G NR system.

This section starts with an introduction of some typical NoMA schemes, which are under discussion in 3GPP 5G NR standardization, followed by some theoretical analysis of a NoMA system. Then a flexible MA structure is presented to meet the requirements of diversified services.

5.6.1 Potential New Multiple Access Techniques for 5G

NoMA Based on Super Position Coding

Superposition coding-based non-orthogonal multiple access (SPC-NoMA) utilizes a power domain for user multiplexing and can be applied for both DL and UL. Established by network information theory, non-orthogonal access with SIC/DPC can achieve the multiuser capacity region both in UL and DL. SPC-NoMA superposes multiple users in

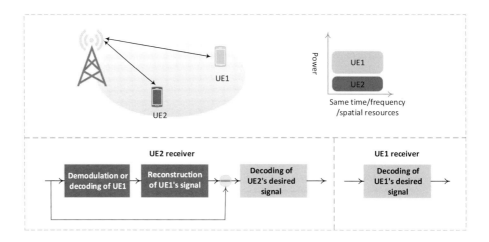

Figure 5.24 Illustration of SPC-based NOMA and transmitter/receiver.

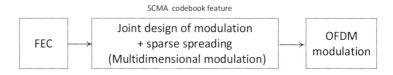

Figure 5.25 Abstracted SCMA transmit procedure for each data layer.

the power domain and exploits channel gain difference between the multiplexed users with the aid of an advanced receiver, e.g., a successive interference cancellation (SIC) receiver, for user separation. Figure 5.24 shows signal transmission and receiving in DL SPC-NoMA system with two users. In the SPC-NoMA system, at the transmitter side, the complex modulated symbols of different UEs are superposed with different transmission power settings. At the receiver side, the symbols of different UEs can be recovered by interference cancellation. With the joint optimization of transmitter and receiver in the SPC-NoMA, multiple layers of data can be simultaneously delivered in the same time, frequency, and spatial resource. SPC-NoMA techniques were discussed in 3GPP under the study item of "study on DL multiuser superposition transmission" in release 13, which was confined to DL data transmission. For the 5G system, more application scenarios of SPC-NoMA techniques, such as UL and control channel, and more advanced SPC-NoMA techniques, such as the combination with MIMO techniques and intercell mitigation schemes, are investigated.

SCMA

SCMA is a novel MA technique. It maps coded bits of a data stream to a sparse codeword of a codebook built based on a multidimensional constellation.

As shown in Fig. 5.25, at each SCMA layer, the SCMA modulator maps input bits to a complex multidimensional codeword selected from a layer-specific SCMA codebook, which has its own sparsity pattern (location of nonzero entries). One or multiple SCMA layers can be assigned to a user/data stream.

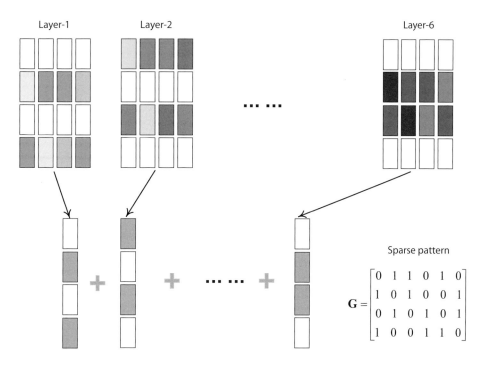

Figure 5.26 SCMA codebook illustration: bit-to-codeword mapping.

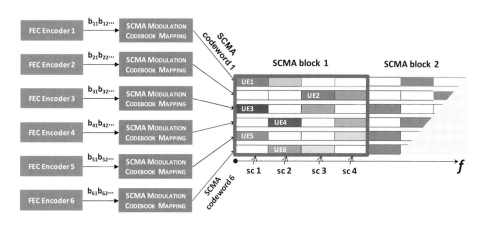

Figure 5.27 Illustrative features for SCMA.

Figure 5.26 shows an example of a codebook set with six data layers [79]. Each codebook has eight multidimensional complex codewords that correspond to eight points of constellation. The length of each codeword is four, which is the same as the spreading length. Upon transmission, the codeword of each layer is selected based on the input bit sequence. The codewords from different layers are overlaid with each other.

Figure 5.27 summarizes the main features of SCMA, and the explanations are listed as follows:

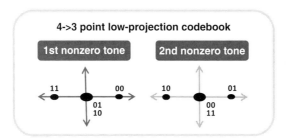

Figure 5.28 Example of low-projection SCMA codebook.

- **Code domain signal superposition**: SCMA allows superposition of multiple symbols from different users on each resource element (RE). For example, on subcarrier 1, symbols from UE1, 3, and 5 are overlapped with each other. The superposition pattern on each RE can be different and is defined in the SCMA codebook.

- **Sparse spreading**: To reduce interlayer interference so that more symbol collisions can be tolerated with low receiver complexity. This allows overloading, meaning that more data layers than the spreading length can be accommodated. For example, in Fig. 5.27, six data layers can be supporting by spreading length 4, resulting in an overloading factor of 150.

- **Multidimensional constellation**: For better SE and low receiver complexity.

- **SCMA codebook design**: Codebook design is the key feature that distinguishes SCMA from other NoMA schemes. The design of the SCMA codebook can be considered as the joint optimization of the sparse spreading pattern design and the multidimensional constellation design. In general, the aim of the codebook design is to provide good distance properties (Euclidean and/or product) among the points in the overall multidimensional constellation to maximize the coding/shaping gain. Another feature of SCMA codebooks is the possibility of having a lower number of projection points over each resource element. This is due to the multi-dimensional nature of the codebooks, which allows two constellation points to collide over some of the nonzero components, as they can still be separated over the other nonzero components. An example is shown in Fig. 5.28, in which the constellation points corresponding to 01 and 10 collide over the first tone, but are separated over the second tone, making three projection points instead of four [79]. This feature can be considered in the design of SCMA codebooks with the goal of reducing the receiver complexity.

- **Receiver**: For a non-orthogonal system like SCMA, there are more than one OFDM symbols overlaid on each RE, so joint multiuser detection algorithms are needed. In general, maximum a posteriori probability (MAP) detection is optimal but with very large complexity. Furthermore, due to the sparsity of the SCMA codeword structure, a message-passing algorithm (MPA) on a factory graph with much lower complexity can be adopted to achieve a suboptimal performance. Although a MPA has significantly lower complexity over MAP detection, it is still

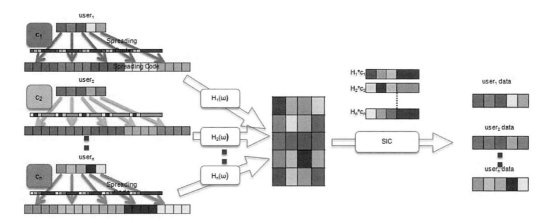

Figure 5.29 Concept of multi-user shared access (MUSA).

suffering from implementation complexity and a wide dynamic range of exponential operations when computing the likelihood function. Then, inspired by the idea of max-log-MAP decoding algorithm for turbo codes, a further simplification can be achieved by implementing MPA in a log domain. This subsection is about the implementation of MPA in a log domain, which is named as log-MPA. By log-domain transform, exponential operations are omitted, multiplication operations are replaced by addition operations, and addition operations are replaced by maximum operations.

MUSA

Multiuser shared access (MUSA) is a NoMA scheme operating in the code domain. Conceptually, each user's modulated data symbols are spread firstly by a specially designed sequence that facilitates robust successive interference cancellation (SIC) implementation compared to the sequences employed by traditional DS-CDMA (direct-sequence CDMA). Then, each user's spread symbols are transmitted concurrently on the same radio resource by means of shared access, which is essentially a superposition process. Finally, decoding each user's data from a superimposed signal can be performed at the BS side using SIC technology. The major processing blocks of the MUSA transmitter and receiver are illustrated in Fig. 5.29 [79].

The spread sequence design is a key component of MUSA, and it has a direct impact on the system performance and computation complexity of the corresponding SIC implementation. Long pseudo-random spread sequences used by traditional DS-CDMA, such as in IS-95 standard, may exhibit relatively low cross-correlation even if the number of sequences is larger than the length. Thus, those sequences can provide a soft capacity limit on the system rather than a hard capacity limit. This soft capacity limit concept can also be understood as the overloading ability of a system. Long spread sequences may be attractive in terms of soft capacity limit, however, the SIC receiver tends to be less efficient when very large spreading factors are used and the system

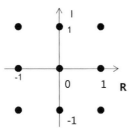

Figure 5.30 Constellation of elements of MUSA sequences.

needs to be operated in extremely overloaded situations to achieve a reasonably good capacity. MUSA relies on a special family of complex spread sequences that can enjoy relatively low cross-correlation even when they are very short, e.g., eight, or even four. In one example of MUSA spread sequence, the real and imaginary parts of the complex spread sequence are from an M-ary real value set. By this method, even the short spread sequences with real and imaginary part selected from a simple 3-value set, $-1,0,1$, can deliver quite impressive performance in terms of overloading. The corresponding trilevel constellation is depicted in Fig. 5.30 [79].

It should be pointed out that the spread sequences used in MUSA are different from the spreading codes in the sense that MUSA spreading does not have the low density property. While the low density codes are more friendly to advanced symbol-level detectors such as using a MPA, the codeword-level SIC can downplay the importance or necessity of using the advanced detectors. Equipped with the well-optimized spreading sequence and state-of-the-art SIC technology, MUSA is capable of decoupling the multiuser mingled data, even if those users are contending to access the system. Potentially a large number of devices are allowed to transmit data at their will (by randomly picking spread sequences) spread the data, and send them. In another words, MUSA is suitable for the scenario where the UL transmissions are not tightly scheduled, and the grants for transmission are not signaled on a per-user basis, and with a high overloading. The relaxed UL synchronization requirement for MUSA allows simple derivation of UL time from a DL synchronization process, which can greatly cut down on battery consumption. Lastly, the code domain superposition nature of MUSA can turn the near–far problem into a near–far advantage. The disparity in the received signal-to-noise ratio (SNR) across simultaneously transmitting users can be exploited in MUSA to facilitate SIC. Tight transmit power control is no longer needed, which can further lower the device cost and its power consumption.

PDMA

PDMA (pattern division multiple access) is a kind of NoMA technology based on the principle of the introduced reasonable diversity between multiuser to promote the capacity, which can obtain higher multiuser multiplexing and diversity gain by designing a multiuser diversity PDMA pattern matrix to implement non-orthogonal signals transmission in such domains as time, frequency, power, and space. PDMA can design patterns for specific users in time, frequency, and space resources. Figure 5.31 [79]

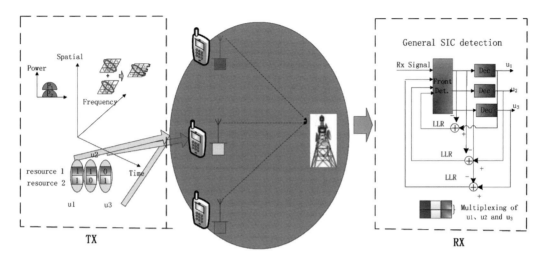

Figure 5.31 The technical framework of the PDMA UL application.

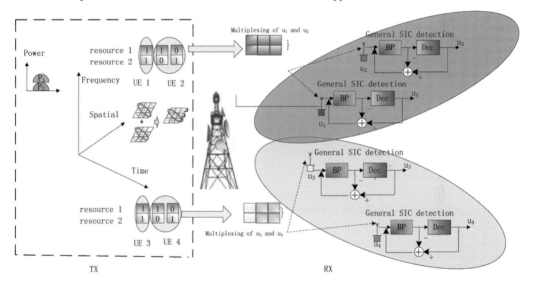

Figure 5.32 The technical framework of the PDMA DL application.

shows the technical framework of the PDMA UL application, Fig. 5.32 [79] shows that of the PDMA DL application.

As shown in Figs. 5.31 and 5.32, the PDMA technical framework includes two parts, the transmitter and the receiver, which reflects that the PDMA technology consider the joint design of the transmitter and the receiver, based on the optimization point of view for a multiuser communication system. On the transmitter side, we distinguish users by using the non-orthogonal characteristic pattern based on the multiple signals domain (including time, frequency and the space domain, etc.). On the receiver side, we can realize suboptimal multiuser detection by general SIC, based on the features of the user pattern.

5.6.2 Theoretical Analysis of a NoMA System

Constellation-Constrained (CC) Capacity

In this section, to provide insight into the achievable sum rate for NoMA in the UL, we analyze the constellation-constrained (CC) capacity [80] of NoMA schemes in the multiple access channel (MAC). The CC capacity is measured by the mutual information between the input and the output in a Rayleigh fading channel, where modulated symbols of each user are constrained to a finite set of constellation points with a uniform distribution. For a K-user MAC channel, the received symbol vector $\mathbf{y}$ at the base station is

$$\mathbf{y} = \mathbf{H} \odot \mathbf{S}\mathbf{x} + \mathbf{n} = \mathbf{H}_{eff}\mathbf{x} + \mathbf{n} \tag{5.21}$$

where $\mathbf{H} = [\mathbf{h}_1, \mathbf{h}_2, \cdots, \mathbf{h}_K]$ of size $N \times K$ denotes the channel matrix for all K users, where the (i, j)-th entry $h_{i,j}, \forall i \in \{0, 1, \cdots, N-1\}$, is assumed to be an independent and identically distributed (i.i.d) complex Gaussian random variable with zero mean and unit variance, $\mathbf{H}_{eff} = \mathbf{H} \odot \mathbf{s}$ denotes the effective channel matrix for all k users, $\mathbf{x} = [x_1, x_2, \cdots, x_K]^T$ refers to the transmitted symbol vector of all k users with normalized power $E\left[|x_i|^2\right] = 1$, $\odot$ denotes the element-wise Hadamard product of two matrices, and finally $\mathbf{n} \sim (0, \sigma^2\mathbf{I})$ is the noise vector.

For illustrative purposes, we utilize sparse code-based NoMA to exemplify the NoMA scheme. We write the received signal vector y for the sparse code-based NoMA scheme with sparse pattern matrix $\mathbf{P}_{2 \times 3} = \begin{bmatrix} 1 & \sqrt{2} & 0 \\ 1 & 0 & \sqrt{2} \end{bmatrix}$ as follows:

$$\mathbf{y} = \begin{bmatrix} h_{11} & h_{12} & h_{13} \\ h_{21} & h_{22} & h_{23} \end{bmatrix} \odot \begin{bmatrix} 1 & \sqrt{2} & 0 \\ 1 & 0 & \sqrt{2} \end{bmatrix} \begin{bmatrix} x_1 \\ x_2 \\ x_3 \end{bmatrix} + \mathbf{n} = \mathbf{H}_{eff}\mathbf{x} + \mathbf{n}. \tag{5.22}$$

Using the chain rule from the information theory, we can express the sum of the CC capacity as follows:

$$\mathrm{I}(\mathbf{x}, \mathbf{y}) = \mathrm{I}(x_1; \mathbf{y}) + \mathrm{I}(x_2; \mathbf{y} | x_1) + \mathrm{I}(x_3; \mathbf{y} | x_1, x_2). \tag{5.23}$$

The term $\mathrm{I}(\mathbf{a}; \mathbf{b})$ denotes the mutual information between the variables $\mathbf{a}$ and $\mathbf{b}$, whereas the term $\mathrm{I}(\mathbf{a}; \mathbf{b} | \mathbf{c})$ denotes the mutual information between the variables $\mathbf{a}$ and $\mathbf{b}$, conditioned on the knowledge of the variable $\mathbf{c}$. Where $\mathrm{I}(x_1; \mathbf{y}) = \mathrm{H}(\mathbf{y}) - \mathrm{H}(\mathbf{y} | x_1)$, $\mathrm{H}(\mathbf{y}) = -\int p(\mathbf{y})\log_2(p(\mathbf{y}))d\mathbf{y}$, $p(\mathbf{y}) = \frac{1}{\Pi_{i=1}^3 |\chi_i|} \sum_{\chi} p(\mathbf{y} | \mathbf{x})$, and χ_i denotes the size of modulation order of the k-th user, which is assumed to be 4 (i.e., QPSK constellation is considered) for all users in this paper without loss of generality. To realize 150% overloading, without loss of generality, we can also utilize the following sparse pattern matrix for an example and others are not precluded.

$$\mathbf{P}_{4 \times 6} = \begin{bmatrix} \sqrt{2} & \sqrt{2} & \sqrt{2} & 0 & 0 & 0 \\ \sqrt{2} & 0 & 0 & \sqrt{2} & 0 & \sqrt{2} \\ 0 & \sqrt{2} & 0 & \sqrt{2} & \sqrt{2} & 0 \\ 0 & 0 & \sqrt{2} & 0 & \sqrt{2} & \sqrt{2} \end{bmatrix} \tag{5.24}$$

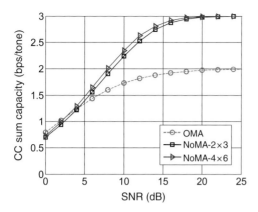

Figure 5.33 Numerical results of constellation constraint capacity for UL sparse code-based NoMA overloading 150%.

Then, we can compute the CC sum capacities of these different NoMA schemes with different sparse pattern matrices in Fig. 5.33, where the CC sum capacity of OMA is also shown for comparison.

Here we want to emphasize that the goal of our comparison for the UL MA study is the UL access user number for a given system target spectrum efficiency (the range of operation point under the given modulation order), not the single-user throughput. So the same modulation order (QPSK) is assumed for each user both in the OMA case and NoMA case in the above numerical simulation.

5.6.3 A Unified Framework of Multiple Access Schemes

Advanced MA technology has been envisioned as one of key enablers of 5G communications. The signals from different users will be superposed into the same time and frequency resource and demodulated by an advanced receiver algorithm to provide higher spectrum efficiency and system capability. Grant-free transmission will be allowed to significantly reduce signaling overhead, shorten access latency, and decrease terminal power consumption. The MA techniques as introduced in the literature are summarized in Table 5.6.

The just-discussed advanced MA schemes, as well as the traditional OMA scheme, e.g., OFDMA, are both identified as potential candidates for 5G. Based on the diverse deployment scenarios and traffic requirements of 5G, flexible MA can be utilized to meet the verified demands. For example, in the case of massive connections, the question of how to accommodate more users with limited resources has become a critical problem for next-generation access networks. With NoMA schemes, e.g., SCMA, MUSA, PDMA, or RSMA [81], the same resources are shared and reused by multiple users, thus the number of connections increases. To support the traffic with low latency requirement, NoMA schemes help to realize grant-free MA, with which the latency is much lower, and the power consumption of the devices can be reduced. In other scenarios, such as DL machine-type traffic, the simple OMA schemes are better, due

Table 5.6 Summary of multiple access techniques.

	BDM	MUSA	SPC-NOMA	PDMA	RSMA	SCMA
Scenario	DL eMBB	UL MMC, DL eMBB	eMBB, MMC, URC	eMBB, MMC, URC	UL MMC/ UL URC	eMBB, MMC, URC
Multiplexing domain	Code/ Power	Code/ Power	Power	Code/Power /Spatial	Code/ Power	Code/ Power
Transmitter Overloading	High	High	Medium	High	High	High
Transmitter Spreading	No	Yes	No	Yes	Yes	Yes
Transmitter multidimensional constellation	No	No	No	No	No	Yes
Receiver	MMSE/SIC	SIC	SIC	SIC/MPA	SIC	MPA/SIC
Receiver Complexity	Low (SSD), Medium (MSD)	Medium	Medium	Medium	Medium	Medium

to device cost and implementation complexity. A study of scenarios and requirements for next-generation access technologies has been made. The requirement of support for wide a range of services to be deployable on a single continuous block of spectrum in an efficient manner is proposed in the document. To support this operational requirement, we propose a compatible MA structure as depicted in Fig. 5.34 [82]. By the unified structure, we can flexibly configure different MA schemes according to various 5G scenarios.

Without loss of generality, here we take the DL transmission as an example, the unified MA framework can also be used in the UL transmission. As depicted in Fig. 5.34, the differences among various MA schemes lie in the different realization of bit-level and symbol-level's operations, e.g., cell-specific/user-specific interleaver design in the

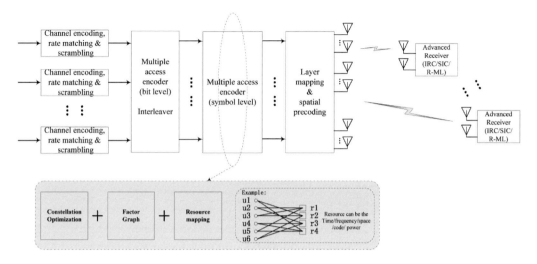

Figure 5.34 Unified framework of MA schemes.

bit-level processing, constellation optimization, factor graph, and multiplexing domain in the symbol-level operations. The detailed explanations are listed in Table 5.7.

Note 1: The identity/permutation matrix means independent/dependent mapping method respectively. For example, like the MUST Cat-1 interleaver, the coded bits of each user is mapped to symbols without considering the bit/symbol of the other co-scheduled user, such mapping is independent mapping. Otherwise it is dependent mapping (permutation matrix in interleaver).

Note 2: The constraint permutation matrix means that it is not a totally dependent mapping method. The bits or symbols that are mapped should satisfy certain conditions, e.g., bits from the same user should be adjacent.

Note 3: The element of the sparse matrix maybe different among various schemes, e.g., the element of the sparse matrix can only be "0" or "1" in the SCMA and PDMA , and it can also be "-1" in the MUSA.

5.6.4   Summary

NoMA is an attractive solution to boost system capacity by accommodating more users at the same time/frequency resource, and reducing system latency caused by scheduling and queuing, as well as relaxing the dependency on precise channel state information and feedback quality. In particular, for UL, NoMA-enabled grant-free is a competitive solution for small packet transmission in many scenarios, including mMTC, URLLC, and eMBB.

In the coming study of 3GPP NoMA SI, more works will be dedicated to the comprehensive evaluations of the various candidate schemes based on the unified framework to better understand the commonality and differentiation of different schemes, to find the recommended configurations for different target scenarios. Moreover, as other technologies are evolving in parallel in 3GPP, the study of how these radio technologies can be

Table 5.7 Configuration method of different MA schemes based on the unified framework.

		Interleaver	Constellation mapping	Factor graph	Resource mapping (multiplexing domain)
OMA		Identity matrix[1]	Gray-mapped legacy constellation	Identity matrix	time/frequency/code/space
NOMA	MUST Cat 1	Identity matrix	non-Gray-mapped superposed constellation	Identity matrix	power
	MUST Cat 2	Constraint permutation matrix[2]	Gray-mapped superposed constellation	Identity matrix	power/bit
	MUST Cat 3	Permutation matrix	Gray-mapped legacy constellation	Identity matrix	bit
	SCMA	Identity matrix	joint optimization (multidimensional modulation + Sparse matrix[3])		code/power
	PDMA	Identity matrix	legacy modulation	Sparse matrix	code/power/space
	… …	… …	… …	… …	… …

integrated with NoMA shall be carried out. As one example, the integration of NoMA with (massive) MIMO has been raised in the literature. In addition, the efficient and flexible MA adaptation schemes, the MA codebook and RS design, and low-complexity receiver design are also key to the commercialization of NoMA in the future. The EE performance at the UE side is an important KPI. Therefore, the EE–SE codesign of various MA schemes needs to be investigated in-depth [1].

5.7   Full Duplex

Out of the potential technologies for 5G and beyond, full duplex has drawn much attention because it may bring a revolution of the duplex mode in future wireless communications. The current TDD mode is half duplex, since its DL and UL are on the same frequency but not simultaneous. The FDD mode is a full duplex mode in time domain by transmitting and receiving simultaneously, but operates on different frequencies in the DL and UL. Transmitting and receiving essentially at the same time and frequency, full duplex may maximally achieve doubled SE compared with either TDD or FDD. Much progress has been made so far in the research and development of full duplex, with multiple successful demos of the feasibility of short-range wireless connection in either relay or single access point scenarios [83–90].

One fundamental challenge to full duplex is self-interference cancellation, the self-interference with the following techniques prescribed to solve this challenge: antenna cancellation, analog cancellation, and digital cancellation. As the antenna number increases at the access point, the self-interference cancellation will become more complicated, since an analog cancellation circuit is generally needed for each Tx and Rx antenna pair, to cancel possible multipath self-interference signals. Currently the analog cancellation circuit is large, e.g., the prototype board measures 10×10cm for a single Tx and single Rx full duplex transceiver [90]. More efforts are still required in the design of efficient and space-compact interference cancellation circuits.

Some solutions have been proposed to realize full duplex with multiple antennas. One solution is to utilize antenna cancellation via symmetric placement of multiple Tx and Rx antennas where the performance was better than conventional MIMO, but required N radios and 2N antennas [87]. Digital signal processing techniques can further mitigate self-interference in MIMO full duplex systems, e.g., the time-domain transmit beam-forming method [88]. When the antenna array becomes a massive array with over 64 elements, the excessive degrees of freedom in massive MIMO full duplex systems can be leveraged [89]. When the antenna size in the MIMO full duplex system is a problem, reducing the Tx and Rx antenna number can significantly reduce the array size. A full duplex implementation using a single antenna was presented in [90], where novel analog and digital cancellation techniques were utilized to cancel the self-interference to the receiver noise floor. In addition to the self-interference cancellation, MAC mechanisms were also investigated. For example, the full duplex physical layer was designed in [91] with a MAC protocol backward compatible with current IEEE 802.11 systems.

In the case of a multicell full duplex network, interference management becomes even more complex [92]. Recently, full duplex networking issues were investigated when full duplex is considered for a wireless network. An interference management strategy was proposed in [93] to handle the intercell interference to achieve gains in data rates over half duplex systems, with the assumption that all BSs in the network have instantaneous access to the global CSI. Extensive system-level simulations were carried out in [94] to evaluate the throughput of full duplex cellular systems, where a suboptimal resource allocation scheme was considered. The results showed that full duplex could significantly increase the aggregate throughput of current cellular systems in both DL and UL. However, these analyses and simulations may be overly optimistic, since instantaneous availability of the CSI and perfect interference mitigation are assumed.

The EE performance of a full duplex system has also been investigated. In [95], the authors analyzed SE–EE trade-off with a full-duplex BS and half-duplex UE under two different kinds of residual self-interference. Corresponding optimization algorithms were proposed to achieve EE maximization. [96] studied four types of power control schemes in full-duplex networks, which shows remarkable gains on EE compared to half-duplex networks. In [97], joint beamforming and time allocation algorithms were proposed to maximize the sum rate and EE, taking account of energy harvest-enabled UE. EE maximization in full-duplex networks with MIMO techniques were studied in [98, 99]. In [98], precoding in the context of full-duplex MU-MIMO was studied, and in [99] power allocation algorithm was studied in the context of full-duplex same cell with massive MIMO. In [100], the authors investigated critical EE challenges in implementing full-duplex relaying in mm-Wave systems and outlined a number of promising EE-oriented solutions for designing full-duplex relaying-enabled systems.

In a practical system deployment, severe intercell and intracell interferences due to simultaneous transmission and reception in each cell make the deployment of full duplex networks very difficult. For example, DL/UL channel measurement and estimation may not be easily achieved due to mutual interference from both the same cell and adjacent cells. In addition, proper scheduling of DL and UL users requires inter-user channel information, which also may incur heavy signaling overhead. Though tremendous progress has been made in the study and implementation of full duplex technologies, there exist many open issues for successful deployment of full duplex network in 5G and beyond, including, for example: a desirable full duplex frame structure and the required modification of the current standards, DL and UL reference signals, efficient intracell/inter-cell interference mitigation, transceiver structures, the extension of TDD and FDD to full duplex, and implementation of MIMO full duplex.

This section aims to shed some insights on how a full duplex cellular network should be designed, from the perspectives of interference mitigation techniques, frame structure design, and TDD/FDD extension to full duplex. So far, 5G NR has started to standardize the flexible frame structure with configurable DL and UL transmissions, and the mechanism of cross-link interference mitigation. There are still many open issues pending in the research and standardization. The work in this section is expected to provide some reference designs.

5.7.1   Interference Mitigation in Full Duplex Networks

In the current TDD or FDD system, the DL-to-DL interference received at the UE and the UL-to-UL interference received at the BS have been extensively studied in literature and standardization bodies. For example, the CoMP technologies were standardized in 4G LTE-A and IEEE 802.16m to counteract these co-channel interferences. In a multicell full duplex network, the interference condition is more severe. A two-cell full duplex network is shown in Fig. 5.35a, where BS1 and BS2 are transmitting to UE1 and UE3 in the DL, respectively, while UE2 and UE4 are transmitting to BS1 and BS2 in the UL, respectively. In addition to the self-interference from the Tx to Rx at each BS, there are intracell UL-to-DL interferences from UE2 to UE1 and from UE4 to UE3, intercell UL-to-DL interferences from UE2 to UE3, and DL-to-UL interferences from BS1 Tx

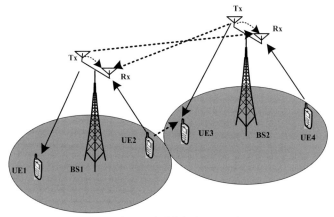

(a) Interferences in full duplex network

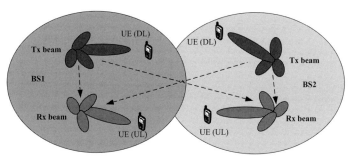

(b) Interference mitigation via beamforming

Figure 5.35 Interferences mitigation in a full duplex system.

to BS2 Rx and from BS2 Tx to BS1 Rx. These interferences have significant impact on whether a full duplex system works, and must be mitigated properly, in addition to the already existing DL-to-DL and UL-to-UL interferences.

In the multicell scenario, due to the interferences explained above, the DL and UL in each cell should be jointly scheduled for an optimized performance (e.g., maximum sum rate of both DL and UL). Self-interference mitigation, which is the predominant issue in the point-to-point application of full duplex, should be jointly considered with many other issues, like DL and UL user pairing, DL and UL power control, DL-to-UL and UL-to-DL (either intracell or intercell) interference mitigation, and the DL and UL QoS.

UL-to-DL Interference Channel Measurement and Feedback

To maximize the sum rate of both DL and UL in a full duplex system, minimizing UL-to-DL interference is very important. This requires that the BS schedules proper DL and UL users with negligible inter-user interference. In the following, an intracell

UL-to-DL interference channel measurement and feedback scheme is presented, which can be extended to the intercell case easily. The key features of this scheme are listed below:

- Simultaneous transmission of orthogonal DL and UL RS: The DL and UL RSs should be orthogonal (e.g., in code or in frequency domain) and transmitted simultaneously, such that the DL UEs can measure the DL and UL RSs simultaneously and calculate the SINR accurately.
- Orthogonal UL RSs: The BS allocates orthogonal RSs resource to each prescheduled UL UE, then each UL UE will transmit its own RSs with a given transmit power, e.g., the max power. The number (N) of the UL UEs, and the UL RSs are broadcasted.
- Feedback mechanism: The DL UEs feedback the lowest M ($M \leq N$, determined by the scheduler) interference power and the corresponding UL RS indexes. If some UL RSs are not received due to large propagation attenuation between the DL and UL UEs, the associated interference power is assumed to be 0. The indexes of these un-received UL RSs are also known to the DL UEs since N and all the UL RSs are broadcasted.

It should be pointed out that large-scale, fading-based approaches can be utilized alternatively to reduce the possibly increased overhead of the above scheme, e.g., the DL and UL UEs with a sufficiently large gap between their large-scale fading to the BS can be scheduled.

Joint Interference Mitigation

In addition to the UL-to-DL interference information, the BSs also need to know all the channel information: the DL and UL channels of each UE with the adjacent BSs, the self-interference channels at BSs, and the DL-to-UL interference channels. Note that the self-interference channels and the DL-to-UL interference channels are semi-static, and hence allow much lower overhead in channel estimation.

As shown in Fig. 5.35b, one feasible approach to maximizing the sum rate is via joint Tx and Rx beamforming [92, 101] when multiple antennas are available at the BSs. The DL data of each BS is precoded such that DL data transmission is improved while the self-interference, the DL-to-UL interferences, and the DL-to-DL interferences are mitigated, e.g., nulls are formed toward the corresponding Rx antennas. To further improve performance, UL beamforming can be adopted similarly.

The above joint interference mitigation approach requires instantaneous CSI at the BSs and a central controller responsible for all the corresponding signal processing. This can be significantly facilitated via the C-RAN [102]. With well-designed DL and UL RSs for various measurement purposes, full or partial CSI of all users are possibly available at the C-RAN baseband pool. System level optimization can be made possible in a real sense.

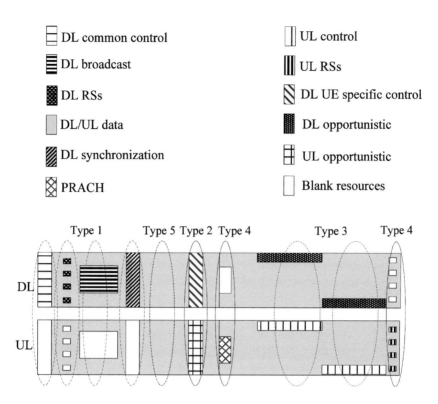

Figure 5.36 Frame structure for full duplex.

5.7.2   Full Duplex Frame Structure Design

Obviously, the TDD frame structure is not applicable directly to full duplex, since it does not support simultaneous DL and UL transmissions. Neither is the FDD frame structure, since applying the DL and UL frame structures of the FDD system (e.g., FDD-LTE) on the same frequency may bring severe interference between DL and UL signals, e.g., between the UL PRACH and the PDCCH (which is a kind of DL common control channel). Therefore, the full duplex frame structure is required to both support simultaneous DL and UL transmissions and to have the capability of DL and UL interference mitigation. This is a dramatic departure from conventional TDD and FDD systems and brings unique technical challenges.

As a virtual bridge between the cellular network and the mobile users, OTA signaling and control information are essential to the stability, reliability, and operation efficiency of the system. Naturally, the basic principle of full duplex frame structure design is that the DL (UL) signaling and control information have a higher priority than data and hence should not be interfered by UL (DL) data, signaling, and control. As one potential example, high-level DL and UL frame structures are illustrated in Fig. 5.36, with the following five types of regions to cover almost all the important DL and UL transmissions of signaling, control, and data.

Type 1: DL Signaling and Control with UL Being Muted

In this region, the essential DL signaling and control information are transmitted, the transmission in the corresponding UL resources need to be muted to ensure that no DL UE may fail to detect these DL signals due to interferences from UL UEs. This information includes the following:

- Primary synchronization sequence (PSS) and secondary synchronization sequence (SSS), which are DL synchronization signals to enable UEs to identify the presence of a network, and acquire the time and frequency of the network.
- The main information block (MIB) in physical broadcast channel (PBCH), which contains the most essential physical-layer information like system bandwidth, frame number, etc., and must be detected successfully before further access to more information in the system information block (SIB).
- Cell common control signals, which deliver control messages to support radio resource management and data transmissions. A UE needs to decode the following three DL control channels before it receives and decodes the data on the PDSCH allocated to it.
- The PDCCH, which provides physical-layer signaling to support MAC-layer operations. The common PDCCH, which carries information such as paging, PRACH response, and system information, belongs to region type 1. On the contrary, the user-specific PDCCH does not belong to this region.
- The physical hybrid-ARQ indicator channel (PHICH), which includes the information of HARQ ACK/NACK feedback for UL data transmissions on the PUSCH.
- The physical control format indicator channel (PCFICH), which indicates the number of OFDM symbols designated for PDCCHs in the current subframe.
- Cell common DL RSs, which are designed for CSI measurement or estimation.

Type 2: DL Signaling and Control with Opportunistic UL

In this region, the UE-specific control information, or the non-delay-sensitive radio resource management signaling is transmitted on the PDSCH. The UL transmission is possible only if its interference to the DL UEs is tolerable. The UL is therefore opportunistic.

Type 3: UL Signaling and Control with Opportunistic DL

In this region, the UL SRS, physical UL control channel (PUCCH) and DMRS are transmitted. The transmission of the DL data or UE specific control information is possible if the UL-to-DL interference can be managed by scheduling proper DL and UL users. The DL is therefore opportunistic.

Type 4: UL Signaling and Control with DL Being Muted

- When the UE transmits the random access signal on the PRACH for network entry, the DL transmission is difficult to be scheduled since the UL UEs transmitting PRACH are not known to the BS yet. Therefore, the DL is suggested to be muted.

- When some special UL control signals are transmitted, the DL transmission needs to be muted. For example, when the UL channels of one UE to adjacent BSs need to be measured accurately for intercell interference mitigation, DL transmissions at the adjacent BSs should be muted to avoid interference to the UL reception at each BS (e.g., the Tx of BS2 is muted when UE2 transmits RSs for channel measurement to BS1 and BS2, as shown in Figure 5.35a.

Type 5: Opportunistic DL/UL Data

For data transmission, the BS can freely schedule DL and UL users.

In addition to the interference-aware considerations above, another issue that needs to be addressed is the time gap between DL and UL in a traditional TDD system. This time gap was designed to be sufficiently large to provide enough time for the DL-to-UL switch of the BS circuitry, and to mitigate intracell and intercell DL and UL interferences. In a full duplex system, this gap can be removed because the BS is not expected to quickly switch from DL to UL, since Tx and Rx have separate radios. In addition, fast switching in the UE from DL to UL is not necessary, because the UE's DL and UL can be scheduled with a time gap.

With the above design principles, the full duplex frame structure can be devised, e.g., based on either a TDD-LTE or FDD-LTE frame structure. More efforts are needed to elaborately design each type of region.

5.7.3 Extension of FDD and TDD to Full Duplex

Potential standardization of full duplex key technologies in LTE-A or IEEE 802.11ax may start in the near future. The focus may lie in leveraging full duplex capabilities at infrastructure nodes to support half duplex UEs, since full duplex UEs still seem impractical due to, e.g., complexity and the large size of the current analog cancellation circuit.

Extending the traditional FDD system (e.g., operating on the carrier frequency f1 in the UL and f2 in the DL) to a full duplex system generally requires doubled transceiver number for both f1 and f2. As shown in Fig. 5.37a, the full duplex BS (with f1 and f2 in both DL and UL) transmits to FDD UE1 on f2, and receives from UE1 on f1 simultaneously. This is exactly the FDD mode to UE1. Meanwhile, the BS transmits to FDD UE2 on f1 and receives from UE2 on f2. This is also the FDD mode to UE2. However, with the full duplex frame structure in Fig. 5.36 on both f1 and f2, the current FDD UEs cannot be supported directly, thus mandating corresponding changes to the UE design. Moreover, the design of UE1 and UE2 should be different, since UE1 is operating on f1 in the UL and f2 in the DL, while UE2 is operating on f1 in the DL and f2 in the UL.

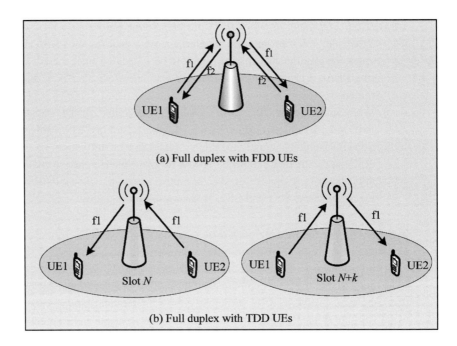

Figure 5.37 TDD and FDD extension to full duplex.

The extension of the traditional TDD system (e.g., operating on the carrier frequency f1 in both DL and UL) to full duplex is shown in Fig. 5.37b. The BS schedules TDD UE1 in the DL and TDD UE2 in the UL in time slot N, and schedules UE1 in the UL and UE2 in the DL in time slot $N + k$ (k is an integer determined by the BS scheduler). The current TDD UEs still operate in TDD mode, while the whole cell operates in the full duplex mode. Different from the FDD case, the current TDD UEs can be supported directly by the full duplex BS with frame structure in Fig. 5.36.

5.7.4   Summary

In this section, several key design issues for full duplex networks were investigated, with potential solutions proposed. Firstly, the interference situation was analyzed and potential interference mitigation techniques were discussed. Design principles of the full duplex frame structure were then presented, aiming to provide efficient means to mitigate the severe interferences in full duplex networks. The extension of traditional TDD and FDD systems to full duplex was further addressed.

Full duplex technologies are expected to reduce both the control and data-plane latency and to double the link and system capacity maximally via simultaneous DL and UL operation. The application of full duplex in future wireless communication systems like 5G is able to remove the clear distinction between TDD and FDD and to better utilize the unpaired spectrum. With growing interest and efforts on the research and development of full duplex technologies, especially on fundamental issues like efficient intracell/intercell interference mitigation, full duplex frame structure, DL and UL reference signals, implementation of full duplex with multiple antennas, etc., full duplex network can be feasibly deployed in the future.

Recall that three types of frame structures are discussed in Subsection 5.3.1 for the DL, UL, and bidirectional transmission, respectively. They are fundamentally different from the full duplex frame structure presented in this section. The bidirectional transmission within one subframe of the self-contained frame structure is time duplexed, not full duplex in essence. The framework of SDAI allows flexible configuration of frame structures to enable real full duplex. However, more work is to be done in the future standardization of full duplex, based on the 5G NR specification on frame structure.

5.8   Flexible Signaling, Control, and Protocol

5.8.1   Introduction

Considering the intensified diversity of user traffic, conventional network signaling, control and protocol (SCP) are facing tremendous challenges, especially in scenarios with critical performance targets in 5G networks.

The conventional "one-size-fits-all" mechanism offers undifferentiated network signaling, control, and protocol toward all kinds of mobile traffic. Take LTE for example: there are two RRC states for the user, i.e., RRC_IDLE and RRC_CONNECTED. A UE in the RRC_IDLE state maintains no connection with the network. It is free from frequent interactions with the network, thus is energy efficient. On the other hand, a UE in the RRC_CONNECTED state is connected to the network and is required to maintain the connection. When a RRC_IDLE UE wants to transmit data, it is required to first establish an RRC connection by executing the RRC connection setup procedure. Then, after expiry of a data-inactive timer, the RRC connection between the UE and the network is released for terminal power savings. As demonstrated in table 5.2.1–2 in [103], each RRC connection setup/release procedure requires about 18 signaling interactions with a 265-bytes payload. Such RRC procedure incurs very heavy signaling overhead in small packet transmission [104], and therefore is not energy efficient. Meanwhile, by the LTE procedure, the control-plane latency, which refers to the time to move from a battery efficient state (e.g., IDLE) to start of continuous data transfer (e.g., ACTIVE), can hardly be satisfied, considering its 10ms target in 5G. As a result, to optimize network performance and efficiency for each individual 5G scenario, scenario- and service-aware flexible SCP design is motivated.

With the flexible SCP framework, different SCP function components (e.g., UE states and scheduling mechanisms) can be orchestrated by the RRC and radio resource management (RRM) entity for different traffic scenarios. For example, for eMBB, in the control plane, novel RRC state and RRC procedures could be designed to enable fast connection establishment; in the user plane, more dedicated treatment could be applied to differentiate user traffic and enhance the user experience. For mMTC of massive small data connectivity, slim signaling can be introduced in both the control plane and user plane to reduce signaling consumption. For URLLC, radio resources could be pre-allocated to reduce data latency incurred by resource grant.

The flexible SCP is an element of the SDAI slice. In the later part of this chapter, new SCP function components are introduced: firstly, a new RRC state with lean signaling design is introduced for mMTC and eMBB for control-plane latency reduction and signaling overhead reduction; Secondly, grant-free MAC scheduling is introduced for mMTC signaling reduction and URLLC low-latency data transmission; Thirdly, the concept of smart RAN is introduced to enhance users' experience with optimized cross-layer processing for different services.

5.8.2   New SCP Function Components

Service-Oriented New UE State: RRC_KEEP_ALIVE / RRC_INACTIVE

In [104] it is revealed that small data bursts result in orders of magnitude higher OTA signaling overhead than the streaming services, by a metric termed as data-signaling ratio (DSR). As a solution, a new UE state, RRC_KEEP_ALIVE is proposed to support both low signaling overhead and small packet transmission. UE behaviors of RRC_KEEP_ALIVE are characterized as follows:

1.   No RRC connection maintenance;
2.   No handover;
3.   Context reservation and slim signaling before data transmission;
4.   Small data transmission;

Since the RRC_KEEP_ALIVE state requires no RRC connection maintenance, the UE in RRC_KEEP_ALIVE behaves like an RRC_IDLE UE and will consume much reduced power. Meanwhile, since only slim or little signaling is needed before data transmission, state transition signaling can be saved and fast data transmission is enabled.

It is evaluated that, for a UE with periodical keep-alive data transmission, application of RRC_KEEP_ALIVE achieves a 6-fold DSR gain compared with the conventional RRC_IDLE/CONNECTED approach. This is because the RRC_KEEP_ALIVE state saves RRC maintenance signaling compared to the RRC_CONNECTED state, and it requires less signaling for activation of data transmission compared to the RRC_IDLE state.

In 3GPP Release 15, a similar concept, RRC_INACTIVE, is introduced. This standardized new UE state is mainly motivated by the requirements for fast RRC state transition and reduced control plane latency. It is clearly characterized in [105] by the following features:

1.   Broadcast of system information;
2.   Cell reselection mobility;
3.   5G core and NG-RAN connection (both control and user planes) is established for UE;
4.   The UE access stratum context is stored in at least one gNB and the UE;

5. Paging is initiated by NG-RAN;
6. Discontinuous reception (DRX) for NG-RAN paging configured by NG-RAN;
7. RAN-based notification area (RNA) is managed by NG-RAN;

Obviously, though the two states are somehow different in motivations, they do share most behaviors and merits. For both of the states: UE context is reserved and slim signaling can be enabled; UE behaves similar as in energy saving RRC_IDLE state. RRC_INACTIVE complements the RRC_KEEP_ALIVE state with the feature of accurate paging support in RAN. And RRC_KEEP_ALIVE is keener on optimization of small data users, which supports small data transmission without RRC state transition.

Service- and Load-Aware MAC Scheduling: From Grant-Based to Grant-Free

To further resolve the issues of low signaling efficiency (e.g., low DSR) and long data transmission latency in small data transmission, grant-free scheduling is investigated in 5G NR. By pre-allocating radio resources to grant-free users via RRC signaling, grant-free scheduling enables users to transmit UL data without further resource request and grant. In this way, grant signaling in the MAC layer can be saved. Data transmission latency could possibly be reduced, since the user data could be transmitted before a dedicated grant is possibly received by the UE.

The performance difference between the grant-free and the grant-based scheduling is shown in Fig. 5.38, where the system bandwidth is 1.08MHz, and the basic scheduling resource unit is 180kHz in frequency and 0.5ms in time. Suppose in every 10 TTIs, there is a dedicated TTI configured for the small data. Then there are N=1200 basic scheduling units. Assume the system serves 300,000 static mMTC users, each with a traffic arrival rate of 1 packet per 5 minutes. In this simulation, the grant-free transmission is free of any signaling interaction, while the grant-based transmission involves signaling of scheduling request and scheduling grant. For the grant-based approach, the total bandwidth is divided into two parts: one for scheduling request, and the other for data transmission and accompanying grant signaling. The resource of scheduling request is configured similar to PRACH in LTE, i.e., 12 basic resource units for 64 preambles and each preamble conveys a scheduling request. For each grant, overhead of 3/7 scheduling resource unit is consumed. It is assumed that the SE is 1.8 and 1 bps/Hz for the grant-based and grant-free UEs, respectively.

A metric termed "effective spectral efficiency" is introduced for evaluation. It takes into account the signaling impacts on data transmission efficiency and is calculated as the total transmitted data payload bits divided by total radio resources consumed by both data payload and supporting signaling per Hz per second.

As can be seen in Fig. 5.38, efficiency of the grant-based scheduling is quite resistant to load increase, and it improves with increased packet sizes. This explains why grant-based scheduling is widely adopted in existing cellular systems. However, it is also observed that efficiency of grant-based scheduling is low in the case of small data packets, while the grant-free solution is the other way around. Obviously, grant-free is especially efficient to small data transmission and may serve as an ideal solution for the

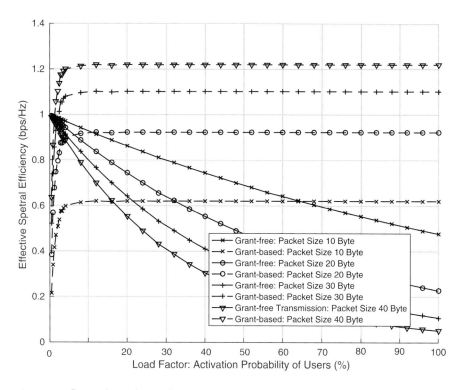

Figure 5.38 Comparison of grant-free and grant-based MAC scheduling.

5G mMTC scenario for signaling and energy consumption reduction. Besides, it can be also applied to the latency stringent URLLC services. Therefore, grant-free scheduling can be taken as a new and important component of the 5G MAC scheduler.

In ongoing 3GPP standardization, the 3GPP Release 15 is devoted to study grant-free solutions by RRC preconfiguration of grant-free resources. The pre-allocated grant-free resources may be shared by one or a group of UEs.

Context-Aware Service Delivery: Smart RAN

Given the ever-increasing mobile traffic load and limited radio resources, it would be very challenging to meet the requirements of mobile applications in 5G. To make it worse, the limited resources in even the current mobile networks are far from being fully utilized. One major cause lies in the isolated design of the mobile network and mobile applications. The mobile network radio resource allocation is performed based on rapidly varying radio conditions, while rate adjustment of the application is based on relatively long-term statistical E2E throughput observations. Therefore, there is a clear mismatch between the air interface data rate and the application data rate, which may lead to inefficient resource usage and degraded user experience. Operators would be highly motivated to remove the separation wall between the mobile network and mobile application.

Smart RAN is a novel concept proposed by CMCC. It is dedicated to enable more real-time coordination between RAN and the applications. On the one hand, smart RAN is capable of mobile application awareness, and is able to configure RAN strategy accordingly. Its awareness of applications can be assisted by the application layer. For example, considering the DL video, the video application server can mark the video packets with different labels, e.g., primary video segment packets or supplementary packets. Upon detection of different labels, RAN configures differentiated radio protocol stack for different packets. For example, a higher MAC scheduling priority may be applied to packets with a "primary" label. Meanwhile, the video client in UE can report to RAN its application buffer status, based on which RAN allocates preemptive radio resource to the buffer-hungry UEs. On the other hand, smart RAN is capable of exposing the RAN condition (e.g., the available radio bandwidth) as required by the application server via e.g., application programmable interfaces (APIs) to the video server. Then, the video server can adjust the video bit rate according to the radio bandwidth. By alignment of the network and the application, the network resources can be more fully utilized and the UE experience can be improved. Furthermore, to enable more timely coordination between RAN and the applications, edge deployment of the application server can be considered in smart RAN.

Both the industry and academia have been devoting efforts to smart RAN case studies. In a Release 14 study item led by CMCC, "Context-aware service delivery," application use cases of smart RAN are discussed, including data caching, TCP data transmission, and video transmission [106]. The simulation results in its annex show that "video playout buffer-aware scheduling" achieves a 25% capacity gain and significantly reduces the stalling probability. In [107], a cross-layer moving mean algorithm (CMMA) is proposed; it is demonstrated that by radio condition-aware application rate adjustments, the CMMA can boost throughput of the typical video protocol "dynamic adaptive streaming" over HTTP (DASH) by up to 30%. The reason is that the video client in DASH conventionally chooses an optimum video code rate by estimation of available bandwidth based on throughput observations on the client side. Nevertheless, since time granularity of the client observations is much larger than the millisecond-level radio variations, video transmission can hardly fully utilize the available radio bandwidth. With assistance of the radio throughput information provided by RAN, the DASH server is capable of estimating the throughput condition more accurately.

5.8.3   Summary

In this section, some examples of new SCP function components were introduced to meet the 5G requirements. First, new RRC_KEEP_ALIVE/RRC_INACTIVE states in the RRC layer were introduced, which can be easily merged for the satisfaction of both signaling reduction for mMTC and control-plane latency reduction for eMBB. Then, grant-free scheduling in the MAC layer was proposed to avoid dedicated resource grant signaling for massive small packets, and to reduce URLLC data latency. Furthermore, smart RAN was proposed to overcome the gap between the network and applications, and to achieve more efficient radio resource usage. It is possible to further distinguish traffic types within both a session and a bearer, and to apply finer optimization accordingly. Flexible SCP, as a soft and green solution, is expected to be implemented in 5G mobile networks and beyond.

References

[1] Q. Sun, S. Han, C.-L. I, and Z. Pan, "Software defined air interface: A framework of 5G air interface," *IEEE WCNC*, 2015.

[2] C.-L. I, S. Han, Z. Xu et al. "New paradigm of 5G wireless internet," *IEEE J. on Sel. Areas in Commun.*, vol. 34, no. 3, pp. 472–482, Mar. 2016.

[3] S. Zhang, G. Wang, and C.-L. I, "Is mmWave ready for cellular deployment," *IEEE Access*, 2017, DOI: 10.1109/ACCESS.2017.2711491.

[4] R. Crane, *Propagation Handbook for Wireless Communication System Design*, CRC Press, 2003.

[5] A. F. Molisch, *Wireless Communications*, Second Edition, John Wiley & Sons, 2012.

[6] T S. Rappaport, *Wireless Communications: Principles and Practice*, Second Edition, Prentice Hall, 2002.

[7] J. Ramakrishna, *Radiowave Propagation and Smart Antennas for Wireless Communications*, Springer, 2001.

[8] T. L. Marzetta, *Fundamentals of Massive MIMO*, Cambridge University Press, 2016.

[9] B. M. Hochwald, T. L. Marzetta, and V. Tarokh, "Multiple-antenna channel hardening and its implications for rate feedback and scheduling," *IEEE Trans. on Inf. Theory*, vol. 50, no. 9, pp. 1893–1909, Sept. 2004.

[10] H. Q. Ngo and E. G. Larsson, "No downlink pilots are needed in TDD massive MIMO," *IEEE Trans. on Wireless Commun.*, vol. 16, no. 5, pp. 2921–2935, May 2017.

[11] K. Zheng, S. Ou, and X. Yin, "Massive MIMO channel models: A survey," *International Journal of Antennas and Propagation*, vol. 2014.

[12] L. Liu et al., "The COST 2100 MIMO channel model," *IEEE Commun. Mag.*, vol. 19, no. 6, pp. 92–99, Dec. 2012.

[13] 3GPP, TR 36.873, "Study on 3D channel model for LTE (Release 12)."

[14] J. Meinilä, P. Kyösti, L. Hentilä et al., "Deliverable 5.3: WINNER+ final channel models," WINNER+/Celtic project CP5-026, Jun. 2010.

[15] S. Salous, "COST IC1004 white paper on channel measurements and modeling for 5G networks in the frequency bands above 6 GHz," 2016. www.ic1004.org/.

[16] ETSI, "New ETSI group on millimetre wave transmission starts work," 2015. www .etsi.org/news-events/news/866-2015-01-press-new-etsigroup-on-millimetre-wave-transmission-starts-work.

[17] 3GPP, TR 38.900, "Study on channel model for frequency spectrum above 6 GHz (Release 14)," 2016.

[18] Various Contributors, "White Paper on 5G channel model for bands up to 100 GHz," white paper, 2015. www.5gworkshops.com/5G_Channel_Model_for_bands_up_to100_GHz(2015-12-6).pdf.

[19] S. Jaeckel, L. Raschkowski, K. Borner, and L. Thiele, "QuaDRiGa: A 3-D multi-cell channel model with time evolution for enabling virtual field trials," *IEEE Trans. Antenna Propag.*, vol. 62, no. 6, pp. 2921–2935, Jun. 2014.

[20] IEEE Wireless LAN medium access control (MAC) and physical layer (PHY) specifications amendment 3: Enhancements for very high throughput in the 60 GHz band," 2012. https://ieeexplore.ieee.org/document/6392842.

[21] MiWEBA, "Channel modeling and characterization," June 2014. www.miweba.eu/wp-content/uploads/2014/07/MiWEBA_D5.1_v1.011.pdf.

[22] METIS, "METIS Channel Models," 2015.

[23] mmMAGIC, "Measurement campaigns and initial channel models for preferred suitable frequency ranges" 2016. https://5g-mmmagic.eu/results/deliverables.

[24] M. K. Samimi and T. S. Rappaport, "3-D millimeter-wave statistical channel model for 5G wireless system design," *IEEE Trans. on Microwave Theory and Techniques*, vol. 64, no. 7, pp. 2207–2225, Jul. 2016.

[25] X. Lin, J Andrews et.al, "An overview of 3GPP device-to-device proximity services," *IEEE Commun. Mag.*, vol. 52, no. 4, pp. 40–48, Apr. 2014.

[26] X. Cheng, Y. Li, B. Ai, et al., "Device-to-device channel measurements and models: a survey," *IET Communications*, 2014.

[27] R. He, et al., "High-Speed Railway Communications: From GSM-R to LTE-R," *IEEE Veh. Technol. Mag.*, vol. 11, no. 3, pp. 49–58, Sept. 2016.

[28] B. Ai, et.al., "Challenges toward wireless communications for high-speed railway," *IEEE Trans. on Intell. Transp. Syst.*, vol. 15, no. 5, pp. 2143–2158, Oct. 2014.

[29] R.. He et al., "A measurement based stochastic model for high speed railway channels," *IEEE Trans. on Intell. Transp. Syst.,* vol. 16, no. 3, pp. 851–854, Jun. 2015.

[30] G. Li, B. Ai, K. Guan et al., "Path loss modeling and fading analysis for channels with various antenna setups in tunnels at 30 GHz band," *IEEE EuCAP*, 2016.

[31] T. Zhou, C. Tao, S. Salous, and L. Liu, "Measurements and analysis of angular characteristics and spatial correlation for high-speed railway channels," *IEEE Trans. on Intell. Transp. Syst.*, vol. 19, no. 2, pp. 357–367, Feb. 2018.

[32] 3GPP RAN1 R1-162156, "Scenario & design criteria on flexible numerologies," Huawei, HiSilicon.

[33] Q. Li, G. Li, W. Lee et al., "MIMO techniques in WiMAX and LTE: A feature overview," *IEEE Commun. Mag.*, vol. 48, no. 5, pp. 86–92, 2010.

[34] 3GPP, TR 38.913, "Study on scenarios and requirements for next generation access technologies."

[35] F. Rusek, D. Persson, B. K. Lau et al., "Scaling up Mimo: Opportunities and challenges with very large arrays," *IEEE Signal Process. Mag.*, vol. 30, no. 1, pp. 40–60, Jan 2013.

[36] S. Han, C.-L. I, Z. Xu, and C. Rowell, "Large-scale antenna systems with hybrid analog and digital beamforming for millimeter wave 5G," *IEEE Commun. Mag.*, vol. 53, no. 1, pp. 186–194, Jan. 2015.

[37] Z. Xu, S. Han, Z. Pan, and C.-L. I., "Alternating beamforming methods for hybrid analog and digital MIMO transmission," *2015 IEEE Int. Conf. on Commun. (ICC)*, pp. 1595–1600.

[38] W. Roh, J. Seol et al., "Millimeter-wave beamforming as an enabling technology for 5G cellular communications: Theoretical feasibility and prototype results," *IEEE Commun. Mag.*, vol. 52, no. 2, pp. 106–113, Feb. 2014.

[39] Z. Pi and F. Khan, "An introduction to millimeter-wave mobile broadband systems," *IEEE Commun. Mag.*, vol. 49, no. 6, pp. 101–107, Jun. 2011.

[40] F. Khan, Z. Pi, and S. Rajagopal, "Millimeter-wave mobile broadband with large scale spatial processing for 5G mobile communication," *50th Annual Allerton Conf. on Commun., Control, and Comput. (Allerton)*, 2012, pp. 1517–1523.

[41] O. E. Ayach, R. W. Heath, S. Rajagopal, and Z. Pi, "Multimode precoding in millimeter wave MIMO transmitters with multiple antenna sub-arrays," *2013 IEEE Global Commun. Conf. (GLOBECOM)*, pp. 3476–3480.

[42] O. E. Ayach, S. Rajagopal, S. Abu-Surra, Z. Pi, and R. W. Heath Jr.,"Spatially sparse precoding in millimeter wave MIMO systems," *IEEE Trans. Wireless Commun.*, vol. 13, no. 3, pp. 1499–1513, Mar. 2014.

[43] S. Kutty and D. Sen, "Beamforming for millimeter wave communications: An inclusive survey," *IEEE Commun. Surveys & Tutorials*, vol. 18, no. 2, pp. 949–973, 2016.

[44] X. Gao, L. Dai, S. Han, C.-L. I, and R. W. Heath, "Energy-efficient hybrid analog and digital precoding for mmWave MIMO systems with large antenna arrays," *IEEE J. Sel. Areas Commun.*, vol. 34, no. 4, pp. 998–1009, Apr. 2016.

[45] A. Mezghani, F. Antreich, and J. A. Nossek, "Multiple parameter estimation with quantized channel output," *Smart Antennas (WSA)*, pp. 143–150.

[46] L. Jacques, J. N. Laska, P. T. Boufounos, and R. G. Baraniuk, "Robust 1-Bit compressive sensing via binary stable embeddings of sparse vectors," *IEEE Trans. on Inf. Theory*, vol. 59, no. 4, pp. 2082–2102, Apr. 2013.

[47] J. Mo and R. W. Heath, "Capacity analysis of one-bit quantized MIMO systems with transmitter channel state information," *IEEE Trans. Signal Process.*, vol. 63, no. 20, pp. 5498–5512, Oct. 2015.

[48] A. Liu and V. Lau, "Sum capacity of massive MIMO systems with quantized hybrid beamforming," in *2016 IEEE Int. Symposium on Inf. Theory (ISIT)*, pp. 320–324, 2016.

[49] T. Shin, G. Kim, H. Park, and H. M. Kwon, "Quantization error reduction scheme for hybrid beamforming," *18th Asia-Pacific Conf. on Commun. (APCC)*, pp. 243–247, 2012.

[50] T. Demir and T. E. Tuncer, "Hybrid beamforming with two bit RF phase shifters in single group multicasting," *2016 IEEE Int. Conf. on Acoust., Speech and Signal Process. (ICASSP)*, pp. 3271–3275.

[51] H. Q. Ngo, E. G. Larsson, and T. Marzetta, "Energy and spectral efficiency of very large multiuser MIMO systems," *IEEE Trans. on Commun.*, vol. 61, no. 4, pp. 1436–1449, Apr. 2013.

[52] Z. Xu, Z. Pan, and C.-L. I, "Fundamental properties of the EE-SE relationship," *IEEE Wireless Commun. and Netw. Conf. (WCNC)*, 2014.

[53] G. Y. Li, Z. Xu, C. Xiong et al., "Energy-efficient wireless communications: Tutorial, survey, and open issues," *IEEE Wireless Commun.*, vol. 18, no. 6, pp. 28–35, Dec. 2011.

[54] 3GPP, "Final report of 3GPP TSG RAN WG1 #85," 2016.

[55] 3GPP, "Final report of 3GPP TSG RAN WG1 #86," 2016.

[56] 3GPP, "Final report of 3GPP TSG RAN WG1 #86b," 2016.

[57] Qualcomm, "OFDM based waveform single user evaluation," R1-164685, Nanjing, China, May 2016.

[58] J. Abdoli, M. Jia, and J. Ma, "Filtered OFDM: A new waveform for future wireless systems," *IEEE 16th Int. Workshop on Signal Process. Advances in Wireless Commun. (SPAWC)*, 2015.

[59] Huawei, "f-OFDM scheme and filter design," R1-164033, Nanjing, China, May 2016.

[60] X. Wang, T. Wild, F. Schaich, A. Santos, and Alcatel-Lucent, "Universal filtered multi-carrier with leakage-based filter optimization," *European Wireless*, 2014.

[61] B. Farhang-Boroujeny, "OFDM versus filter bank multicarrier: development of broadband communication systems," *IEEE Signal Proc. Magazine*, pp. 92–112, May 2011.

[62] M. Bellanger, "FS-FBMC: an alternative scheme for filter," *ISCCSP*, Rome, Italy, May 2012.

[63] N. Michailow, M. Matth, I. S. Gaspar et al., "Generalized frequency division multiplexing for 5th generation cellular networks," *IEEE Trans. on Commun.*, vol. 62, no. 9, Sept. 2014.

[64] A. Farhang, N. Marchetti, L. E. Doyle, "Low Complexity transceiver design for GFDM," 2015. arXiv:1501.02940v1.

[65] Cohere Technologies, "OTFS modulation waveform and reference signals for new RAT," R1-162930, Busan, South Korea, Apr. 2016.

[66] A. M. Sayeed and B. Aazhang, "Joint multipath-doppler diversity in mobile wireless communications," *IEEE Trans. on Commun.*, vol. 47, no. 1, pp. 123–132, Jan. 1999.

[67] F. Hasegawa, S. Shinjo, A. Okazaki et al., "Static sequence assisted out-of-band power suppression for DFT-s-OFDM," *PIMRC 2015*, Hong Kong, pp. 61–65, Sept. 2015.

[68] G. Berardinelli, F. M. L. Tavares, T. B. Sorensen et al., "Zero-tail DFT-spread-OFDM signals," *2013 IEEE Global Commun. Conf. (GLOBECOM)*, Sept. 2013.

[69] D. A. Guimaraes, "Contributions to the understanding of the MSK Modulation," *REVISTA Telecommunications*, vol. 11, no. 1, Dec. 2008.

[70] K. Murota and K. Hirade, "GMSK modulation for digital mobile radio telephony," *IEEE Trans. on Commun.*. vol. 29, no. 7, pp. 1044–1050, Jul. 1981.

[71] 3GPP RANI #86 R1-167963, "Way forward on waveform," Huawei, Aug. 2016.

[72] 3GPP #84-BIS R1-163222, "Waveform for NR," Ericsson, Apr. 2016.

[73] 3GPP #37 R1-040642, "Comparison of PAR and cubic metric for power de-rating," Motorola, May 2004.

[74] 3GPP RANI #86-BIS R1-1609929, "Discussion and evaluation of UL waveforms," CMCC, Oct. 2016.

[75] 3GPP RANI #85 R1-166004, "Response LS on realistic power amplifier model for NR waveform evaluation," RAN4, Nokia.

[76] 3GPP TS 36.101, "User Equipment (UE) radio transmission and reception (Release 12)," 2016.

[77] 3GPP TR 36.873, "Study on 3D channel model for LTE," 2014.

[78] 3GPP TR 38.913, "Study on scenarios and requirements for next generation access technologies."

[79] "White paper, v2.0D-5G enabler: Alternative multiple access v1," Nov. 2015. http://www.future-forum.org/2009cn/member.asp.

[80] R1-164889, 3GPP WG1, "Analytical evaluation of multiple access and preliminary LLS results," CMCC. http://www.3gpp.org.

[81] R1-163510, 3GPP WG1, "Candidate NR multiple access schemes," Qualcomm. http://www.3gpp.org.

[82] R1-162870, 3GPP WG1, "On unified framework for multiple access schemes," CMCC. http://www.3gpp.org.

[83] S. Hong et al, "Applications of self-interference cancellation in 5G and beyond," *IEEE Commun. Mag.*, vol. 52, no. 2, pp. 114–121, Feb. 2014.

[84] DUPLO Deliverable D2.1, "Design and measurement report for RF and antenna solutions for self-interference cancellation," Apr. 2014.

[85] M. Duarte and A. Sabharwal, "Full duplex wireless communications using off-the-shelf radios: Feasibility and first results," *Proc. IEEE Asilomar Conf. on Signals, Systems and Computers*, pp. 1558–1562 Nov. 2010.

[86] M. Jain, J. I. Choi, D. Bharadia et al., "Practical, real-time, full duplex wireless," *Proc. ACM 17th Annual Int. Conf. on Mobile Comput. and Networking (MobiCom)*, pp. 301–312, Aug. 2011.

[87] E. Aryafar, M. A. Khojastepour, K. Sundaresan, S. Rangarajan, and M. Chiang, "Midu: enabling MIMO full duplex," in *Proc. ACM 17th Annual Int. Conf. on Mobile Comput. and Networking (MobiCom)*, pp. 257–268, Aug. 2012.

[88] Y. Hua, P. Liang, Y. Ma, A. C. Cirik, and Q. Gao, "A method for broadband full duplex MIMO radio," *IEEE Signal Process. Lett.*, vol. 19, no. 12, pp. 793–796, Dec. 2012.

[89] B. Yin, M. Wu, C. Studer, J. R. Cavallaro, and J. Lilleberg,, "Full duplex in large-scale wireless systems," in *Proc. IEEE Asilomer Conf. on Signals, Systems, and Computers*, Nov. 2013.

[90] D. Bharadia, E. McMilin, and S. Katti, "Full duplex radios," in *Proc. Sigcomm*, Aug. 2013.

[91] B. P. Day, A. R. Margetts, D. W. Bliss, and P. Schniter, "Full duplex bidirectional MIMO: Achievable rates under limited dynamic range," *IEEE Trans. Signal Process.*, vol. 60, no. 7, pp. 3702–3713, Jul. 2012.

[92] DUPLO Deliverable D4.1.1, "Performance of Full duplex systems," Jan. 31, 2014.

[93] A. Sahai, S. Diggavi, and A. Sabharwal, "On uplink/downlink full duplex networks," *Proc. IEEE Asilomer Conf. on Signals, Systems, and Computers*, Nov. 2013.

[94] S. Goyal, P. Liu, S. Hua, and S. Panwar, "Analyzing a full duplex cellular system," *Proc. 47th IEEE Annual Conf. on Inf. Sciences and Systems (CISS)*, pp. 1–6, Mar. 2013.

[95] D. Wen et al., "Results on energy- and spectral-efficiency tradeoff in cellular networks with full-duplex enabled base stations," *IEEE Trans. on Wireless Commun.*, vol. 16, no. 3, pp. 1494–1507, Mar 2017.

[96] C. Feng et al., "Power control in full duplex networks: Area spectrum efficiency and energy efficiency," *IEEE Int. Conf. Computers (ICC)*, 2017.

[97] V. Nguyen et al., "Spectral and energy efficiencies in full-duplex wireless information and power transfer," *IEEE Trans. on Commun.*, vol. 65, no. 6, pp. 2220–2223, May 2017.

[98] D. Nguyen et al., "Precoding for full duplex multiuser MIMO systems: Spectral and energy efficiency maximization," *IEEE Trans. on Signal Process.*, vol. 61, no. 16, pp. 4038–4051, Aug. 2013.

[99] Y. Li et al., "On the spectral and energy efficiency of full-duplex small-cell wireless systems with massive MIMO," *IEEE Trans. on Veh. Technol.*, vol. 66, no. 3, pp. 2339–2353, Mar. 2017.

[100] Z. Wei et al., "Energy-efficiency of millimeter-wave full-duplex relaying systems: Challenges and solutions," *IEEE Access*, vol. 4, pp. 4848–4860, 2016.

[101] D. Nguyen, L.-N. Tran, P. Pirinen, and M. Latva-aho, "On the spectral efficiency of full duplex small cell wireless systems," *IEEE Trans. Wireless Commun.*, vol. 13, no. 9, pp. 4896–4910, Sept. 2011.

[102] C.L. I, J. Huang et al, "Recent progress on C-RAN centralization and cloudification," IEEE Access, vol. 2, pp. 1030–1039, Aug. 2014.

[103] 3GPP TR 36.822, "LTE radio access network (RAN) enhancements for diverse data applications (Release 11)."

[104] Y. Chen, G. Li, Z. Pan, and C.-L. I, "Small data optimized radio access network signaling/control design," *IEEE Int. Conf. on Commun.*, pp. 49–54, 2014.

[105] 3GPP TS 38.300 "NR; Overall description; Stage 2 (Release 15)."

[106] 3GPP TR 36.933 "Study on context aware service delivery in RAN for LTE (Release 14)."

[107] H. Liu, Y. Liu, Z. Chen, L. Sang, and D. Yang, "A cross-layer bandwidth estimation algorithm for DASH services," *Proc. IEEE Int. Carnahan Conf. Security Technol. (ICCST)*, 2015.

6 Energy-Saving Solutions and Practices

Currently, 3GPP is actively progressing in the standardization of 5G NR. The concepts and many solutions proposed in previous chapters have been adopted in the NR technical report, including, for example, service-based network architecture, two-layer (CU and DU) RAN architecture and signaling interfaces, software-defined air interface with flexible frame structure, unified multiple access schemes with orthogonal and non-orthogonal solutions, unified signaling procedure for MIMO with hybrid beamforming structures, multiple waveform support, etc. Toward green and soft networks of 5G and beyond, more efforts are needed in both standardization and practical deployment.

In this chapter, some practices in CMCC's wireless communication networks are discussed, from 2G, 3G, 4G, even to future 5G. The existing energy-saving schemes and practices in cellular networks and WLAN are first surveyed. Then the field trial of C-RAN is presented, with test results discussed in various scenarios. This chapter continues to explore the feasibility of energy-saving in mobile applications running on mobile phones. Finally, a massive MIMO platform named "Invisible BS" is introduced, which is featured by low-power, low-cost UE-grade RF IC for power-saving, and flexibly configurable SmarTiles with integrated RF and antennas.

6.1 Green Wireless Technologies in Cellular Networks

6.1.1 Energy-Saving in GSM

Energy-saving techniques in GSM have been investigated extensively [1–6]. A GSM multi-carrier BS can utilize digital intermediate frequency (IF) and multi-carrier power amplification (MCPA) technology to achieve multi-carrier on a single radio frequency (RF) channel [6]. Attributed to high integration, it is more energy-efficient than a traditional GSM BS using a single-carrier power amplifier (SCPA). According to field test results, the energy efficiency of a multi-carrier BS is as high as 25%, while that of the traditional single carrier BS is only 15%. Additionally, the BS is able to adjust the operating voltage of the MCPA according to the resource utilization of multiple channels. Field test results show that applying this technique alone can help each BS save 438 kWh per year. At present, it has been widely implemented in many GSM networks.

6.1.2   Energy-Saving in TD-SCDMA

There are mainly two schemes for energy-saving in TD-SCDMA systems, i.e., the BBU baseband deactivation technique and the RRU time slot deactivation technique based on power amplification (PA). The former can reallocate or migrate a small amount of traffic to one or a few baseband boards during low traffic load, and deactivate the idle baseband boards to save energy. When the traffic load rises, it will activate the sleeping boards in advance to avoid traffic congestion caused by insufficient resources. The latter scheme takes advantage of the key characteristics of TDD systems, i.e., the UL and DL time slots are utilized alternately. When there is no traffic on the DL, the boards can be turned off in corresponding time slots in real time, thereby reducing power consumption.

6.1.3   Energy-Saving in LTE

Compared with GSM and TD-SCDMA, the PA technology of LTE equipment is more mature and widely used. At present, the energy conversion efficiency for LTE RRU is up to 40%. From a hardware perspective, it is difficult to further improve the conversion efficiency. There are many energy-saving techniques in LTE [7–16]. The most straightforward solution is to turn off the BSs at various granularity, e.g., symbol-level turnoff, channel-level turnoff, and small BS shutdown. Symbol-level and channel-level turnoffs are supported and implemented within eNBs, while small BS turnoff may impact the interface among eNBs. This technique has been standardized by 3GPP. In the following subsections, we will discuss these energy-saving methods and their test results in CMCC's networks.

Symbol-Level Turnoff: PA Shut down on Symbol-Level

When there is no actual data transmission in certain symbols, PA can be powered off so as to reduce power consumption and possibly reduce interference to the adjacent cells. For LTE normal frames, each subframe contains 14 symbols, some of which include reference symbols (RS). When in high traffic load situations, all symbols are carrying data and PA cannot be shut down, whereas in low traffic load situations, PA can be powered off rapidly, such that power consumption can be reduced. In the next symbol or frame duration, PA can be powered on again if there is data to send. The trial results show that approximately 5% of energy consumption of RRU can be saved when the traffic load is 10%. The approximate annual electricity savings in CMCC's networks is a very substantial 30 million kWh.

Channel Shutdown

When under low traffic load, a cell can choose to use only parts of its total transmit channels to transmit data, and the rest can be shut down to save energy. As a result, the transmit/receive chain can be turned off, including PAs and other analog devices, since the LTE BS implements multiple RF chains to support multiple frequency channels. If we shut down some transmit channels without affecting the capacity and coverage of cells, we can achieve the goal of energy-saving effectively. Our test results show that approximately 15% of energy consumption of RRU can be saved when the traffic load is 10%.

Small Base Station Turnoff

As the BS take up a large amount of energy consumption in cellular networks, turning off the BSs according to the traffic patterns is an effective way for energy-saving. We specify three energy-saving scenarios: inter-RAT, inter-eNB, and intra-eNB.

Inter-RAT energy-saving

In the inter-RAT energy-saving scenario, the LTE BS is used for capacity enhancement and overlays with the existing 2G and 3G networks and there are different methods to allow the BS to enter or leave sleep mode:

- The OAM method: e-UTRAN cells get into sleep mode according to the decision by centralized OAM.
- The signaling method: in order to achieve energy-saving gain, the energy-saving target cell and adjacent cells need to exchange specific network condition parameters, such as the traffic load threshold, the length of energy-saving time, and current energy consumption.

Inter-eNB energy-saving

As in Fig. 6.1, in the inter-eNB energy-saving, the E-UTRAN cells A and B provide wide area coverage, and small cells C–G are deployed within the initial coverage area to enhance capacity in some locations. It is also possible that the coverage area of the energy-saving cell and the coverage compensation cell do not overlap completely. When powering off the energy-saving cells, the coverage compensation cell needs to take measures to provide consistent availability to users in the area of interest, as shown in Fig. 6.2.

For the OAM-based method, the cells are configured by OAM as either an energy-saving cell or a coverage compensation cell. Besides the OAM-based method, the cells'

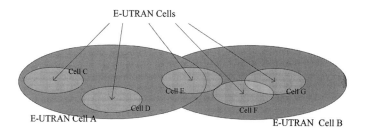

Figure 6.1 Inter-eNB energy-saving, where small cells C–G can be turned off.

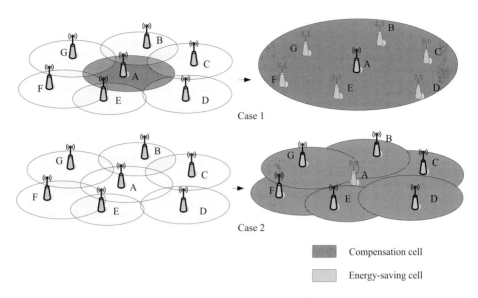

Case 1

Case 2

Compensation cell

Energy-saving cell

Figure 6.2 Illustration of coverage compensation in inter-eNB energy-saving.

role can also be dictated through the exchange of information signaling and/or OAM signaling.

Intra-eNB energy-saving

Intra-eNB energy-saving focuses on analyzing different methods that can be used by a single eNB. Possible solutions are:

- Multicast broadcast single frequency network (MBSFN) subframe configuration: When the normal frames change to MBSFN, power consumption can be reduced, since the overhead of the cell reference signal (CRS) in MBSFN is much reduced.
- The configuration of DwPTS length: for a TDD system, a special subframe includes DwPTS, UpPTS, and the guard period. A shorter DwPTS means less radiated power. Therefore, we can accomplish energy-saving by configuring fewer DwPTS when the traffic load is low.

6.2   Multi-RAT Cooperation Energy-Saving System (MCES)

In the previous section, we briefly introduced the concept and technique of energy-saving in a multi-RAT scenario. This section will put forth a novel system called MCES for effective improvement on the overall energy efficiency of mobile networks. MCES interacts with the RAN in real time, manner and can support multi-vendor equipments in 2G/3G/4G RAN, as shown in Fig. 6.3.

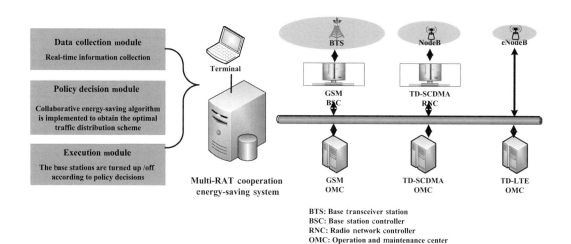

Figure 6.3 Topology of MCES.

6.2.1   Basic Principle

In recent years, with the rapid expansion of the customer base and soaring data traffic, the CMCC has deployed more than 1,500,000 TD-LTE BSs nationwide above the original TD-SCDMA and GSM BSs [17]. Current cellular wireless networks include GSM in 1,800/900 MHz, TD-SCDMA, and TD-LTE in D/F/E frequency bands. Multiple networks with overlapping coverage provide users with better Qos but also bring much-increased energy consumption. It is estimated that the CMCC consumes more than 20 billion kWh every year.

In this section, we first introduce the concept of network-level energy-saving. Network-level energy-saving technology refers to tight cooperation among BSs in single or multiple network topologies in order to improve the energy efficiency of the network [18]. To be specific, the saved power comes from putting some cells (identified as energy-saving cells) into idle mode and diverting its traffic to other cells (identified as compensation cells) [19]. An energy-saving cell usually has a traffic pattern with obvious tidal effect, i.e., the volume of traffic is very large during some time of the day and very small during the rest of day. The compensation cell provides basic coverage over the energy-saving cell, thus its coverage is generally slightly larger than the energy-saving cell. The relationship between the energy-saving cell and compensation cell is obtained by calculating the intercell coverage correlation, which is demonstrated in Fig. 6.4.

MCES is an application of the network-level energy-saving technology in a heterogeneous network consisting of GSM, TD-SCDMA, and TD-LTE. By collecting and analyzing the measurement report (MR) and the basic information on the 2G/3G/4G BSs, we have found that there are four types of overlapping scenarios in the network, which are listed in Table 6.1.

The technical precedure of MCES can be summarized as three sequential steps: energy-saving cell detection, energy-saving cell deactivation, and energy-saving cell activation:

Table 6.1 Four types of overlapped scenarios.

	Energy-saving cell	Compensation cell
Type 1	GSM 1800 MHz band	GSM 900 MHz band
Type 2	TD-SCDMA	GSM 1800 MHz/900 MHz band
Type 3	TD-SCDMA	TD-LTE cell D/F band
Type 4	TD-LTE cell D band	TD-LTE cell F band

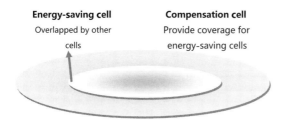

Figure 6.4 The coverage relationship between energy-saving cell and compensation cell.

1. Energy-saving cell detection: MCES collects the latitude, longitude, and MRs of all the BSs within the network of interest, and assigns the energy-saving cell-compensation cell pair. At the same time, MCES will continuously monitor the real-time traffic load of the energy-saving cell.

2. Energy-saving cell deactivation: When the energy-saving cell goes into a low-traffic load state, MCES can transfer its data traffic to the the compensation cells and deactivate the energy-saving cell.

3. Energy-saving cell activation: Thanks to the real-time monitoring function, MCES can wake up the sleeping energy-saving cell before high-data traffic arrives to ensure the QoS in the network.

6.2.2   Functional Architecture

MCES is divided into four modules: collection module, strategy module, execution module, and monitoring module, as illustrated in Fig. 6.5.

Collection module: It automatically collects all 2G/3G/4G cell performance and configuration information, MRs, system alarm information, etc.

Strategy module: A massive amount of data from the collection module is used as the input to the core algorithm for the purpose of guaranteeing the QoS in the network. First of all, according to different dimensions of data obtained, one of the two algorithms is selected to judge the relevance of cell coverage. Then, the combination of historical data and real-time data will be used to predict the traffic load of the energy-saving cells. In addition, the 2G/3G/4G network energy-saving parameters can be configured flexibly to meet the specific needs of different regions.

Execution module: According to the scheduling scheme generated by the strategy module, MCES can automatically complete the task of scheduling. Based on the specification of BS equipment, MCES can automatically send the instructions in parallel

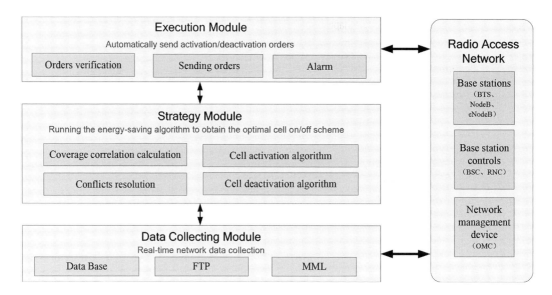

Figure 6.5 Functional architecture of MCES.

to the BS to shorten the execution time, and the BSs can automatically execute the deactivation and activation orders. Afterwards, MCES will keep tracks of the execution efficiency of such orders.

Monitoring module: MCES not only monitors processes, such as data collection, traffic prediction, and order execution, but also supervises performance statistics and alarm information in real time. If an abnormal event occurs, MCES will quickly provide the appropriate solution/action and notify the responsible parties.

6.2.3   Technical Characteristics

MCES is an energy-saving system based on real-time interaction with the RAN and is designed to support multi-vendor RAN equipments. MCES can turn off some of the overlapping cells in a timely manner to achieve network energy-saving in the scenario of significant coverage overlapping. Through analyzing massive MR data and the traffic profile, MCES can find the energy-saving cell and its corresponding compensating cell, and predict their traffic load trends. As for the energy-saving cell with low traffic load, MCES will migrate its traffic to the compensating cell and schedule the energy-saving cell to sleep mode. To ensure the reliability of the network, MCES also has a real-time monitoring function to turn on the sleeping cell when sudden high traffic occurs. Specifically, the MCES system has three major technical characteristics, which will be discussed next.

Network-Level Energy-Saving

Since traditional energy-saving technologies (e.g., micro DTX) are limited to single-network and single-cell scenarios, they cannot achieve optimal energy efficiency. MCES can oversee the wireless communication equipment that belongs to different networks

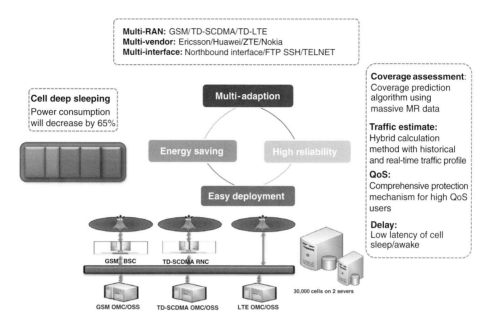

Figure 6.6 Technical characteristics of MCES.

and/or different vendors. Taking into consideration the overlay networks' respective status and the dynamic variations of traffic load, MCES can find the best actions for multiple networks and cells by analyzing the MR data.

Energy-Saving Cell Prediction Technologies Based on Big Data

MCES has access to cell configuration information, network performance statistics, and cell MRs, and could calculate intercell coverage correlation while locating the energy-saving cells and their corresponding cells in a multi-RAT network. The ML algorithm is used to make the accurate network traffic forecast for both types of cells through analyzing the historical traffic data and real-time traffic data.

Real-Time Cell On/Off Control

MCES can perform minimal periodic interaction with wireless network devices, obtain network performance data from the BS controller and OMC-R, and send instructions to the BS every 15 minutes. When the energy-saving cell goes into deep sleeping mode, its power consumption will immediately decrease by 65%.

6.2.4 Deployment Progress in China

Until now, MCES (see Fig. 6.6) has been deployed across about 7,000 cells in six provinces in China. The average annual electricity savings is 400,000 kWh per 10,000 cells.

In Nanning, for example, all 40,746 cells of CMCC's communication network can be accessed by MCES, where 6,704 cells could be turned off as energy-saving cells, and

4,880.2 kWh can be saved daily on average. It is estimated that MCES will help to save 1,781,300 kWh every year, which is equivalent to 1,780,000 yuan. In the deployment process, both drive test results and the network performance KPIs were monitored. The results indicated that the cell throughput has not been affected by using MCES.

In Shanghai, the trial exceeds 20,000 cells, where 6,108 cells were identified as energy-saving cells, and 1,243 kWh could be saved daily on average. If MCES can reach the whole network of Shanghai, it is expected to achieve a daily energy-saving of 6,173 kWh and an annual cost savings of 2,250,000 yuan.

6.3   WLAN Energy-Saving Technology

With the development of mobile internet and the popularization of smart terminals, the scale and number of active WLAN networks have been rapidly increasing. Therefore, the soaring energy consumption of the WLAN access system has become increasingly prominent and power-saving solutions are in urgent need [20–25].

In the WLAN access system, the energy consumption of the access point (AP) takes a large proportion of the total energy consumption, and the radio frequency (RF) module accounts for the largest proportion of the total device energy consumption (up to 80%). Therefore, WLAN access system's energy-saving strategy mainly focused on how to reduce the idle energy consumption of the AP RF module so as to reduce energy waste in idle status. Several simple energy-saving mechanisms are briefly discussed in the following subsections.

6.3.1   AP Device Shutdown

An AP device could be shut down by configuring the on/off switch during a specific time window in a specific area based on the tidal effects of the traffic load. The function can be realized by upgrading the WLAN network management system, improving the power supply switch structure, or introducing a third-party control module. Nowadays, the last method has been deployed over a small scale of current network, aiming to maintain the original ONU/PoE architecture unchanged. It can implement power supply control and remote power failure monitoring, with the advantages of easy realization, low cost, and easy deployment. Energy-saving expectation depends on the ratio of off time and on time, which is estimated theoretically as about 30%.

6.3.2   AP RF Channel Shutdown

As the RF module accounts for the largest proportion of AP energy consumption, AP RF channels could be switched on/off to provide flexible and dynamic control strategy. Its implementation is relatively simple, such as by adjusting the PA bias voltage. The energy-saving expectation depends on the ratio of the AP RF module power consumption and shutdown time, which generally is about 50%.

6.3.3   AP Single/Dual Band Selection

An AP single/dual band selection also helps to save energy in scenarios with tidal effects or intermittent service needs. Existing AP devices generally support 2.4 GHz and 5.8 GHz dual bands. According to a user's connection and signal strength, the 5.8 GHz band can be turned off or on automatically. The power consumption of the single-band APs could be reduced by at least 10% (depending on the different RF channel architectures) compared to that of the dual-band APs.

6.4   C-RAN Field Trials

6.4.1   Introduction

It has been pointed out that the key features of C-RAN include centralization, cooperation, and cloudification, as well as a clean system. With these features, C-RAN claims such advantages as cost reduction in terms of total cost of ownership (TCO), fast network deployment, improved system performance with support for cooperative technologies, on-demand and highly efficient resource utilization, and so on.

In order to demonstrate the features and verify the advantages of C-RAN, CMCC has been devoted to various field trials and prototype development since 2012.

The first step toward C-RAN is BBU centralization, which is relatively easy to implement and can be tested with existing 2G, 3G, and 4G systems. In the past few years, extensive field trials have been carried out in more than 10 cities in China using commercial 2G, 3G, and pre-commercial TD-LTE networks with different centralization scales. The main objective of C-RAN deployment in 2G and 3G is to demonstrate deployment benefits by centralization, including speedup of site construction and power consumption reduction. One example of the trial is in the city of Changchun, where 506 2G BSs in five counties were upgraded to a C-RAN-type architecture, i.e., centralized to several physical sites. In the largest central site, 21 BSs were aggregated to support 101 RRUs with a total of 312 carriers. It was observed that power consumption was reduced by 41% due to shared air conditioning. In addition, system performance in terms of call drop rate as well as DL data rate was enhanced using multiple-RRU-co-cell technologies. For detailed results and benefits demonstrated by centralization in 2G and 3G trials, readers can refer to [26, 27].

Similar results have also been verified in 4G LTE systems. However, a critical challenge for LTE C-RAN centralization in the past has been fronthaul (FH) transportation. FH is the link between BBU and RRUs. A typical FH interface includes CPRI and OBSAI, with CPRI being the most frequently adopted interface. FH transportation for 2G and 3G is not an issue, but becomes very challenging for LTE given the increasing number of antennas and bandwidth. Take CMCC's networks as an example. The RRUs are usually equipped with eight antennas with 20MHz of LTE bandwidth. As a result, the CPRI data rate is around 9.8 Gbps per carrier. It is clear that the larger scale of centralization, the higher the FH bandwidth is required. Therefore, fiber cores are required to enable C-RAN large-scale centralization, which unfortunately is rather expensive to most of the operators in the world due to the scarcity of fiber resources. FH transport had been an issue preventing C-RAN centralization.

Fortunately, with the advance in the transport area, low-cost FH transport solutions have gradually become feasible. In particular, WDM solutions have become more and more mature as mainstream FH solutions for 4G C-RAN centralization.

The basic idea of WDM is to use different wavelengths to transmit different signals within a single fiber core. There are two kinds of WDM solutions: passive WDM (p-WDM) and active WDM (a-WDM). Compared to the a-WDM solution, the passive solution does not need any power supply for the equipment. Therefore, it is more convenient for deployment; yet it usually offers less capacity than the active solution. For example, 12 wavelengths are typically supported by a set of p-WDM equipment while the capacity of active solutions is usually 48 wavelengths. In addition, compared to a-WDM, p-WDM is more stable and cost-effective.

6.4.2   Demonstration of WDM-Based FH Solutions

The test results for a-WDM-based C-RAN systems could be found in [28]. In this section, we will only provide the test results to demonstrate the applicability and the performance of the p-WDM solution for FH transportation. One of the reasons is that the p-WDM solution is the one widely adopted in CMCC's current 4G C-RAN networks. In addition, the tests are performed based on CMCC's commercial network, which could give the readers a more direct sense of how the real networks operate.

In the trial network, for each BS site, a pair of p-WDM boxes, called multiplexer/de-multiplexer, is needed. One box is put in the central office, together with the BBU pool, while the other is deployed on the remote site close to the RRUs. Instead of traditional gray optics, with the p-WDM solution, colored optical modules are needed on the BBU, RRU, and the multiplexer/de-multiplexer. Taking the UL as an example, the colored optics transform the radio signals to different colored wavelengths and multiplex them via the multiplexer. Then the multiplexed signal is transmitted via the single fiber core from the remote to the central site. In the central office, the wavelengths are de-multiplexed and distributed to different colored optics on the BBU where the signal will be processed. In CMCC's network, the data rate for each wavelength is 10 Gbps and each BS site consists of 3 RRUs with each covering one sector.

6.4.3   Test Methodology

The network in the test covers around 10 districts. In order to have a comprehensive test and evaluation of the p-WDM solution, in this test we consider and compare four scenarios: highway, suburbs, residential area, and enterprise buildings. For each scenario, the network consists of three types of deployment method: C-RAN, equipment room-based deployment, and outdoor cabinet-based deployment. The number of BSs for each deployment method and scenario is shown in Table 6.2. A set of performance indexes were tested, including RRC connection success rate, E-RAB connection success

Table 6.2 Number of BSs with different deployment methods for different scenarios.

	DI	D2	D3	D4	D5	D6	D7	D8	D9	D10	Total
Highway	1	3	6	0	2	2	0	1	0	2	17
C-RAN	0	2	2	0	1	0	0	1	0	0	6
Outdoor cabinet	0	0	2	0	0	1	0	0	0	0	3
Traditional equipment room-based deployment	1	1	2	0	1	1	0	0	0	2	8
Suburb area	1	1	0	2	0	1	0	1	0	0	6
C-RAN	1	1	0	0	0	0	0	0	0	0	2
Outdoor cabinet	0	0	0	1	0	1	0	1	0	0	3
Traditional equipment room-based deployment	0	0	0	1	0	0	0	0	0	0	1
Residential area	8	10	4	5	13	10	9	16	13	6	94
C-RAN	3	3	1	4	5	5	1	7	3	2	34
Outdoor cabinet	3	4	2	1	5	3	4	4	5	4	35
Traditional equipment room-based deployment	2	3	1	0	3	2	4	5	5	0	25
Enterprise	5	2	4	7	3	3	2	0	1	6	33
C-RAN	1	0	1	0	2	1	0	0	1	2	8
Outdoor cabinet	2	1	1	3	0	0	1	0	0	1	9
Traditional equipment room-based deployment	2	1	2	4	1	2	1	0	0	3	16
Total	15	16	14	14	18	16	11	18	14	14	150

rate, average number of RRC connection, the maximum number of RRC connections, throughput, packet loss rate, delay, RRC connection establishment delay, block error rate, and so on.

Test Results
Network Fault Analysis
We tested the network for five months and continuously collected the statistics, as shown in Table 6.3.

From that table, the C-RAN network acted similar to traditional deployment methods (i.e., equipment room-based deployment) and was better than the outdoor cabinet solution in terms of network breakdown. Note that network breakdown mainly happens due to power failure or BBU equipment malfunction. Since the BBUs in a C-RAN network were centralized in the same office, it was thus easier for the engineers to pin down the faulty parts. As a result, the average recovery time for C-RAN was much shorter. Furthermore, there is no malfunction caused by high temperature of C-RAN. This is mainly because in the current C-RAN deployment the number of centralized BBUs is fewer than five (i.e., five sites centralized), and the air conditioning in the facilities in the central office in particular were much better than traditional BS sites.

Table 6.3 Test results of the p-WDM-based C-RAN.

	Test item	Traditional equipment room-based deployment	Outdoor cabinet	C-RAN
Wireless	# of breakdown	79	16	73
	# of malfunction due to high temperature	4	1	0
	# of power failure	55	8	51
	# of BBU breakdown	20	7	22
	Average recovery time (hour)	19.9	4.2	12.4
Transport	# of breakdown	22	30	45
	# of optical link failure	19	29	42
	# of equipment breakdown	3	1	3
	Average recovery time (hour)	16.7	19.9	13.6

From the transport perspective, link errors happened much more frequent in C-RAN than on traditional sites. This is expected since in a C-RAN network the distance between the central office and the remote RRUs was much longer, leading to the increased possibility of link error. However, it turns out that the fault points were limited to some certain locations, which decreased the difficulty for fault detection and thus reduced the recovery time.

Performance Analysis

Through the five-month operation, we also collected extensive data to verify the performance of the p-WDM-based C-RAN networks. The data is shown in Table 6.4. Various indexes were collected from different perspectives. Note that there are two vendors involved in the test network by providing their C-RAN equipment. Both vendors' network was tested from a performance perspective.

From the test data in Table 6.4, the performance of p-WDM-based C-RAN is almost the same as the traditional deployment solutions.

6.4.4   C-RAN-Based UL CoMP Test

This trial was to demonstrate the performance gain of UL CoMP facilitated by the C-RAN architecture. The test was performed in a commercial network. The difference between inter-site and intra-site CoMP was also tested.

Test Environment and Network Configuration

The network in the test was a typical campus scenario that consisted of six 8T8R macro sites and one 2T2R small site. Here, "nTmR" means "n" transmission antennas and "m" receiving antennas. Each macro site contained three sectors while the small site was a single cell. In the central office, there were three BBUs, each controlling two macro

Table 6.4 System performance for the C-RAN network with p-WDM as fronthaul solution.

Vendor 1		C-RAN	Outdoor cabinet	Traditional deployment	Total
Network completion rate-related	# of RRC connection request	4,989	14,878	14,571	11,765
	# of successful RRC connections	4,984	14,865	14,556	11,754
	RRC connection success ratio	99.91	99.91	99.91	99.91
	# of E-RAB request	4,555	13,926	14,347	11,227
	# of successful E-RAB connection	4,551	13,914	14,336	11,217
	E-RAB connection success rate	99.88	99.91	99.91	99.90
	Network completion ratio	99.78	99.82	99.83	99.81
Service-related	Average # of RRC connection	4.76	15.73	14.79	12.06
	Maximum number of RRC connections	16.44	40.65	38.26	32.45
	Total U-plane cell throughput (DL+UL) (bytes)	4,98,869	13,35,139	13,23,356	10,76,453
User experience-related	Cell-level DL packet loss rate	0.02	0.03	0.05	0.02
	Cell-level average DL delay (ms)	22.47	27.64	26.72	25.75
	MAC-layer UL block error rate	0.40	0.33	0.33	0.35
	MAC-layer DL block error rate	0.08	0.09	0.07	0.08
	Average time for RRC connection establishment (ms)	25.01	26.34	25.52	25.64
Vendor 2		C-RAN	Outdoor cabinet	Traditional deployment	Total
Network completion rate-related	# of RRC connection request	5,072	10,107	22,102	13,286
	# of successful RRC connections	5,056	10,091	22,025	13,246
	RRC connection success ratio	99.73	99.79	99.80	99.78
	# of E-RAB request	4,591	8,634	20,979	12,213
	# of successful E-RAB connection	4,587	8,631	20,893	12,178
	E-RAB connection success rate	99.93	99.94	99.87	99.91
	Network completion ratio	99.66	99.74	99.66	99.69
Service-related	Average # of RRC connection	5.35	10.07	26.09	14.86
	Maximum number of RRC connections	21.95	27.23	53.24	35.65
	Total U-plane cell throughput (DL+UL) (bytes)	4,48,195	7,82,517	21,79,566	12,20,804
User experience-related	Cell-level DL packet loss rate	0.05	0.01	0.02	0.02
	Cell-level average DL delay (ms)	31.78	33.19	35.36	33.63
	MAC-layer UL block error rate	0.23	0.34	0.22	0.26
	MAC-layer DL block error rate	0.29	0.26	0.21	0.25
	Average time for RRC connection establishment	24.79	26.66	29.27	27.14

Table 6.5 System configuration.

Parameters	Value
Frequency	F and D band
Bandwidth	20MHz
Frame structure	• Configuration 2, Normal CP • Subframe configuration: DSUDDDSUDD
CFI	3
Antenna mode	DL：TM3/8 self-adaptive UL：SIMO
UL power control	On
HARQ	On
AMC	On
BS transmit power	$8 \times 5W$
Handover	Contention-based

sites. In addition, there was a high-speed switch in the central office to connect the BBUs together for the sake of timely information exchange. Each macro site consisted of three sectors, with each sector having three cells on different frequency bands. As a result, the total number of macro cells was $6 * 3 * 3 = 54$. Together with a small cell, there were 55 cells in total in the trial network.

To have a better observation of the CoMP performance gain, the data was collected and analyzed during busy hours of the day. The system configuration is as shown in Table 6.5.

The test methodology is shown as follows.

• First, turn off the switch and any CoMP algorithms, then collect the statistics for one week;
• Second, turn on the intra-site CoMP while keeping the switch off and collect the statistics for one week;
• Third, turn on the inter-site CoMP and the switch to interconnect the BBUs, then collect the data for one week.

Note that as mentioned earlier, the data was collected during busy hours, which typically lasts for two hours.

Test Results

Without enabling any CoMP algorithms, the baseline performance is shown in Table 6.6. The performance gain with intra-site CoMP and inter-site CoMP open is shown in Table 6.7. It could be seen that CoMP has brought slight gain in terms of average UL cell spectral efficiency and average UL user experience rate. However, compared to intra-site CoMP, the performance by inter-site CoMP, which is enabled by the C-RAN architecture, is much higher.

When it comes to the performance gain for edge users, as shown in Table 6.8, the intra-site CoMP still brings a small gain. In comparison, C-RAN-enabled inter-site CoMP has much higher performance gain.

Table 6.6 System performance with CoMP function off.

Average UL interference (dBm)	Average UL MCS	Average cell user number	Edge user (%)	Average utility rate for UL PRB (%)	Average throughput per E-RAB (KB)	Average UL throughput per PRB (bits)	Average UL throughput per PRB for edge users (bits)
-114.28	20.78	34.15	19.85	5.69	325	306	201

Table 6.7 Cell average UL CoMP gain.

UL CoMP	Average gain on cell UL spectral efficiency	Average gain on cell UL user experience rate
Intra-site UL COMP	2.42%	0.89%
Inter-site UL COMP	4.78%	10.91%

Table 6.8 UL CoMP gain for edge users.

ULCOMP	Average gain on UL spectral efficiency for edge users	Average gain on UL user experience rate for edge users
Intra-site UL COMP	2.90%	3.47%
Inter-site UL COMP	8.84%	17.25%

6.5 Green Application

As consumers rely more and more on mobile devices and applications (apps), their expectation of network connectivity is very high. Background services, software, and accessing data over the network are all part of the user experience. Both for smartphones and tablets, there is one common user focus: better battery life on mobile devices. In order to ensure that consumers get the longest battery life from their devices, feedback must be provided from all involving parties (network integrators, consumers, developers) (Fig. 6.7).

Consumers expect a device to feature outstanding battery life, as promised by the vendors. A typical claim is at least eight hours' usage on Wi-Fi, or four hours on LTE. Developers expect their application not to drain the battery too fast. Meanwhile, streaming media applications must meet users' expectations for performance. Network integrators focus on how much network activity is used by an application and how much signaling is generated. Research indicates that smartphones consume the same amount of data as feature terminals, while network signaling consumption is more than 10 times that.

To achieve better user experience, we believe a green application should not only be functionally complete, stable, and attractive but also power efficient. A green application will prolong battery life and generate less heat.

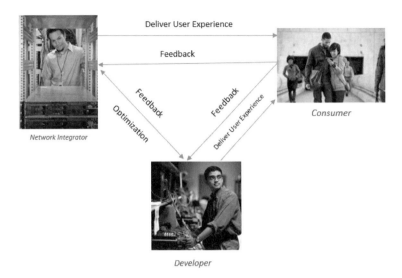

Figure 6.7 Consumer, developer, and network integrator dependencies.

This section describes an E2E system for green applications and intelligent network design, in order to optimize performance and minimize power consumption. The vision of this chapter is to construct an app-friendly ecosystem and a green community to help the whole industry understand how terminals, batteries, networks, and apps work together so as to guide developers to use a more efficient and friendly development mode. Consequently, both user experience and network efficiency can be improved.

6.5.1 Key Factors for App Power Consumption

The hardware architecture of a smartphone can be generally divided into three areas: application processor, communication processor, and peripherals. The application processor generally includes a central processing unit (CPU)/graphics processing unit (GPU), memory, etc. The communication processor mainly refers to network-communication-related modules. Peripherals are the screen, speaker, camera, sensors, and other parts.

Different types of smartphones in different operation modes have different power consumption ratios for the application processor, communication processor, and peripherals. For casual game playing, due to the short interaction time and few data exchanges with the server during the run time, the application processor plays an important role in power consumption. Therefore, for a smartphone running a casual game, the main performance optimization object is the application processor part. However, for apps such as online video, which requires constant high-rate network connection, the optimization focus should be shifted to the communication processor.

In the following subsections, we will separately discuss the impacts of the application processor and communication processor in terms of power consumption.

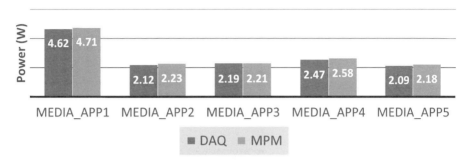

Figure 6.8 Streaming media power consumption in watts.

Application Processor

Consumers want smartphones to provide a great user experience with excellent battery life. Research indicates that battery life has become the single biggest reason people choose smartphones [29]. It is common that the performance of a best-in-class device is degraded by the application running on it. A poorly optimized application could destroy a good user experience. To understand how an application can impact the battery life, we have profiled various categories of applications for power consumption, including media streaming, casual games, and camera-based applications. For measurement, we use National Instrument's data acquisition (DAQ) and Intel's MPM. The goal of combining both hardware and software tools is to allow us to correlate power between the two different collection modes: application and hardware.

Category 1: Streaming Media
Media playback on smartphones or tablets is one of the top usages by consumers. Nearly 90% of usage on tablets is from media content [30]. To understand the impact of power consumption for media we tested similar video clips running at 1080p quality. There was a slight difference in quality, depending on the service provider who was streaming the video. Figure 6.8 shows the power consumption of several anonymous applications during media playback. Lower power indicates a longer expected battery life. MEDIA_APP2 can perform streaming and playback for 10 hours on a 20 Whr-battery while MEDIA_APP1 can run for less than 5 hours.

Category 2: Casual Game Play
Compared to streaming media, games are more computationally intensive. To understand the impact, we tested games with similar action-oriented game play. Figure 6.9 shows the power consumption of arcade games during normal usage.

Category 3: Camera Applications
More and more customers use smartphones and tablets for taking photos or videos. Even though the click takes hardly any time, the user might spend more time in the preview.

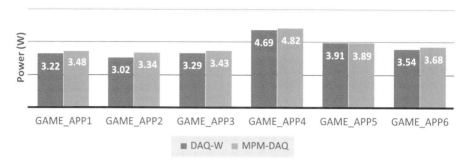

Figure 6.9 Casual game power consumption in watts.

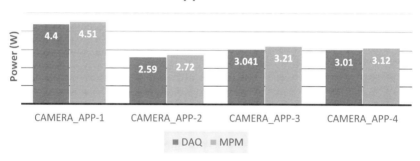

Figure 6.10 Camera application power consumption in watts.

Some applications add more editing features while others have predefined settings. We used a black-and-white effect to keep our data consistent between different applications. Figure 6.10 shows the power consumption of camera apps.

Communication Processor

Many mobile app developers design mobile communication similarly to how they would design for a wireless network, resulting in unfriendly network behaviors, such as higher power consumption, sustained mobile network signaling interaction, frequent small data transaction, and so on. They lack an appropriate understanding of the mobile network mechanism and the specific behavior of apps in a mobile network [31].

The key factor affecting application performance and network energy efficiency is the radio resource control (RRC) state machine whose purpose is to efficiently utilize limited radio resources and to improve smartphone's battery life. Application traffic patterns trigger RRC state transitions, which in turn affect radio resource utilization, smartphone power consumption, and user experience. Developers need to be aware of the reason and effect behind RRC state transition behavior [32–36].

6.5.2 Optimization

There is no specific manual for optimizing applications, but there are guidelines that a developer should follow and then implement necessary changes in their applications. Note that different categories of applications will require different optimization. Different optimization methods can be adapted according to different application categories, as the following list shows.

Media: (Including streaming media, local media, audio streaming, photography, and video editing)

- Using the graphics and media pipeline; offloading to the GPU provides significant benefits for power and performance.
- Buffering during playback, record, or audio streaming; combining frames during processing saves significant power on the CPU and other components, such as storage and network access.
- Running simultaneously between CPU and graphics; overlapping calls helps to keep the CPU and GPU active at the same time.
- Using proper codecs and reducing total bandwidth; this results in running at a lower frequency and thus will make for lower power consumption.
- No overdrawing the user interface (UI) for the application; re-rendering the UI can cause unnecessary GPU usage.

Browsing: (Including watching media and casual browsing)

- No rendering the text or photo on the browser; this impacts graphics performance.
- No updating JavaScript animation during browsing.
- No buffering data over the network for advertisements and background processing.

Casual Gaming

- Using graphics for rendering: including UI rendering.
- No using the texture of very large sizes or unsuitable sizes (such as 13x19).
- Arranging draw call sequence with depth from near to far.
- No drawing regions invisible to end users.
- No doing alpha blending if unnecessary.
- Avoiding using branches in shading language if unnecessary.
- No using native screen resolution if up-sampling is acceptable since higher resolution can impact GPU usage.

Social, sharing, and emails

- Push notifications and background services causes high network usage and power.

Generic optimizations

- Optimizing timers, sleep, and spin to save power and increase performance.
- Minimizing background services updates to use less cache memory.
- No re-rendering UI that will impact frame-per-second and graphics utilization.

6.6   "Invisible BS"

6.6.1   Motivation

In current cellular networks, there are a large quantities of BSs, including legacy 2G, 3G, and current 4G BSs. From the perspective of a network operator, a large portion of its capital expenditure (CAPEX) and operating expense (OPEX) lies in the purchase, deployment, and maintenance of such BS equipment. Moving forward with new BS technology, such as large scale antenna systems (LSASs), we can see capacity and power consumption being improved. However, the physical form factor and weight of a LSAS BS are of particular concern. The commercial deployment of new BSs faces resistance from the public, especially commercial property owners, regarding the aesthetics and potential/perceived health issues due to constant exposure to electromagnetic waves. The increasing footprint of the BSs will not only bring significant tower construction challenges but also greater confrontation. Moreover, it is more difficult to change the deployment site once chosen, which does not fit well with the flexibility requirement of next-generation communication networks.

In order to tackle these challenges, the concept of "invisible BS" has been proposed. By disguising the antenna system, the BSs can be made invisible to the public. A previous approach was to construct towers as fake trees, which are actually quite obvious. We modularize the BS, and design multiple antenna elements in the form of "tiles." By separating the LSAS panel into multiple tiles, one can flexibly deploy antennas in an irregular shape and as a part of the building facade or signage, thus blending into the surrounding environment. Another distinguishing feature is that low-power, low-cost RFIC is adopted in the "invisible BS" for further RF power-saving.

Figure 6.11 shows the system architecture of CMCC's hardware platform for an invisible BS. This platform most notably consists of a high-capacity software-configurable baseband unit and modular active antenna arrays, which is called SmartTile. The software-configurable baseband platform is responsible for baseband and physical-layer processing, particularly multiuser beamforming. Each SmarTile module contains eight RF chains and eight dual-polarized antenna elements. The baseband unit is connected to the SmartTile via the common public radio interface (CPRI) interface.

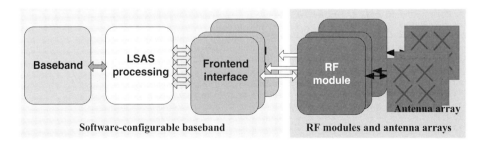

Figure 6.11 System architecture of CMCC's hardware platform for "invisible BS."

The purpose of this system is to investigate the feasibility of invisible BSs and to verify different key 5G technologies, including massive MIMO, ultra wideband operation, RF requirements on large antenna arrays, OTA requirement, as well as highly effective heat dissipation techniques. There are several benefits to designing the LSAS system in this integrated manner:

- Dramatically increase system capacity by deploying multiple SmarTiles, equipped with many antennas and RF chains.
- Compact modular design, small form factor, easy deployment (e.g., on external building walls or logos), blends with surroundings, flexible site selection.
- Reduce power consumption due to the use of highly integrated, low-cost and low-power RFIC.
- Flexible arrangement of multiple modules.

6.6.2 Powerful Baseband Platform with a Unified Design

It is a great challenge for the 5G baseband platform to handle the amount of data generated from a large number of antenna elemnts (128/256/512) and a wide carrier bandwidth (e.g., above 100MHz). It is roughly estimated that tens of Gbits of data needs to be processed per second and several hundred Gbits of data needs to be transmitted per second. Our powerful baseband processing unit is designed with those high requirements in mind, using a universally acceptable design framework. A detailed system diagram is shown in Fig. 6.12.

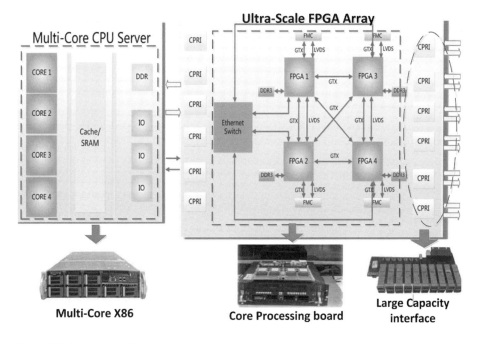

Figure 6.12 Baseband architecture.

For the baseband platform, some other testbeds have also been proposed for massive MIMO verification. Wireless open-access research platform (WARP) [37] is a unified wireless network testbed for education, whose programmability and flexibility excels in implementing wireless communication systems quickly and cost-effectively. It has been used by Rice University in its ArgosV1 and ArgosV2 platforms [40]. However, such massive MIMO implementation becomes diffcult when the number of antenna elements increased significantly since the fronthaul interface and real-time baseband processing are the bottleneck. Lund University, in collaboration with National Instruments, has also developed a testbed named LuMaMi [37] for massive MIMO verification. It has the processing ability to support more than 100 antenna elements/RF chains for real-time wireless transmission [38]. The architecture design of the system is mainly based on FPGA, therefore, its compatibility with other platforms and softwares is rather weak.

Compared to other platforms, the interoperability and processing ability have both been taken into consideration for the design of the baseband platform described in Fig. 6.12. The baseband platform has a two-part architecture. The first level is the common processing architecture and the second is the boosted processing architecture. Based on this, the platform can support over 10 Gbps data processing and 800 Gbps fronthaul transmission. With the powerful hardware processing ability, various 5G key technologies can be verified, such as massive MIMO, new waveform, NoMA, etc.

Firstly, the platform has a multicore x86 server for tasks that are less computationally demanding, such as protocol stack processing and certain simple physical layer functions, since x86 CPU is good for task scheduling and multitask procedures. The obvious benefit here is the ability to place network functions and different implementations of protocol stacks in a flexible way and on the same server to ensure multi-vendor compatibility. Secondly, the boosted processing architecture focuses on highly complex signal processing, fast data exchange, and transmission. Therefore, it is mainly made of high-speed processing chips, such as FPGA, and super fast transmission interfaces, such as quad small form factor pluggable (QSFP+). It is based on the standard design architecture ATCA (advanced telecom computing architecture), and comprises an exchange processing board, a high-performance core processing board, and a high-capacity interface board. The exchange processing board mainly focuses on the scheduling and interactive processing of data; the high-capacity interface board is used at the connection between the RRU with interfaces, such as CPRI and serializer-deserializer (SERDES). The high-performance core processing board is the main computation hub within the BBU and is responsible for digital signal processing in the UL and DL transmission.

The high-performance core processing board, shown in Fig. 6.11, integrates with four high-performance FPGA chips of the Xilinx V7 series. The FPGA has abundant hardware resources and can handle complex baseband signal processing, such as high-order MU-MIMO, matrix decomposition, and channel estimation. For example, the calculation of inversion and singular-value decomposition (SVD) in matrix decomposition involves iterative complex calculation, such as multiplication, division, and extraction of the square root. The amount of computation required will likely increase with the antenna number. Large-scale real-time processing requires 128-IFFT/FFT processors

Figure 6.13 Baseband core processing board.

working at the same time. Each FPGA can provide millions of logical operations and thousands of DSP operations. As such, the core processing board with four such FPGA chips can provide baseband processing for massive MIMO systems. Besides, there is a total of 96 pairs of GTX (gigabit transceivers), providing 800 Gbps fronthaul transmission capability (theoretically equal to the data amount from 512 antennas), and LVDS (low-voltage differential signaling) for connections between the FPGAs and the ATCA back board.

To summarize, on one hand, it can support the complex calculations of massive MIMO. On the other hand, the x86 common server makes it much more flexible and compatible with different software providers and enables convenient software upgrade in the future.

6.6.3 SmarTile

The traditional BS baseband processing unit and antenna systems are highly integrated. However, we are seeing more and more benefits of heavy data processing distribution and BS deployment flexibility. Our SmarTile is an active antenna system that is composed of RF chains, antennas, self-calibration networks, and a baseband processor. It is among the first to realize such a compact modular system in the industry. Each unit can be used as an eight-antenna TD-LTE RRU, or an arbitrary number of units can be combined as a LSAS to suit diverse environments. Such design has been widely adopted in commercial equipment by vendors. The feature of being super compact (only 1.28L) and lightweight makes it rather portable and easy to mount on walls, roofs, and other structures such as lampposts, and thus well blended into the surrounding environment. Additionally, the superior energy efficiency demonstrates our ideal of being "green," which is worth doing not only from an economic sense but also from the environmental sense. Table 6.9 lists some of the most important parameters of the SmarTile.

The SmarTile is so highly integrated that all on board components are assembled by hardware connections. The hardware structure of the SmarTile is shown in Fig. 6.14.

Table 6.9 Detailed specifications of the SmarTile.

Frequency band	1880–1920MHz
Channel band	20MHz
RF chain number	8
Antenna number	8
Output power per chain	21dBm
Antenna gain	6dBi
Size	$16 \times 16 \times 5$cm
Power consumption	25 W

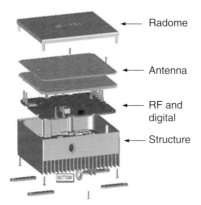

Figure 6.14 Structure of the SmarTile.

Hardware Design

The SmarTile's hardware is designed to reduce power consumption as much as possible. In this section, we look at a couple of design features of the SmarTile and explain in detail why it is unique.

RF Chain

To reduce the complexity of RF design, we choose a zero intermediate frequency (ZIF) structure to build the RF chain. A highly integrated transceiver is adopted, which consists of ADC, DAC, LO, IF amplifier, IF filter, and modulator. The ZIF structure can sample the signal with a lower rate, which can dramatically reduce the power consumption of ADC and DAC. Due to the phase and amplitude imbalance of the signal, the side image and DC offset are the most disadvantageous aspect of ZIF structure RF design, as they cannot be removed by filtering and must be calibrated out by a closed-loop algorithm from the baseband. This transceiver has the ability to remove the side image and DC offset, which makes the baseband design much easier. Each transceiver supports two RF chains with lower power consumption performance. The power amplifier and low-noise amplifier used in this platform are commercial components that are originally designed for user equipment, hence they have better power efficiency compared to BS amplifiers due to the power consumption limitation on user terminals.

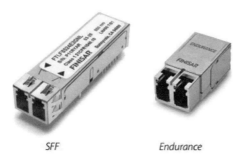

SFF   Endurance

Figure 6.15 Compact SFP optical module.

Digital Hardware

The FPGA is used to process base-band data and is integrated with an ARM core; therefore no extra MCU is needed. The FPGA also contains two 10 Gpbs SERDES, which are responsible of transforming serial data to parallel data and vice visa. The interface between BBU and AAS is optical fiber via a 10 Gbps SERDES.

The majority of the functions at the intermediate frequency, such as DDC and DUC, are performed by RF transceivers, which also helps reduce power consumption.

Again for the purpose of size reduction, we have chosen a compact SFP optical module (shown in Fig. 6.13). Compared to the typical SFP module, the size is reduced from 56.5 mm to 26.41 mm.

Antenna and Calibration Network

The SmarTile has eight RF chains and eight antenna elements, as in shown Fig. 6.16. One patch antenna has two logical elements and is +/− 45-degree dual-polarized. Each element has a 6-dBi antenna gain with an 85-degree HPBW in both horizontal and vertical direction. The distance between patches is half the operating wavelength. We have measured the overall RF performance of this platform, and such a compact array has good array gain and side lobe suppression.

A calibration network is used to make sure all the RF chains have the same phase and amplitude output. The calibration network couples DL/UL signals from/to each antenna element. For better beamforming performance, the phase error should be less than five degrees, and the amplitude error should be less than 0.5dB.

Power Consumption Comparison

SmarTile is designed to be low-power and of small form factor. To demonstrate our superior power-saving performance, Tables 6.10 and 6.11 compare to a commercial eight-antenna BS.

Due to the lower power consumption, the size of the heat sink can be reduced to make the BS even smaller and lighter. It also reduces the cost of electricity and ultimately improve the carbon emissions.

Table 6.10 The RF power consumption for commercial RRU.

Component	Power consumption	Number
Amplifier	100 mW	8
Mixer	100 mW	4
IQ modulator	200 mW	8
ADC	1.5 W	2
DAC	1 W	4
Total	9.8 W	

Table 6.11 The RF power consumption for SmarTile.

Component	Power consumption	Number
Transceiver	1.1 W	4
Total	4.4 W	

Table 6.12 RF performance for a SmarTile.

Test item	Specification
Output power	21.005dBm
ACLR upper	−47.25dB
ACLR lower	−48.56dB
EVM 64QAM	2.6%
EVM QPSK	1.13%

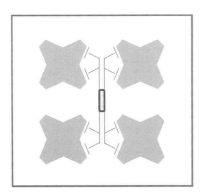

Figure 6.16 Antenna element and calibration network.

SmarTile RF Performance

The DL RF parameter is measured through the antenna ports by a spectrum analyzer. The test mode was ETM 3.1. The test results are shown in Table 6.12 and listed as follows.

Figure 6.17 shows the output power and the adjacent channel leakage ratio (ACLR) of the SmarTile, which is a key performance indicator (KPI) of the BS's transmitter quality. The output power of each RF chain is 21dBm with a bandwidth of 20 MHz,

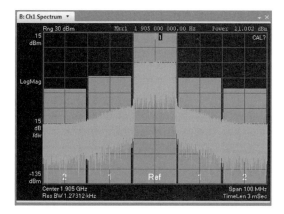

Figure 6.17 The output power and ACLR for a SmarTile.

and the theoretical output power of the whole BS is 30dBm. The measured ACLR was 47.25dBc, better than the 3GPP standardized requirement, which is −45dBc.

6.6.4   Flexible Over-the-Air Calibration Scheme

In the LSAS, the DL beamforming weights are derived from the UL channel estimation. For TDD systems, the channel information of UL can be used for DL precoding calculation because of UL and DL channel reciprocity. This is considered as the most significant advantage over FDD systems, where training pilots waste radio resource on the DL. Furthermore, the computation and overhead become prohibitive as the BS antenna elements grow large. However, for TDD systems, the RF paths of transmitter and receiver are asymmetric, and such discrepancies in the frequency and magnitude response leads to unacceptable channel mismatch between UL and DL. Moreover, the beamforming weights calculated based on the UL channel will no longer match the DL actual channel anymore, resulting in performance degradation. Therefore, TDD large-scale antenna system requires UL and DL calibration. The most common method is to estimate such difference beforehand and, periodically, to compensate the deviation in the DL beamforming parameters.

The mainstream antenna calibration scheme currently uses external measurement devices. The hardware calibration network is implemented physically for all RF paths [39, 40]. By performing predefined measurement procedures, the related parameters of each RF path, such as amplitude, phase, and frequency, can be obtained, which can be used to calculate the calibration coefficients for the BS. The architecture of the calibration network is shown in Fig. 6.18.

However, there are several drawbacks for this method. Firstly, extra hardware devices are needed, which means extra engineering actions, such as installation and debugging. Secondly, it will be a great challenge to calibrate antenna elements that are distributed far from each other physically. Moreover, the cost and complexity of such a hardware

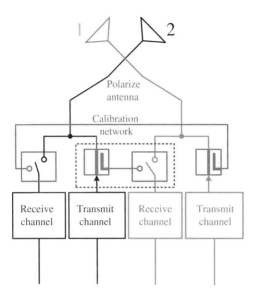

Figure 6.18 Antenna calibration network.

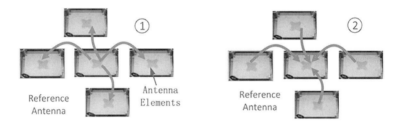

Figure 6.19 Current over-the-air calibration scheme.

network will increase with the number of antennas. Therefore, the old calibration network approach is no longer feasible for the distributed massive MIMO systems, which are envisioned as the most likely deployment scenarios for invisible BS.

Calibration over the air is the most promising scheme and was originally proposed in [41]. In this method, the reference signal for calibration interacts with wireless media other than the complex cabled network. The main step is to choose a reference antenna in advance, and the calibration signal will be transmitted and received between the reference antenna and all other antennas sequentially. Then, the calibration factor can be computed and compensated in the DL beamforming procedure. As in the Fig. 6.19, the antenna in the center is the reference element and the other four antennas around it are the to-be-calibrated antennas.

However, this OTA method still has some problems. The reference antenna is usually chosen as the center element of the array. The performance of the calibration is closely

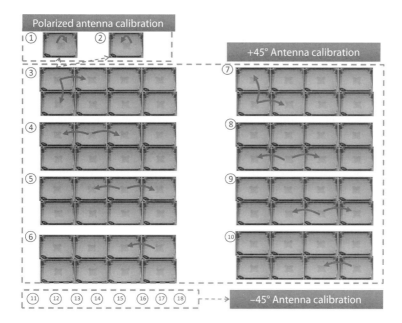

Figure 6.20 The procedure for flexible calibration scheme.

correlated with the location of the reference antenna. However, for a large antenna array, the distance to the reference antenna can be quite different for different elements. Besides, for the invisible BS, the distribution of antennas is irregular and the central element cannot not always be found.

To combat this issue, a flexible OTA calibration scheme is proposed for the massive MIMO system with an irregular antenna array in the invisible BS use case. The method is first proposed and published in [42]. The basic idea of this scheme is that the reference antenna is not fixed. Each RF path will send the calibration signal consecutively and the nearby antenna will receive it. Then the reciprocity factor can be calculated in each transmit-and-receive procedure. By a series of calculations, the calibration factor can be deduced. An example of eight SmarTiles (each SmarTile has a set of dual-polarized antennas) is given in Fig. 6.20.

In this example, several procedures are implemented and the whole process is introduced in detail:

1. The two polarized antenna calibration inside one SmarTile described as ① − ②;
2. The antenna calibration with 45 degrees polar, described as ③ − ⑩;
3. The antenna calibration with −45 degree polar, described as ⑪ − ⑱;
4. The calibration factor can be calculated and deducted;
5. The compensating factor is added in the DL beamforming.

The flexible calibration scheme has been verified to be very effective. For the different shapes of the antenna arrays, the calibration scheme performs well. The results

Table 6.13 The performance of flexible air calibration.

Amplitude deviation	Phase deviation
0.0272dB	3.8842°

in Table 6.13 come from the real test in the system. It indicates that the performance of the proposed OTA calibration method can meet the requirement of LTE well. It can be used in a regular and irregular distributed antenna array without an external calibration network and reference antenna. It is first proposed and verified in the real testbed system and can be a potential scheme for commercialized devices. Moreover, for the standardization of 5G in 3GPP, lots of discussions[43–45] have been made for calibration in the Release 15 agenda item 7.1.2.5, and this scheme provides some reference and support for the discussions, which is very meaningful for communication systems in the future.

6.6.5   High-Efficiency Heat Dissipation Testbed

The designed power consumption of a SmarTile is 25 watts. Due to its small form factor, it is hard to dissipate the heat generated by operating with such high ouput power. Therefore, we have used a SmarTile and built a prototype to verify some new emerging heat-dissipation techniques.

Heat Dissipation Testbed

The prototype is composed of the SmarTile and temperature monitor system described in Fig. 6.21. Different thermal dissipation technologies can be verified on this prototype by looking at the temperature changes of the key components on the SmarTile, such as FPGA and clock modules.

The data processing board is a baseband prototype, which controls transmission and reception of data to/from the SmarTile. The reference signal source on the processing board supplies the whole system clock. Thermal couple wires were used to measure the temperature of each IC component.

For an outdoor BS, the main method for thermal dissipation is through natural heat radiation. There are three main types of material used to aid thermal radiation design: interface material, including thermal conductive adhesive, thermal pad, and phase change material; thermal-averaging material including graphite sheets and flat plate heat pipe; and a thermal radiation shell. Our prototype is used to perform thermal design research and test these three materials, as described in Fig. 6.22.

Interface Material

Interface material transfer/conducts thermal heat from the contact interface to the outside. In an ideal situation, the thermal source and the interface of the radiator will have the same temperature; that way, heat radiated by the source can be dissipated

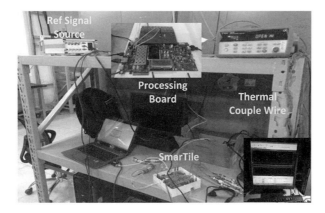

Figure 6.21 Heat dissipation testbed.

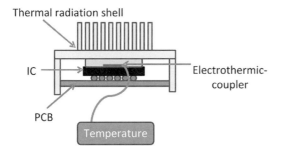

Figure 6.22 The thermal design of the prototype.

quickly. For better thermal dissipation, the interface material should have higher thermal conductivity.

An ordinary silicon-based thermal pad has a thermal conductivity of 2.0 W/m-k. In order to enhance thermal conductivity, we adopt a new interface material based on graphene with a thermal conductivity of 15.0 W/m-k as of now, which could significantly improve thermal performance. The lab test in Table 6.14 also verified such improvement.

Thermal Radiation Shell

The heat produced in the BS should be conducted to the outside shell effectively, and then heat can dissipate in the air through thermal exchange. Hence, the thermal resistance should be as low as possible. One way to lower the thermal resistance is to enlarge the size of the shell, and the other method is to increase the thermal conductivity of the shell. Because the size of the SmarTile is fixed, we strive to find a metal material that has higher thermal conductivity. Table 6.15 shows two types of aluminum alloy that have different thermal conductivities.

Table 6.14 Thermal performance comparisons of different materials.

Thermal pad	FPGA (°C)	CLOCK (°C)	Transceiver (°C)
Silicon	61.95	58.24	61.38
Grapheme	53.41	54.90	59.57
Delta	**8.54**	**3.34**	**1.81**

Table 6.15 Thermal conductivity of two types of aluminum alloy.

	thermal conductivity（W/m-K）
Ordinary aluminum alloy	90–110
High-conductivity aluminum alloy	150–180

Through simulation by adopting high-conductivity aluminum alloy, the temperature of FPGA had a reduction of 1.5°C.

Thermal-Averaging Material

The thermal-averaging material is used to eliminate very high temperature points. For example, the graphite sheet is used widely in terminal design. In the future, the thermal-averaging material based on graphene will have even better thermal conductivity. And we hope that this kind of material will be used in the BS to further facilitate better heat dissipation performance.

References

[1] M. S. Ilyas et al., "Low-carb: Reducing energy consumption in operational cellular networks," *2013 IEEE Global Commun. Conf. (GLOBECOM)*, Atlanta, GA, pp. 2568–2573.

[2] T. Bitzer, J. Achard, and A. Pascht, "New energy-saving multicarrier transceivers and their standardization," *Bell Labs Tech. J.*, vol. 15, no. 2, pp. 53–58, Sept. 2010.

[3] C. Lubritto et al., "Telecommunication power systems: Energy saving, renewable sources and environmental monitoring," *IEEE 30th Int. Telecommunications Energy Conf.*, San Diego, CA, pp. 1–4, 2008.

[4] L. Anaya, D. Valerdi, L. Lezhan, X. Wenbo, and M. Torres, "Field validation of smart energy saving features in a GSM network," *2nd IEEE PES Int. Conf. and Exhibition on Innovative Smart Grid Technologies*, Manchester, 2011, pp. 1–5, 2011.

[5] A. Vidal, M. C. Purisima, M. T. Perez et al., "GSM timeslot detection and switching for power amplifier duty cycling in community cellular networks," *2017 Int. Conf. on Computer, Inf. and Telecommunication Systems (CITS)*, Dalian, China, pp. 200–204.

[6] Q. Lv, "A milestone of GSM BTS: Application of multi-carrier technology in GSM," *ZTE Technologies*, no. 2, pp. 1–5, 2010.

[7] F. Li, Y. Zhang, and L. Li, "Enhanced discontinuous reception mechanism for power saving in TD-LTE," *3rd Int. Conf. on Computer Science and Inf. Technol.*, Chengdu, China, pp. 682–686, 2010.

[8] A. Virdis, G. Stea, D. Sabella, and M. Caretti, "A distributed power-Saving framework for LTE HetNets exploiting almost blank subframes," *IEEE Tran. on Green Commun. and Netw.*, vol. 1, no. 3, pp. 235–252, Sept. 2017.

[9] Y. B. Lin, L. C. Wang, and W. C. Chen, "eSES: Enhanced simple energy saving for LTE HeNBs," *IEEE Commun. Lett.*, vol. 21, no. 11, pp.2520–2523, Nov. 2017.

[10] P. K. Wali and D. Das, "Enhanced-power saving semi-persistent scheduler for VoLTE in LTE-Advanced," *IEEE Tran. on Wireless Commun.*, vol. 15, no. 11, pp. 7858–7871, Nov. 2016.

[11] V. J. Kotagi, R. Thakur, S. Mishra, and C. S. R. Murthy, "Breathe to save energy: Assigning downlink transmit power and resource blocks to LTE enabled IoT networks," *IEEE Commun. Lett.*, vol. 20, no. 8, pp. 1607–1610, Aug. 2016.

[12] L. You, L. Lei, and D. Yuan, "Optimizing power and user association for energy saving in load-coupled cooperative LTE," *2016 IEEE Int. Conf. on Commun. (ICC)*, Kuala Lumpur, pp. 1–6.

[13] K. Kanwal and G. A. Safdar, "Reduced early handover for energy saving in LTE networks," *IEEE Commun. Lett.*, vol. 20, no. 1, pp. 153–156, Jan. 2016.

[14] R. Vassoudevan and P. Samundiswary, "Performance analysis of DRX power saving technique for LTE based UE under bursty Web traffic," *2015 Int. Conf. on Commun. and Signal Process. (ICCSP)*, Melmaruvathur, India, pp. 924–928.

[15] A. Prasad and A. Maeder, "Energy saving enhancement for LTE-Advanced heterogeneous networks with dual connectivity," *2014 IEEE 80th Veh. Technol. Conf.*, Vancouver, BC, pp. 1–6.

[16] R. Imran, M. Shukair, N. Zorba, O. Kubbar, and C. Verikoukis, "A novel energy saving MIMO mechanism in LTE systems," *2013 IEEE Int. Conf. on Commun. (ICC)*, Budapest, pp. 2449–2453.

[17] CMCC, "China Mobile sustainable development report 2016." http://10086.cn/download/csrreport/cmcc_2016_csr_report_full_cn.pdf.

[18] S. Klein, E. Kuehn, and W. M. Wajda, "Energy savings in mobile networks based on adaptation to traffic statistics," *Bell Labs Tech. J.*, vol. 15, no. 2, pp. 77–94, Sept. 2010.

[19] Y. H. Hsu and K. Wang, "An adaptive energy saving mechanism for LTE-A self-organizing HetNets," *7th Int. Conf. on Ubiquitous and Future Networks*, Sapporo, Japan, pp. 289–294, 2015.

[20] R. P. Liu, G. J. Sutton, and I. B. Collings, "WLAN power save with offset listen interval for machine-to-machine communications," *IEEE Trans. on Wireless Commun.*, vol. 13, no. 5, pp. 2552–2562, May 2014.

[21] H. Tabrizi, G. Farhadi, and J. Cioffi, "An intelligent power save mode mechanism for IEEE 802.11 WLAN," *2012 IEEE Global Commun. Conf. (GLOBECOM)*, Anaheim, CA, pp. 3460–3464.

[22] F. Zhu and Z. Niu, "Priority based power saving mode in WLAN," in *2008 IEEE Global Commun. Conf. (GLOBECOM 2008)*, New Orleans, LA pp. 1–6.

[23] F. Zhu and Z. Niu, "Load-aware power saving mechanism in WLAN," *2007 IEEE Wireless Commun. and Netw. Conf.*, Kowloon, Hong Kong, pp. 2086–2091.

[24] J. H. Jun, Y. J. Choi, and S. Bahk, "Power-Saving schedulers for a WLAN with task-linking topology awareness," *IEEE Trans. on Veh. Technol.*, vol. 56, no. 3, pp. 1345–1356, May 2007.

[25] H. Kim and D. Cho, "Enhanced power-saving mechanism for broadcast and multicast service in WLAN," *IEEE Commun. Lett.*, vol. 9, no. 6, pp. 520–522, Jun. 2005.

[26] C.-L. I et al, "Recent progress on C-RAN centralization and cloudification," *IEEE Access*, vol. 2, pp.1030–1039, Aug. 2014."

[27] J. Wu et al., *Green Communication*, CRC Press, 2013.

[28] C.-L. I et al., "Recent progress on C-RAN centralization and cloudification," *IEEE Access*, vol. 2, pp. 1030–1039, Aug. 2014.

[29] G. Auer et al., "Energy efficiency analysis of the reference systems, areas of improvements and target breakdown," EARTH_WP2_D2.3_v2, Dec. 2010. www.ict-earth.eu/publications/deliverables/deliverables.html.

[30] G. Auer et al., "How much energy is needed to run a wireless network," *Wireless Commun.*, vol. 18, no. 5, pp. 40–49., Oct. 2011.

[31] 3GPP TS 23.401, "General packet radio service (GPRS) enhancements for evolved universal terrestrial radio access network (E-UTRAN) access (Release 8)."

[32] 3GPP TS 36.331, "Evolved universal terrestrial radio access (E-UTRA); Radio resource control (RRC); Protocol specification (Release 8)."

[33] 3GPP TS 36.104, "Evolved universal terrestrial radio access (E-UTRA); Base station (BS) radio transmission and reception (Release 8)."

[34] S. Sesia, I. Toufik, and M. Baker, *LTE–The UMTS Long Term Evolution: From Theory to Practice*, Wiley, 2nd Edition, 2011.

[35] E. C. Ifeachor and B. W. Jervis, *Digital Signal Processing: A Practical Approach, 2nd Edition*, Prentice Hall, 2002.

[36] ETSI, "Radio broadcasting systems; Digital audio broadcasting (DAB) to mobile, portable and fixed receivers," EN 300 401, 2004.

[37] K. Amiri, Y. Sun, P. Murphy et al., "WARP, A unified wireless network testbed for education and research," 2007 IEEE International Conference on Microelectronic Systems Education, Jun. 2007.

[38] S. Malkowsky, K. Nieman, Z. Miers et al., "A flexible 100-antenna testbed for Massive MIMO," 2014 *IEEE GLOBECOM Workshops*, Dec. 2014.

[39] J. Vieira, C. Shepard et al., "Argos: Practical many-antenna base stations," *Proc. 18th Annu. Int. Conf. Mobile Comput. Netw.*, pp. 53–64, 2012.

[40] M. Petermann et al., "Multi-user pre-processing in multi-antenna OFDM TDD systems with non-reciprocal transceivers," *IEEE Trans. Commun.*, vol. 61, no. 9, pp. 3781–3793, Sept. 2013.

[41] R. Rogalin et al., "Scalable synchronization and reciprocity calibration for distributed multiuser MIMO," *IEEE Trans. Wireless Commun.*, vol. 13, no. 4, pp. 1815–1831, Apr. 2014.

[42] D. Liu, W. Ma, S. Shao, Y. Shen, and Y. Tang, "Performance analysis of TDD reciprocity calibration for massive MU-MIMO systems with ZF beamforming," *IEEE Commun. Lett.*, vol. 20, no. 1, Jan. 2016.

[43] 3GPP RAN1 R1-1608809, "Over the air calibration for channel reciprocity in NR MIMO," Fujitsu.

[44] 3GPP RAN1 R1-1609903, "Partial reciprocity based UL MIMO schemes," InterDigital Communications.

[45] 3GPP RAN1 R1-1702603, "OTA calibration for multi-TRP transmission," Qualcomm.

Index